Applications Based on Symmetry/Asymmetry in Fluid Mechanics

Applications Based on Symmetry/Asymmetry in Fluid Mechanics

Guest Editor

Xi Chen

MDPI

Basel • Beijing • Wuhan • Barcelona • Belgrade • Novi Sad • Cluj • Manchester

Guest Editor
Xi Chen
Institute of Fluids
Beihang University
Beijing
China

Editorial Office
MDPI AG
Grosspeteranlage 5
4052 Basel, Switzerland

This is a reprint of the Special Issue, published open access by the journal *Symmetry* (ISSN 2073-8994), freely accessible at: https://www.mdpi.com/journal/symmetry/special_issues/V3UP1Q4W63.

For citation purposes, cite each article independently as indicated on the article page online and as indicated below:

Lastname, A.A.; Lastname, B.B. Article Title. *Journal Name* **Year**, *Volume Number*, Page Range.

ISBN 978-3-7258-6328-0 (Hbk)
ISBN 978-3-7258-6329-7 (PDF)
https://doi.org/10.3390/books978-3-7258-6329-7

Contents

symmetry

MDPI

Editorial

Applications Based on Symmetry/Asymmetry in Fluid Mechanics

Xi Chen

Key Laboratory of Fluid Mechanics of Ministry of Education, Beihang University (Beijing University of Aeronautics and Astronautics), Beijing 100191, China; chenxi97@outlook.com

This Special Issue of Symmetry is devoted to recent advances in the analysis and applications of fluid mechanics based on Symmetry/Asymmetry.

In recent years, the growing significance of symmetry analysis and its applications has been realized in fluid mechanics, due not only to theoretical achievements in this area, but also because of its numerous applications. This symmetry can be either on the space–time level or on the time level. The applications of symmetry in fluid mechanics are usually interdisciplinary, such as mechanical, aerospace, chemical, and process engineering. Therefore, their exploration is crucial for many real-life applications. This Special Issue focuses on the following topics, but is not limited to the following: the importance of symmetry in a variety of fluid flows, heat transfer and its applications, including heat exchangers, thermal storage, heat pipes, etc.

The Special Issue contains twelve papers contributed by researchers from China, USA, Romania, and India, covering a wide spectrum of important problems and topics of current research interest. These topics include the following: the mean velocity distribution in atmospheric surface layers [1]; velocity gradient tensors in two-dimensional isotropic turbulence [2]; tip leakage vortex in axial compressor rotors [3]; axisymmetric and weakly non-axisymmetric supersonic jets [4]; trailing-edge tonal noise with symmetry spanwise source regions [5]; decomposition methods for wall friction and heat flux [6]; blood damage analysis at laminar conditions [7]; asymmetry of two-dimensional thermal convection at high Rayleigh numbers [8]; horizontal divergence asymmetry in the Gulf of Mexico [9]; porous and magnetic effects on axial Couette flows of second grade fluids in cylindrical domains [10]; assessment of the measured mixing time in a water model of asymmetrical gas-stirred ladle [11]; and the influence of fewer strand casting on the symmetry breaking of flow [12].

We hope that this Special Issue comes to serve as a source of ideas for many mechanists, mathematical physicists, and engineers interested in pursuing recent developments in the Symmetry/Asymmetry phenomena of fluids.

Funding: This research received no external funding.

Conflicts of Interest: The author declares no conflicts of interest.

References

1. Ji, Y.; Chen, X. Symmetry Analysis of Mean Velocity Distribution in Stratified Atmospheric Surface Layers. *Symmetry* **2023**, *15*, 1951. [CrossRef]
2. Luo, C.S.; Yang, P.F.; Fang, L. Low-Order Moments of Velocity Gradient Tensors in Two-Dimensional Isotropic Turbulence. *Symmetry* **2024**, *16*, 175. [CrossRef]
3. Hou, J.X.; Liu, Y.W.; Tang, Y.M. A Lagrangian Analysis of Tip Leakage Vortex in a Low-Speed Axial Compressor Rotor. *Symmetry* **2024**, *16*, 344. [CrossRef]
4. Jia, T.; Li, J.W.; Wu, J.; Xiong, Y. Tomographic Background-Oriented Schlieren for Axisymmetric and Weakly Non-Axisymmetric Supersonic Jets. *Symmetry* **2024**, *16*, 596. [CrossRef]
5. Song, Z.C.; Liu, P.Q.; Guo, H.; Sun, Y.F.; Jiang, S.J. The Influence of Low-Frequency Oscillations on Trailing-Edge Tonal Noise with Symmetry Spanwise Source Regions. *Symmetry* **2024**, *16*, 710. [CrossRef]
6. Tang, M.Z.; Zhou, W.F.; Hu, Y.C.; Wang, G.; Yang, Y.G. Local-Energy-Conservation-Based Decomposition Method for Wall Friction and Heat Flux. *Symmetry* **2024**, *16*, 1147. [CrossRef]
7. Krishnamoorthy, G.; Gholizadeh, N. Blood Damage Analysis within the FDA Benchmark Nozzle Geometry at Laminar Conditions: Prediction Sensitivities to Software and Non-Newtonian Viscosity Models. *Symmetry* **2024**, *16*, 1165. [CrossRef]
8. He, J.C.; Bao, Y.; Chen, X. Asymmetry of Two-Dimensional Thermal Convection at High Rayleigh Numbers. *Symmetry* **2024**, *16*, 1583. [CrossRef]
9. Zhou, T.S.; Xie, J.H.; Balwada, D. On the Horizontal Divergence Asymmetry in the Gulf of Mexico. *Symmetry* **2025**, *17*, 136. [CrossRef]
10. Fetecau, C.; Vieru, D. Porous and Magnetic Effects on Axial Couette Flows of Second Grade Fluids in Cylindrical Domains. *Symmetry* **2025**, *17*, 706. [CrossRef]
11. Zhao, Y.S.; Tao, X.; Li, L.B.; Guo, Z.J.; Qi, H.Y.; Wang, J.; Yang, K.; Lin, W.M.; Fan, J.P.; Chen, C. Assessment of the Measured Mixing Time in a Water Model of Asymmetrical Gas-Stirred Ladle with a Low Gas Flowrate Part II: Effect of the Salt Solution Tracer Volume and Concentration. *Symmetry* **2025**, *17*, 802. [CrossRef]
12. Rong, Z.R.; Song, J.T.; Chen, C.; Guo, Z.J.; Wang, H.Z.; Geng, M.J.; Wang, T.Y.; Lin, W.M.; Wang, J.; Sun, Y.H. Influence of Fewer Strand Casting on the Symmetry Breaking of Flow, Temperature Fields, and Transition Billets in a Symmetrical Double Six-Strand Tundish. *Symmetry* **2025**, *17*, 850. [CrossRef]

symmetry

MDPI

Article

Influence of Fewer Strand Casting on the Symmetry Breaking of Flow, Temperature Fields, and Transition Billets in a Symmetrical Double Six-Strand Tundish

Zhiren Rong [1,†], Jintao Song [1,†], Chao Chen [1,*], Zhijie Guo [2], Haozheng Wang [1], Mengjiao Geng [1], Tianyang Wang [1], Wanming Lin [1], Jia Wang [3] and Yanhui Sun [2]

1 College of Materials Science and Engineering, Taiyuan University of Technology, Taiyuan 030024, China; 2023520469@link.tyut.edu.cn (Z.R.); songjintao0348@link.tyut.edu.cn (J.S.); wanghaozheng0676@link.tyut.edu.cn (H.W.); 2023510308@link.tyut.edu.cn (M.G.); wangtianyang0309@link.tyut.edu.cn (T.W.); linwanming@tyut.edu.cn (W.L.)
2 Collaborative Innovation Center of Steel Technology, University of Science and Technology Beijing, Beijing 100083, China; d202210640@xs.ustb.edu.cn (Z.G.); sunyanhui_ustb@163.com (Y.S.)
3 College of Architecture and Arts, Taiyuan University of Technology, Taiyuan 030024, China; wangjia01@tyut.edu.cn
* Correspondence: chenchao@tyut.edu.cn
† These authors contributed equally to this work.

Abstract: In continuous casting, fewer strand operations are sometimes required to match production schedules. However, the study of flow behavior and temperature distribution under fewer strand casting conditions remains insufficiently systematic, especially with regard to the grade casting process, which has not yet been explored. This study presents an innovative investigation of the grade transition process in a symmetrical 12-strand tundish under fewer strand casting conditions. Seven operational cases were analyzed: standard casting (the normal symmetric Case 0), individual closure of strands 1–6 (the asymmetric Cases 1–6), and simultaneous closure of strands 1–2 (the asymmetric Case 7). Notably, strand closures in Cases 5 and 6 significantly impair flow characteristics in their respective strands. The impact area temperature reaches approximately 1844 K (new heat) after 30 min of continuous casting. However, Case 6 exhibits persistent low-temperature regions near strands 5 and 6. The average transition billet lengths for Cases 0 to 7 are 72.41 m, 70.16 m, 70.30 m, 71.68 m, 72.95 m, 72.12 m, 76.35 m, and 65.45 m, respectively. Based on a comprehensive evaluation of flow dynamics, temperature uniformity, and transition billet length, Case 1 emerges as the most favorable single-strand closure strategy. Operational recommendations suggest avoiding strand closure patterns implemented in Cases 5 and 6 during reduced strand casting operations.

Keywords: symmetrical double six-strand tundish; fewer strand casting; transition billet; computational fluid dynamics (CFDs); flow field

1. Introduction

Molten steel flows from the ladle, through the ladle shroud, into the tundish, and finally enters the mold—a process extensively studied by metallurgists and commonly referred to as tundish metallurgy [1–3]. At present, extensive research primarily focuses on single-strand [4] or double-strand [5,6] tundishes. By optimizing internal flow control devices [7–10], applying tundish gas injection technology [11–14], and improving the structure of ladle shroud [15,16], the aim is to reduce the turbulent kinetic energy of molten steel. This approach enhances flow patterns, prolongs residence time [17–19], homogenizes

https://doi.org/10.3390/sym17060850

temperature distribution across strands, and improves inclusion removal efficiency [20–25]. During unsteady casting processes, efforts are primarily directed at optimizing casting flow rate, tundish level, and turbulence inhibitors to reduce slag entrapment [26,27] and re-oxidation [28,29].

The symmetrical double six-strand tundish consists of a single ladle connected to two ladle shrouds [27,30,31], each inserted into two independent six-strand tundishes. The inlet area (impact area) is offset to one side of the tundish, forming a biased flow casting pattern [32–34]. This type of tundish is mainly used for small billet production. In practice, due to constraints on the annual production capacity and casting rhythm, one or two strands in a symmetrical double six-strand tundish are often shut off for fewer strand casting [35]. In multi-heat sequence continuous casting operations, the symmetrical double six-strand tundish bridges molten steel from the preceding and following heats, enabling a continuous casting process. However, if there are compositional differences between the two heats, a segment of billet with inconsistent composition—meeting neither the previous nor the next heat's specifications—may form during the transition. This segment is referred to as a transition billet [36].

Researchers have investigated the effects of fewer strand casting on molten steel flow, temperature distribution, and inclusion removal in tundishes. In terms of steel flow, C. Bruch and P. Valentin [37] conducted numerical simulations on strand closure in a six-strand tundish and found that the streamlines around the stopper rod were significantly sparser compared to normal conditions. T. Merder [38] also studied a six-strand tundish and reported that when outlets far from the ladle shroud were closed, the dead zone volume increased markedly; however, for outlets near the ladle shroud, flow characteristics in a 12-strand tundish improved after strand closure, concluding that closure near the strand resulted in better flow behavior. In terms of temperature field, A. Braun et al. [39] performed a simulations of a two-strand tundish and showed that closing one strand caused the return flow point to shift closer to the ladle shroud, leading to significant temperature drops in the tundish. S. K. Mishra et al. [40] found that two asymmetrically close strands in a six-strand tundish could cause a maximum strand temperature difference of up to 10 °C. In terms of inclusion removal, L. Zhang et al. [41] simulated inclusion behavior in a four-strand tundish and found that strand closure not only enhanced the surface removal of inclusions, but also led to notable temperature drops. P. Zhao et al. [42] modeled inclusion behavior after outlet closure in a six-strand tundish and noted that closing one or two strands, especially those closer to the impact zone, improved flow characteristics and increased inclusion removal rates. X. Wang et al. [35] investigated inclusion removal under fewer strand casting conditions and found that shutting down strands in a six-strand tundish negatively affected the removal of inclusions sized 30–70 μm. In terms of flow control device optimization, L. Zhong et al. [43] optimized baffles by introducing rectangular openings, showing that closing the central strand extended the response time of other strands and reduced the dead zone volume. C. Yao et al. [44] replaced a V-shaped baffle with a U-shaped one, improving the tundish flow field after strand closure. The V-shaped baffle produced a 5 K temperature difference between the two sides of the tundish. J. Song and J. Fan et al. [45,46] compared different flow control setups for a four-strand tundish under fewer strand operation and found that the U-shaped baffle provided better consistency across strands due to its interconnecting structure, compared to the double-baffle configuration.

Regarding transition billets, they are typically either scrapped or downgraded, leading to economic losses. To address this, researchers have widely studied methods to reduce the length of transition billets and accurately predict their start and end positions. These approaches include optimizing tundish operating practices, improving flow control devices and internal structures, analyzing temperature differences between successive heats, and developing mathematical prediction models. In terms of operational practices during grade transition, A. Diener et al. [36] were the first to study the transition casting of different steel grades in the tundish in 1990. They found that reducing the amount of residual steel in the tundish could effectively shorten the length of the transition billet. Later, M.C. Tsai and M.J. Green [47] carried out numerical simulations of grade transitions in a 30-ton single-strand slab caster tundish and discovered that increasing the casting speed and lowering the tundish level could reduce the transition time. M.T. Burns et al. [48] conducted full-scale physical water model experiments of single-strand tundishes and found that, in addition to confirming the relationship between residual steel and transition billet length, transitioning from narrow to wide specification resulted in shorter transition billets. Regarding flow control device optimization, H.S. Chen and R.D. Pehlke [49] used numerical simulations to compare the effects of baffles and dam-weir combinations in double-strand tundish during grade transition. They concluded that baffles were not beneficial for reducing transition billet length. T.J. Piccone et al. [50] conducted industrial trials and water model experiments on a 45-ton single-strand tundish and demonstrated that tundishes equipped with turbulence inhibitors and dams performed better than traditional designs with perforated baffles. S. Kant et al. [51] explored the effect of dam placement on transition billet length through water modeling of a six-strand tundish and found that positioning the dam near the inlet area reduced the transition length. J. Guarneros et al. [52] found that an inhibitor, a baffle, and vortex killer yields a smaller amount of intermixed liquid. S. Chakraborty et al. [53] designed a step-shaped tundish bottom that maintained steel cleanliness while achieving a high liquid level with reduced transition billet. D. Xu et al. [54] elevated the tundish bottom (trapezoidal: higher in the middle, lower at the strands) and optimized the turbulence inhibitor structure, which helped reduce steel level fluctuations, increased plug flow ratio, and decreased the mixing between old and new steel, thereby lowering the transition billet volume. During grade transitions, the temperature difference between the new and residual steel in the tundish results in thermal buoyancy, which affects flow behavior. L.C. Amorim et al. [55] studied this in six-strand tundishes and found that hotter new steel created thermal buoyancy when poured into colder residual steel, altering the flow field. The temperature difference between steel grades significantly influenced mixing behavior. A. Cwudziński [56] performed both water model experiments and numerical simulations on a double-strand tundish and observed that due to density differences at varying temperatures, high-temperature steel layered above the colder, denser residual steel during grade transitions. In recent years, S. Song et al. [57,58] have conducted extensive studies on transition casting of different steel grades. In [57], multi-physics coupling models were used to simulate the transition process in a four-strand tundish. It was found that although reducing the casting speed lowered centerline segregation, it also increased the transition billet length from 10.88 m to 12.41 m. Furthermore, by studying the entire process from the ladle shroud to the solidification end, it was discovered that pouring low-density, high-solute steel into high-density, low-solute steel resulted in a 24.70 m transition billet, whereas the reverse case produced a 33.44 m billet.

To ensure flow consistency across strands and meet the requirements of large-scale production, the tundish is typically designed with symmetrically arranged ladle shrouds [23,24,33–37]. T. Merder [38] found that closing a single outlet leads to an asymmetric nozzle configuration, resulting in increased dead zone volume and reduced strand consistency. S. K. Mishra et al. [40] reported that closing two symmetrically located strands yields better performance than asymmetric closure, with the maximum temperature difference at the outlets increasing by up to 10 °C in the asymmetric case. These findings indicate that fewer strand operation causes asymmetry in the tundish nozzle layout, which deteriorates flow characteristics and temperature uniformity.

Numerous studies have been conducted on fewer strand casting and grade transition processes, primarily focusing on single-strand, double-strand, as well as four-strand and six-strand tundishes. However, these two phenomena are quite common in the production process of a symmetrical double six-strand tundish. At present, research on fewer strand operation and grade transition casting in a symmetrical double six-strand tundish under asymmetric casting conditions remains limited. Moreover, the impact of fewer strand casting on the grade transition process has not yet been reported, and the molten steel flow patterns, temperature distribution, and transition billet length under such conditions are still unclear.

Therefore, this study takes the symmetrical double six-strand tundish as the research object and establishes a numerical model of grade transition casting under fewer strand operation. Using the normal casting condition as a reference, the effects of various strand closure cases on flow field, temperature field, and the start and end positions of the transition billet are compared. The goal is to propose a reasonable strand closure strategy to optimize grade transition casting during the continuous casting process and reduce the length of the transition billet.

2. Materials and Methods

2.1. Geometric Model

A full-scale 1:1 geometric model was established based on a symmetrical double six-strand tundish (capacity: 50 t) from a steel plant as shown in Figure 1. The ladle shroud is located on one side of the tundish, and the strands are numbered from 1 to 6 from the inlet area (impact area) toward the short wall of the tundish. The symmetrical double six-strand tundish is used for casting 150 × 150 mm square billets, with a casting speed of 2.8 m/min and a steel throughput of 2.56 t/min. The tundish liquid level is 820 mm, and the insertion depth of the ladle shroud is 300 mm. The center-to-center spacing between adjacent strands is 1200 mm, and each strand has a length of 710 mm and a diameter of 25 mm.

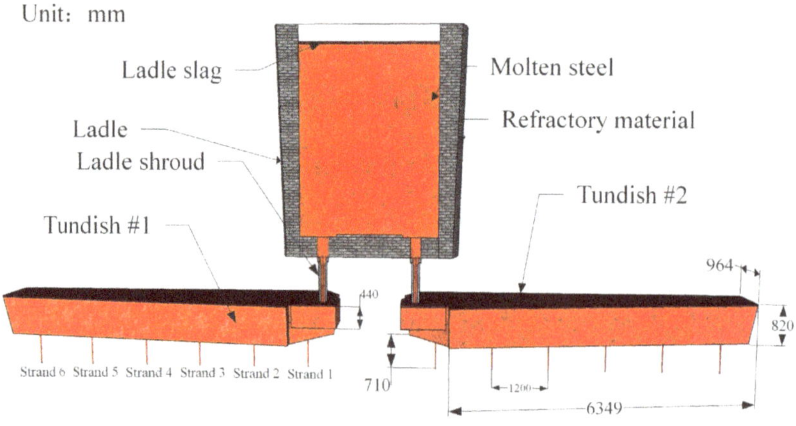

Figure 1. A 12-strand tundish-ladle system.

2.2. Model Assumptions

The following assumptions are made for the 3D mathematical model of the tundish [45]:

1. The molten steel is assumed to be an incompressible Newtonian fluid, with only the flow and temperature changes in the molten steel in the tundish considered;
2. The Realizable k-ε Two-Layer (RKE-2L) turbulence model is used to describe the turbulence phenomena within the tundish;
3. The influence of the slag layer on the molten steel flow and heat transfer in the tundish is neglected;
4. Chemical reactions in the tundish are not considered;
5. Changes in the tundish liquid level are not considered.

2.3. Control Equations

According to Patankar [59], the continuity and momentum equations are unified into a single form. The flow of molten steel inside the tundish is described by the unified differential Equation (1) as follows:

$$\rho\frac{\partial\phi}{\partial t}+\rho\mathbf{u}\frac{\partial\phi}{\partial x}=\frac{\partial}{\partial x}\left[\Gamma_{\phi,eff}\frac{\partial\phi}{\partial x}\right]+S_{\phi},\tag{1}$$

where ρ is the density, kg/m^3; φ represents the solved variables such as velocity, concentration, turbulent kinetic energy, and turbulent dissipation rate; $\mathbf{u}$ is the velocity vector, m/s; t is time, s; $\Gamma_{\phi,eff}$ is the effective diffusion coefficient of the solved variable, m^2/s; and $S\varphi$ is the source term.

2.4. Tracer Transport Model

The tracer transport model adopts a passive scalar approach. The passive scalar is a virtual tracer without physical properties, which is transported passively along with the molten steel flow. In the numerical simulation, the governing equation for passive scalar transport is solved, as shown in Equation (2) [45,60,61]:

$$\frac{\partial}{\partial t}(\rho\omega)+\nabla\cdot(\rho\mathbf{u}\omega)=\nabla\cdot\left(\rho D_{eff}\nabla\omega\right)\tag{2}$$

where ω represents the volume fraction of the tracer within the computational domain—that is, the proportion of each grid cell volume occupied by the tracer. A value of $\omega = 1$ indicates that the grid cell is completely filled with the tracer. D_{eff} denotes the effective diffusion coefficient of the passive scalar, m^2/s.

In the pulsed tracer injection scenario, the tracer is introduced by setting the concentration $\omega = 1$ in the injection region while maintaining $\omega = 0$ elsewhere. The injection occurs with a time interval of 1 s. After the injection is completed, the tracer concentration ω at the inlet region is reset to 0. By tracking the transport of the tracer through the tundish, the variation of ω at the strands is monitored to obtain the residence time distribution (RTD) curves, which reflect the flow characteristics within the tundish.

During grade transition casting, compositional differences between consecutive heats can produce a billet segment with inconsistent properties that meets neither the preceding nor subsequent heat's specifications. This segment is called a 'Transition Billet'. Continuous tracer injection simulates the compositional shift between the previous and new heats. The initial tracer concentration ω in the tundish is set to 0. After $t = 0$, the tracer concentration ω in the inlet region is set to 1. As casting progresses, the tracer concentration within the tundish gradually increases from 0 to 1. The concentration variation at each strand is monitored, where the strand tracer concentration ω represents the proportion of the concentration difference between the previous and the new heats during the grade transition process, as shown in Equation (3):

$$\omega = \frac{w_{\mathrm{O}} - w_{\mathrm{P}}}{w_{\mathrm{N}} - w_{\mathrm{P}}}, \tag{3}$$

where w_{O}, w_{N}, and w_{P} represent the mass fractions of a given element at the strand, in the new heat, and in the previous heat, respectively.

2.5. Heat Transfer Model

The influence of temperature variation on molten steel density is considered using the Boussinesq model. The relationship between density and temperature is given by the following Equation (4):

$$\rho = \rho_{st}(1 - \beta\Delta T), \tag{4}$$

where ρ_{st} is the molten steel density, kg/m^3; β is the thermal expansion coefficient β = 9.72 × 10^{-5} K^{-1}; and ΔT is the instantaneous temperature difference, °C.

2.6. Mesh Generation

The geometric model of the tundish was meshed using a polyhedral mesh generator [62]. The mesh was locally refined in the inlet area, as shown in Figure 2. The total number of mesh elements was 662,278.

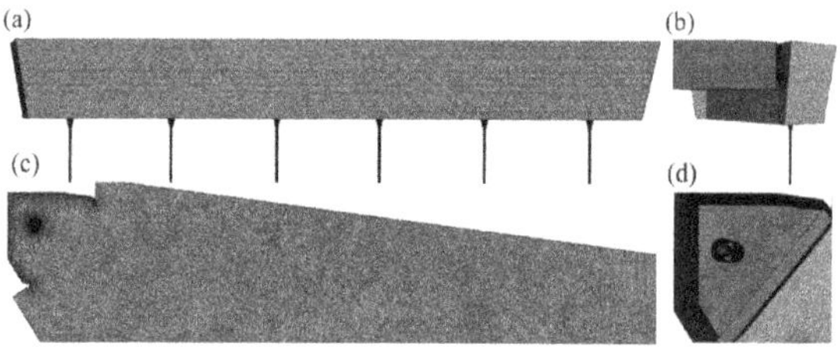

Figure 2. Mesh division of tundish with impact pad. (**a**) Front view; (**b**) left view; (**c**) top view; (**d**) impact pad.

2.7. Boundary Conditions

The boundary conditions were defined as follows: solid surfaces were treated as no-slip, rough walls with a roughness height of 0.00027. The inlet was defined as a velocity inlet. The boundary conditions of the numerical simulation in this study are similar to those used in our previous work [27]. In combination with actual production conditions, the detailed computational parameters and boundary conditions are determined, as shown in Table 1. The turbulence kinetic energy (k) and turbulence dissipation rate (ε) at the inlet were calculated as follows:

$$k = \frac{3}{2}(I\mathbf{v})^2, \tag{5}$$

$$\varepsilon = \frac{C_\mu^{3/4} k^{3/2}}{L} \tag{6}$$

where I is the turbulence intensity, I = 0.02; v is the inlet velocity; C_μ is the model constant, C_μ = 0.09; L is the characteristic length, L = 0.07 D; and D represents the diameter of the ladle shroud.

Table 1. Numerical simulation parameters and wall boundary conditions.

Parameter	Value
Density of molten steel (kg·m^{-3})	7020
Dynamic viscosity (Pa·s)	0.005967
Inlet velocity (m·s^{-1})	1.558
Turbulent kinetic energy (J·kg^{-1})	0.001456
Turbulent dissipation rate (m^2·s^{-3})	0.001807
Free surface heat flux (kW·m^{-2})	15
Bottom wall heat flux (kW·m^{-2})	1.4
Long wall heat flux (kW·m^{-2})	3.2
Short wall heat flux (kW·m^{-2})	3.8
Internal wall (dam, weir, stopper) heat flux (kW·m^{-2})	1.75
Initial temperature (K)	1793
New heat temperature (K)	1844

2.8. Solution Procedure

The governing equations were solved using Simcenter STAR-CCM+ V2021.3 [63], a software based on the finite volume method. The Realizable k-ε Two-Layer (RKE-2L) model was used to iteratively compute the steady-state flow field. The results from the steady-state simulation were then used as the initial condition for the transient simulation. Pressure–velocity coupling was solved using the semi-implicit method for pressure-linked equations (SIMPLE) algorithm. The convergence criterion required the residuals of all variables to fall below 1×10^{-4}. For the transient simulation, the time step was gradually increased: the initial time step was 0.002 s, with a growth factor of less than 1.25, and the maximum time step was set to 1 s, with 20 iterations per time step.

In terms of the temperature field, the simulation modeled the continuous casting process where higher-temperature molten steel from the new heat (1844 K) was poured into the tundish containing lower-temperature molten steel from the previous heat (1793 K). The study did not account for the temperature drop of molten steel in either heat. The temperature calculation process was as follows: the initial temperature of the tundish was set to 1793 K, and wall heat flux boundary conditions were applied. The simulation iterated until all residuals dropped below 1×10^{-4}, at which point the steady-state wall temperature distribution was obtained. This steady-state result served as the initial condition for the transient simulation, during which 1844 K molten steel was continuously poured into the tundish over time.

3. Results and Analysis

3.1. Computational Fluid Dynamic (CFD) Model Verification and Validation

Figure 3 presents a comparison between PIV measurements and CFD-predicted velocity vectors. As shown in Figure 3a, a strong counterclockwise vortex is observed across all liquid heights between the ladle shroud and the right wall, while a clockwise vortex appears at the lower region on the left side of the ladle shroud. The velocity vector directions in the CFD results, shown in Figure 3b, are in good agreement with the PIV measurements, confirming the reliability of the CFD model for simulating the flow field in the 12-strand tundish.

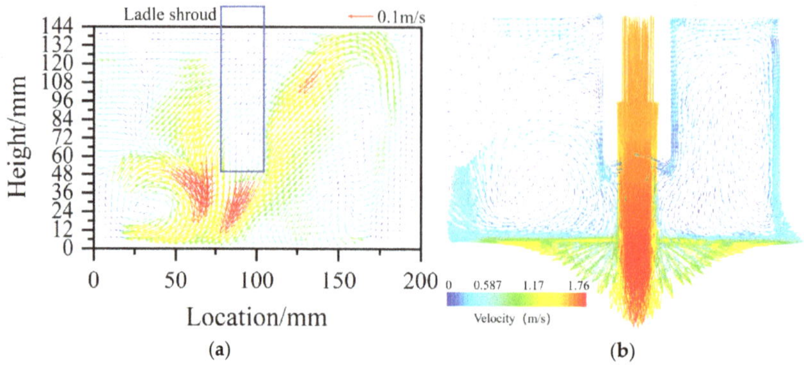

Figure 3. Velocity vectors at the tundish inlet longitudinal section. (**a**) PIV measurement; (**b**) CFD simulation.

3.2. *Influence of Fewer Strand Casting in Tundish on Transition Billet*

3.2.1. Research Plan for Fewer Strand Casting

The normal symmetric Case 0 serves as the control group, in which all 12 strands are fully open and the casting process proceeds symmetrically, representing the ideal symmetric casting condition. In contrast, the reduced-flow casting scenarios focus on the effects of shutting down a single strand—thus disrupting the symmetry—on the molten steel flow characteristics, temperature distribution within the tundish, and the resulting transition billet length. Additionally, for the case of shutting down two strands, strands 1 and 2 are closed, following actual production conditions in the plant. The specific research cases are shown in Table 2.

Table 2. Research plan for fewer strand casting.

Case	Case 0	Case 1	Case 2	Case 3	Case 4	Case 5	Case 6	Case 7
Strand Closure Condition	All Strands Open	Strand 1	Strand 2	Strand 3	Strand 4	Strand 5	Strand 6	Strands 1 and 2

3.2.2. Tundish Flow Field

The streamlines of the tundish flow field for each case are shown in Figure 4. The tundish can be divided into different regions. The inlet region exhibits substantially elevated molten steel velocities (>0.1 m/s), surpassing other flow domains by an order of magnitude. A small and a large recirculation zone are formed near strand 1 and between strands 2 and 3, with velocities around 0.05 m/s. Under both the ideal symmetric condition and the symmetry-breaking scenarios, the flow patterns in the inlet region and near strands 1–3 remain largely consistent. The main flow stream passes through strands 3 and 4, forming an upward flow between strands 4 and 6. In Case 2 (strand 2 closed), the molten steel has a relatively low velocity between strands 4 and 5, falling below 0.02 m/s, as shown in Figure 4e. In Case 6 (strand 6 closed), the upward flow between strands 5 and 6 disappears, as shown in Figure 4g.

According to the definition of slow-flow region (dead zone) in the tundish by F. He et al. [64], the slow-flow region inside the tundish is visualized. The slow-flow regions for each case are shown in Figure 5, where the molten steel flow velocity is $\leq$0.0025 m/s. The volume and percentage of the slow-flow regions in each case are shown in Figure 6. In the normal symmetric Case 0, the volume of the slow-flow region is 0.878 m^3, accounting for 12.2% of the total tundish volume. Under fewer strand casting conditions, all scenarios lead to an expansion of the slow-flow region within the tundish, thereby disrupting the ideal symmetric flow pattern observed in the normal symmetric Case 0. In the asymmetric Cases 1 to 6, the volume of the slow-flow region gradually increases from 1.135 m^3 to 1.731 m^3, with the percentage of the total tundish volume rising from 15.76% to 24.04%. According to the flow streamlines in the tundish shown in Figure 3, strands 1 to 4 are located within the main flow region of the tundish. Therefore, closing strands 1 and 2 or strands 3 and 4 has a similar effect on the volume of the slow-flow region. As a result, the differences in the slow-flow region volume between Case 1 and Case 2, as well as between Case 3 and Case 4, are minimal. In the asymmetric Cases 5 and 6, the slow-flow region occupies more than 20% of the tundish volume, making these conditions more detrimental to the molten steel flow compared to other cases. In the asymmetric Case 7, the volume of the slow-flow region is 1.398 m^3, accounting for 19.42% of the total tundish volume.

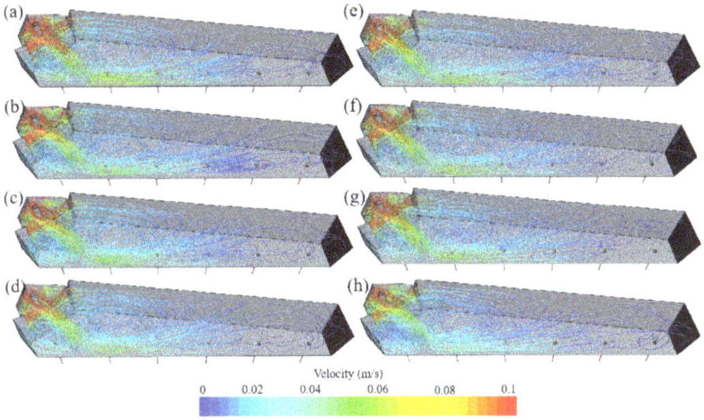

Figure 4. Streamline of the tundish for each case. (**a**) Case 0; (**b–g**) Cases 1–6; (**h**) Case 7.

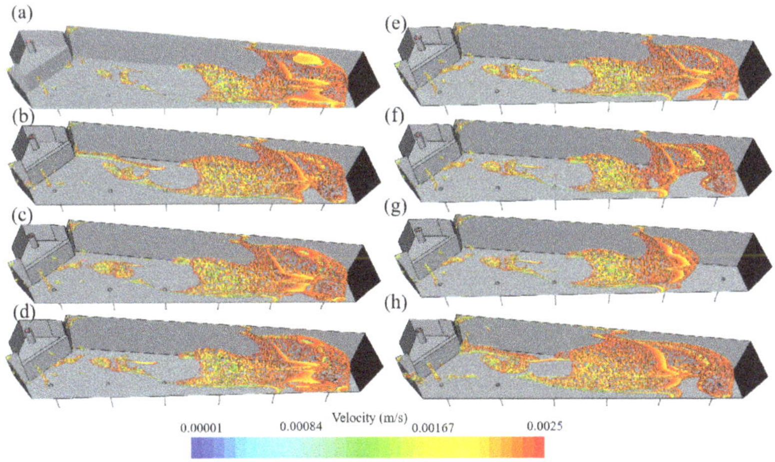

Figure 5. Slow-flow regions in the tundish for each case. (**a**) Case 0; (**b–g**) Cases 1–6; (**h**) Case 7.

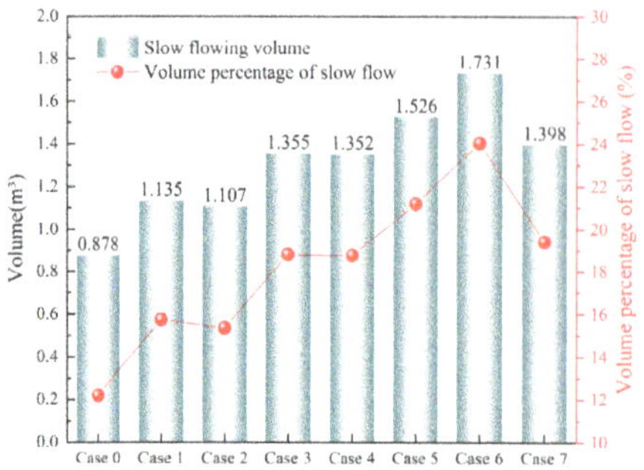

Figure 6. Slow-flow volume and percentage in the tundish for each case.

3.2.3. Residence Time Distribution (RTD) Curves for Each Case Under Fewer Strands Operation

The RTD curves for each strand provide further insights into the fluidity and consistency of the tundish. This study systematically investigates the influence of different strand closures on the tundish flow field by comparing the normal symmetric Case 0 (all strands open) with the asymmetric fewer strand operation cases (Cases 1–7). The RTD curves of different strands under each case are presented in Figures 7–14. The flow characteristic parameters for each strand, calculated based on the RTD curves, are summarized in Table 3 for clarity.

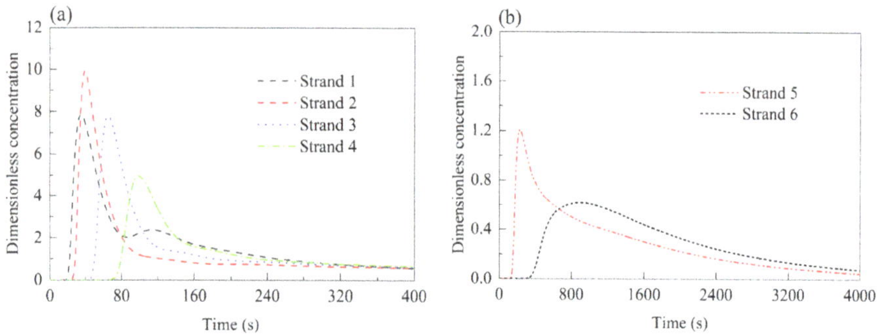

Figure 7. RTD curves of each strand in Case 0 (all strands open). (**a**) Strands 1–4; (**b**) strands 5 and 6.

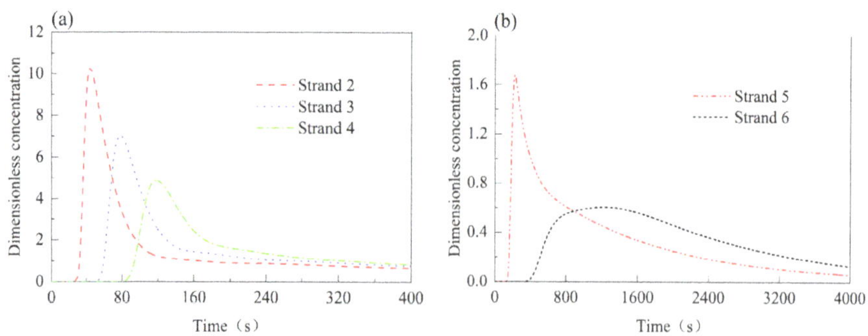

Figure 8. RTD curves of each strand in Case 1 (strand 1 closed). (**a**) Strands 2–4; (**b**) strands 5 and 6.

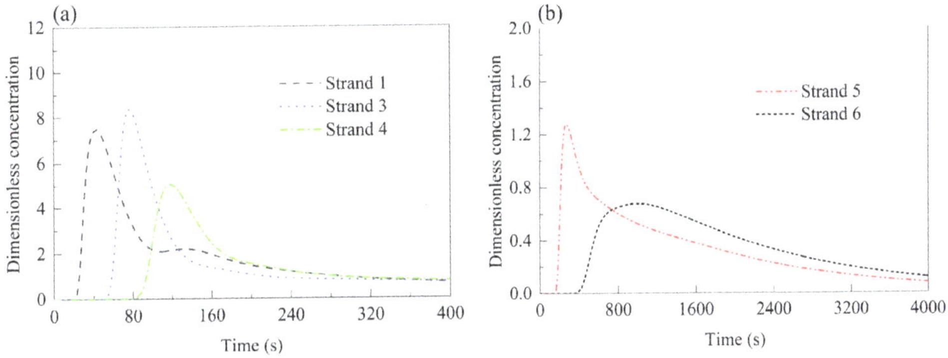

Figure 9. RTD curves of each strand in Case 2 (strand 2 closed). (**a**) Strands 1, 3, and 4; (**b**) strands 5 and 6.

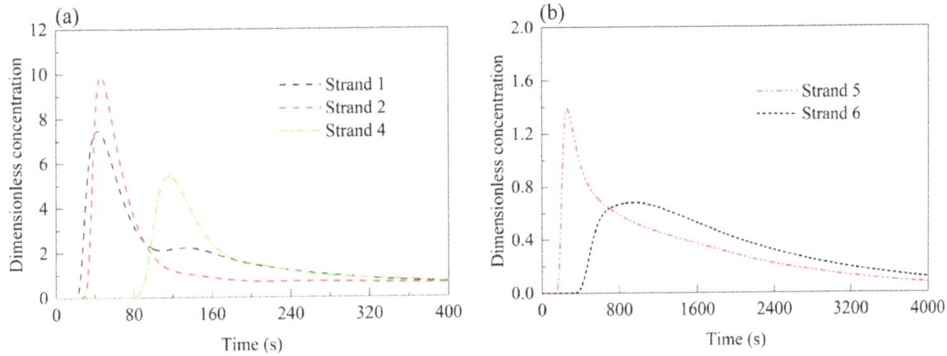

Figure 10. RTD curves of each strand in Case 3 (strand 3 closed). (**a**) Strands 1, 2, and 4; (**b**) strands 5 and 6.

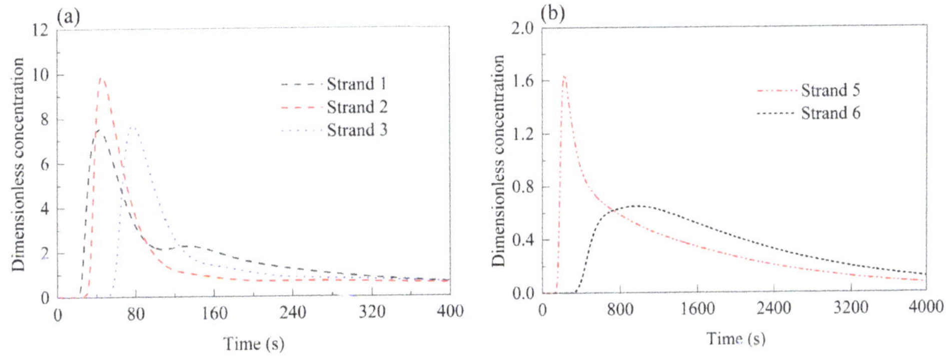

Figure 11. RTD curves of each strand in Case 4 (strand 4 closed). (**a**) Strands 1, 2, and 3; (**b**) strands 5 and 6.

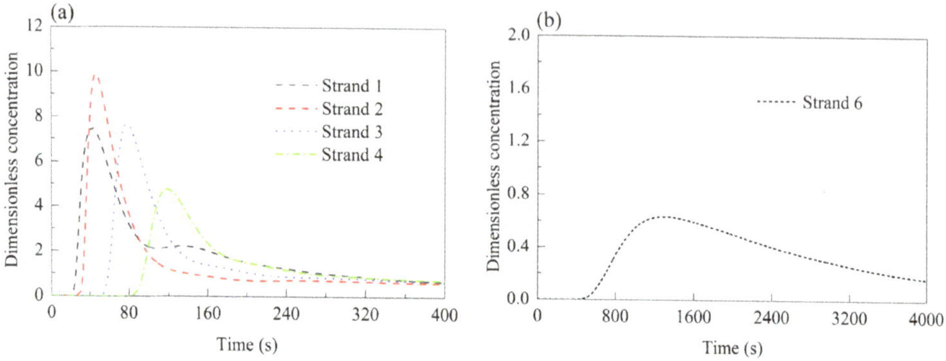

Figure 12. RTD curves of each strand in Case 5 (strand 5 closed). (**a**) Strands 1–4; (**b**) strand 6.

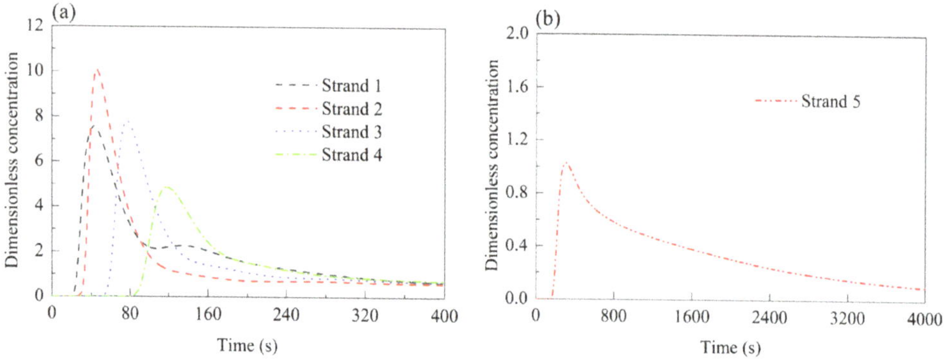

Figure 13. RTD curves of each strand in Case 6 (strand 6 closed). (**a**) Strands 1–4; (**b**) strand 5.

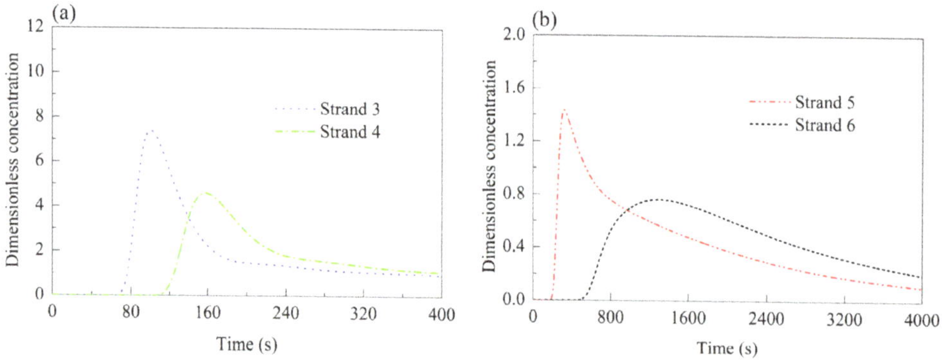

Figure 14. RTD curves of each strand in Case 7 (strands 1 and 2 closed). (**a**) Strands 3 and 4; (**b**) strands 5 and 6.

In the normal symmetric Case 0 (Figure 7), the minimum response times of strand 1 to strand 6 are 16.7 s, 23.3 s, 43.0 s, 68.3 s, 130 s, and 339.5 s, respectively. The peak times follow a similar increasing trend, recorded as 34.7 s, 39.0 s, 65.1 s, 98.8 s, 226.7 s, and 888.8 s. The peak concentration reaches its maximum at strand 2 (9.93), with strand 1 at 7.85, and then decreases gradually to 0.62 at strand 6. The average residence times increase steadily from 200.4 s (strand 1) to 1063.9 s (strand 6). The tundish forms a biased flow casting pattern, with the impact zone equipped with a triangle-like impact pad. As a result, molten steel flows directly toward the area near strand 2 through the open section of the impact pad, leading to the highest peak concentration at strand 2.

Table 3. Flow characteristic parameters of each case.

Case	Parameters	Strand 1	2	3	4	5	6
Case 0	Minimum response time/s	16.7	23.3	43.0	68.3	130	339.5
	Peak concentration time/s	34.7	39.0	65.1	98.8	226.7	888.8
	Peak concentration	7.85	9.93	7.80	4.97	1.21	0.62
	Mean residence time/s	200.4	256.0	353.1	507.3	999.3	1063.9
Case 1	Minimum response time/s		27.0	50.8	80.7	138.6	377.5
	Peak concentration time/s		44.6	78	118	223.2	1177.5
	Peak concentration		10.28	7.02	4.89	1.68	0.61
	Mean residence time/s		300.8	442.4	576.5	971.5	1314.3
Case 2	Minimum response time/s	19.6		49.9	80.1	154.8	387.6
	Peak concentration time/s	42.4		76.2	117.6	272.9	958.5
	Peak concentration	7.51		8.37	5.07	1.28	0.68
	Mean residence time/s	276.8		436.1	605.5	1094.1	1341.5
Case 3	Minimum response time/s	19.6	27.4		78.8	149.5	368.5
	Peak concentration time/s	42.5	45.92		115.8	258.2	909.8
	Peak concentration	7.49	9.9		5.47	1.39	0.682
	Mean residence time/s	261.0	315.2		577.5	1102.2	1374.8
Case 4	Minimum response time/s	19.6	27.5	50.5		139.6	338.5
	Peak concentration time/s	42.5	46.0	77.7		230.6	947.3
	Peak concentration	7.52	9.91	7.69		1.64	0.65
	Mean residence time/s	258.7	319.6	442.7		1028.9	1386.1
Case 5	Minimum response time/s	19.7	27.5	50.5	80.9		487.9
	Peak concentration time/s	42.4	45.7	77.6	118.4		1246
	Peak concentration	7.49	9.91	7.70	4.79		0.63
	Mean residence time/s	259.6	319.0	443.4	621.5		1432.8
Case 6	Minimum response time/s	19.5	27.2	50.0	80.0	158.0	
	Peak concentration time/s	42.4	46.1	77.6	118.6	298	
	Peak concentration	7.57	10.07	7.82	4.89	1.03	
	Mean residence time/s	256.7	314.2	436.2	612.5	1253.5	
Case 7	Minimum response time/s			64.4	105.2	191.4	503
	Peak concentration time/s			100.9	157.9	320	1245.9
	Peak concentration			7.423	4.614	1.44	0.76
	Mean residence time/s			422.8	578.8	993.9	1615.1

In the asymmetric Case 1 (Figure 8), strands 2–4 exhibit slightly longer response and peak times compared to the normal symmetric Case 0, while their peak concentrations decline. Strand 5 shows an earlier peak time than in the normal symmetric Case 0, with a higher peak concentration of 1.68, indicating enhanced fluidity. For strand 6, the peak time extends significantly to 1177.5 s, and the average residence times at strands 5 and 6 increase by approximately 315 s and 250 s, reaching 971.5 s and 1314.3 s, respectively.

Under the asymmetric Case 2 (Figure 9), the flow patterns of strands 1, 3, and 4 remain similar to the asymmetric Case 1, with a slight increase in peak concentration at strand 3. Strand 5, however, experiences an increase in response time to 154.8 s and a sharp rise in peak concentration time to 272.9 s, while its peak concentration drops to 1.28. A sharply rising and falling RTD curve indicates a short-circuit flow at the outlet, while a gradually increasing (sinusoidal) RTD curve suggests slow flow from the outlet. The delayed peak concentration time and the reduced peak concentration at strand 5 indicate a decrease in molten steel flow velocity. These flow characteristics demonstrate that the closure of strand 2 in Case 2 has an adverse effect on strand 5. At strand 6, the peak time decreases slightly to 958.5 s compared to the asymmetric Case 1, and its average residence time increases by 280 s relative to the normal symmetric Case 0.

In the asymmetric Cases 3 and 4 (Figures 10 and 11), the minimum response times of strands 1–5 are consistently extended. The minimum response time at strand 6 is 368.5 s in the asymmetric Case 3—29 s longer than in the normal symmetric Case 0—and 338.5 s in Case 4, slightly shorter than in the normal symmetric Case 0. Peak concentration times at strands 5 and 6 are 258.2 s and 909.8 s in Case 3, and 230.6 s and 947.3 s in the asymmetric Case 4. The peak concentration at strand 5 increases to 1.39 and 1.64 in the asymmetric Cases 3 and 4, respectively, while strand 4 reaches 5.47 in the asymmetric Case 3. Other strands show peak concentrations comparable to the normal symmetric Case 0. Overall, the average residence times in these two cases are notably higher than in the normal symmetric Case 0.

The asymmetric Cases 5 and 6 (Figures 12 and 13) exhibit nearly identical trends. The minimum response times of strands 1–4 range from 20 s to 80 s, and their peak concentration times fall within 40 s to 120 s—both metrics slightly extended from the normal symmetric Case 0. Peak concentrations at strands 1–4 remain similar to those in the normal symmetric Case 0, with strand 2 registering the highest values: 9.91 in the asymmetric Case 5 and 10.02 in the asymmetric Case 6. In Case 5, the minimum response time and peak time at strand 6 are 484.9 s and 1246 s—up by 148.4 s and 357.2 s compared to the normal symmetric Case 0. In the asymmetric Case 6, strand 5 shows response and peak times of 158 s and 298 s, respectively, rising by 28 s and 71.3 s. These observations suggest that closing strands 5 and 6 exerts minimal influence on strands 1–4 but significantly affects flow behavior at the closed or adjacent strands.

Finally, the asymmetric Case 7 (Figure 14) reveals that the minimum response times for strands 4–6 are 64.4 s, 105.2 s, 191.4 s, and 503 s, while their peak concentration times are extended to 100.9 s, 157.9 s, 320 s, and 1245.9 s, respectively. Notably, the peak concentrations at strands 5 and 6 increase to 1.44 and 0.76 compared to the normal symmetric Case 0, indicating altered flow patterns under this closure configuration.

J. Zhang et al. [65] focused only on the flow field and flow behavior under fewer strand casting conditions, which aligns with the perspective of this study. However, they did not investigate the temperature field or the transition billet length under such conditions. This work builds on their findings and extends the analysis to these aspects, contributing to the novelty of the study.

3.3. Temperature Variations in Different Cases of Fewer Strand Casting

When casting a new heat in the tundish, the temperature of the new heat is often higher than that of the present heat in the tundish. When the higher-temperature new heat is cast into the lower-temperature present heat in the tundish, the temperature fields of the tundish under different cases are shown in Figure 15. As shown in Figure 15a, at 120 s after casting the new heat, the impact zone of the tundish and the surface near strand 1 exhibit higher temperatures, ranging from 1797 K to 1820 K. Strands 2–4 are in the temperature transition zone, with a free surface temperature between 1701 K and 1773 K. Strands 5 and 6 experience severe temperature drops on the free surface, forming a low-temperature region. In the asymmetric Case 2, the low-temperature region near strands 5 and 6 is significantly reduced compared to the normal symmetric Case 0. In Case 6, the low-temperature region near strands 5 and 6 tends to expand. In Case 7, after closing strands 1 and 2, the free surface temperature near strands 5 and 6 in the far-flow region is higher than in other cases. At 30 min of casting, as shown in Figure 15b, the overall tundish temperature increases, ranging from 1773 K to 1844 K from the far-flow region to the tundish impact zone, with the impact zone temperature nearly reaching the new heat temperature. In the normal symmetric Case 0, the high-temperature region is significantly larger than in other asymmetric cases. In the asymmetric Case 6, there is still a low-temperature region near the edge strands 5 and 6.

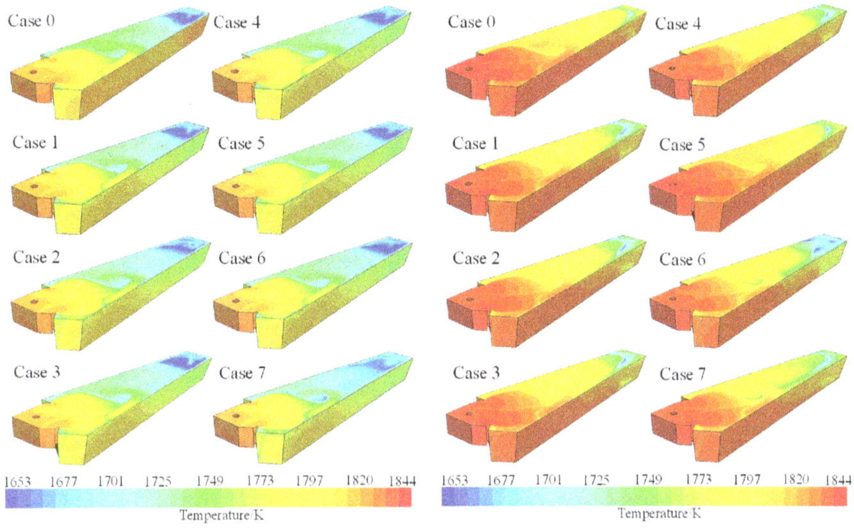

(**a**) (**b**)

Figure 15. Tundish temperature distribution. (**a**) Casting for 2 min; (**b**) casting for 30 min.

When the higher-temperature new heat is cast into the tundish containing the lower-temperature present heat, the outlet temperature variation curves of the tundish for each case are shown in Figure 16. As shown in Figure 16, in the normal symmetric Case 0, the time at which the temperature starts to rise gradually increases from strand 1 to strand 6. The temperature rise at strand 1 is the most significant, while the temperature curves for strands 2–4 are nearly identical. The temperature rise at strands 5 and 6 is relatively slow. The trends of temperature variation in fewer strand casting cases (Cases 1–7) are generally consistent with Case 0, except that the temperature rise at strand 6 in Case 5 is significantly lower than in other cases.

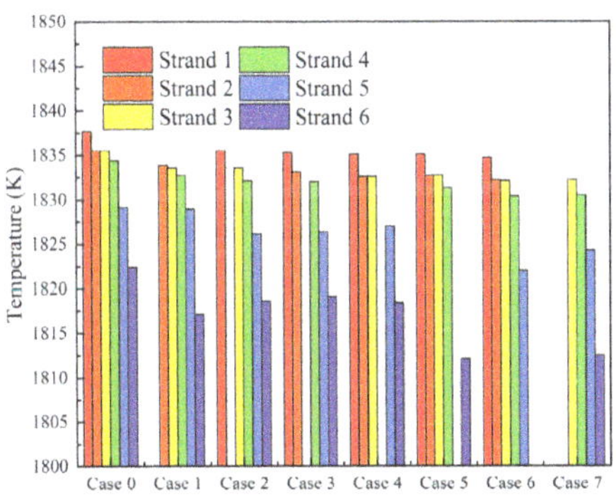

Figure 16. Temperatures at different strands for each case at 1800 s of casting.

As shown in Figure 16, at 1800 s of casting in Case 0, the temperatures at strands 1–6 are 1837.7 K, 1835.5 K, 1835.5 K, 1834.4 K, 1829.1 K, and 1822.4 K, respectively, with a temperature difference of 15.3 K between strand 1 and strand 6. In the asymmetric Case 1, the temperatures at strands 2–4 are 1833.9 K, 1833.6 K, and 1832.7 K, respectively, which are 1.6–1.9 K lower than those in the normal symmetric Case 0. The temperature at strand 5 is 1828.9 K, nearly identical to the normal symmetric Case 0, while the temperature at strand 6 is 1817.1 K, showing a significant decrease of 5.3 K compared to the normal symmetric Case 0. In the asymmetric Cases 2–4, the temperatures at all strands are lower than in the normal symmetric Case 0, with strands 1–5 decreasing by 2–3 K and strand 6 decreasing by 3–4 K. In the asymmetric Cases 5 and 6, the temperatures at strands 1–4 are 2–4 K lower than in the normal symmetric Case 0, while the temperatures at strand 6 in the asymmetric Case 5 and strand 5 in the asymmetric Case 6 decrease by 10.3 K and 7.1 K, respectively. Closing strands 5 and 6 significantly reduces their corresponding temperatures. In the asymmetric Case 7, the temperatures at strands 3–4 decrease by 3–5 K compared to the normal symmetric Case 0, and the temperature at strand 6 decreases by 9.9 K.

Under normal symmetric casting conditions (Case 0), the molten steel temperature shows a decreasing trend from strand 1 to strand 6, indicating that the high-temperature new steel preferentially flows toward strand 1 near the inlet, while strand 6, located farther away, experiences a delayed temperature rise. This suggests significant flow non-uniformity within the tundish. Under asymmetric casting conditions (Cases 1–7), the temperature at each strand decreases compared to the normal casting case, with varying degrees of reduction. Overall, the temperature drop caused by fewer strand casting is closely related to the position of the strand.

3.4. Transition Billet Calculation Results

In the process of grade transition casting, the variation curves of the new heat concentration difference ratio over time for each case, along with the grade transition criteria (0.2–0.8) boundaries, are shown in Figure 17. Overall, the concentration difference ratio of the new heat over time exhibits a rapid change at strand 1, reaching the grade transition boundary first. The curves for strands 2 to 4 show a relatively consistent growth trend, whereas strands 5 and 6 display significant differences, with a noticeable delay in growth. Strand 6, in particular, shows a greater lag compared to strand 5.

The starting positions of the transition billets for different strands (with ω = 20%), representing the length conforming to the present heat, are shown in Figure 18, with detailed data in Table 4. Overall, whether in the normal symmetric state or the asymmetric state, strands 1–4 exhibit shorter billet lengths conforming to the present heat, all less than 10 m. In contrast, strands 5 and 6 have significantly longer billet lengths conforming to the present heat, with strand 5 around 20 m and strand 6 exceeding 35 m. For strand 1, the normal symmetric Case 0 has the shortest billet length conforming to the present heat at 2.82 m, while the asymmetric Cases 2–6 show similar lengths of around 2.88–2.89 m. For strand 2, the billet lengths are relatively close across all cases, ranging from 2.71 m to 2.79 m. Across all cases, strands 1 and 2 exhibit nearly identical billet lengths, with strand 2 being slightly shorter. For strand 3, the billet length conforming to the present heat increases to around 4 m, with the asymmetric Case 2 being the shortest at 4.02 m. Strand 4 sees a further increase to approximately 6–7 m, with the normal symmetric Case 0 being the shortest at 5.29 m. Strand 5 shows a significant increase in billet length compared to strands 1–4, with the shortest being the asymmetric Case 1 (14.82 m) and the longest being the asymmetric Case 6 (21.89 m). Strand 6 has the longest billet length conforming to the present heat, with Case 0 at 41.69 m. Cases 2, 3, and 4 have noticeably shorter lengths at 37.57 m and 37.71 m. The asymmetric Case 5 exhibits a substantial increase compared to the normal symmetric Case 0, reaching 49.47 m. This indicates that during grade transition casting, closing strands 2–4 reduces the billet length conforming to the present heat for strand 6, whereas closing strand 5 significantly increases it.

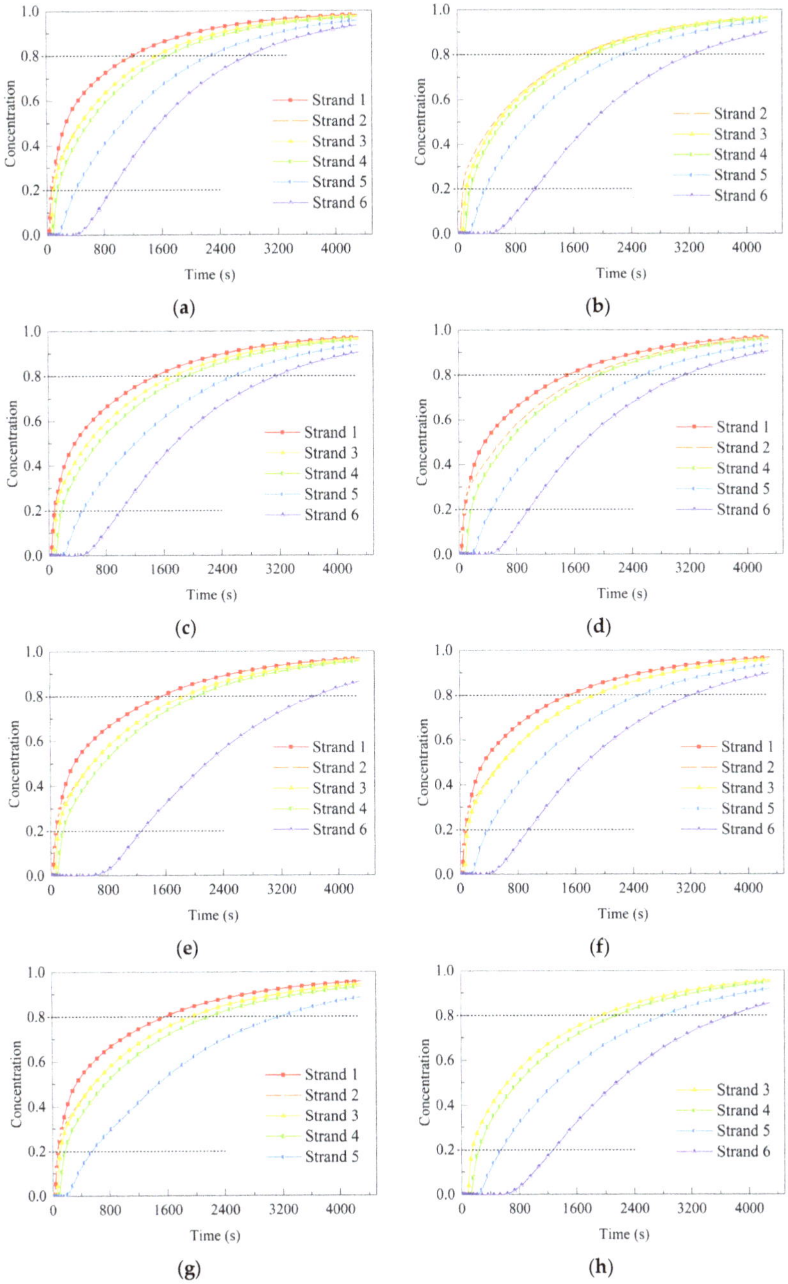

Figure 17. Concentration variation curves over time for each case: (**a**) Case 0; (**b**–**g**) Cases 1–6; (**h**) Case 7.

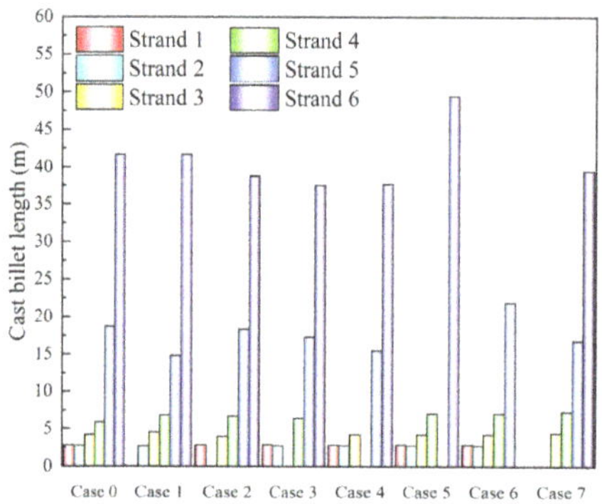

Figure 18. Transition billet starting positions for each case (ω = 20%).

Table 4. Transition billet starting positions (m) for each case.

Case	Strand 1	Strand 2	Strand 3	Strand 4	Strand 5	Strand 6
Case 0	2.82	2.79	4.25	5.92	18.71	41.69
Case 1		2.71	4.55	6.84	14.82	41.71
Case 2	2.88		4.02	6.73	18.36	38.79
Case 3	2.89	2.76		6.47	17.33	37.57
Case 4	2.89	2.78	4.26		15.48	37.71
Case 5	2.89	2.78	4.26	7.05		49.47
Case 6	2.89	2.77	4.26	7.07	21.89	
Case 7			4.44	7.29	16.82	39.49

In actual production, a steel plant designates the first 30 m of each strand's billet during grade transition casting as belonging to the present heat. However, based on the computational results, the billet lengths for strands 1–5 at the 20% concentration difference threshold are significantly shorter than 30 m, whereas strand 6 exceeds 30 m. This suggests that using a uniform 30 m criterion is unreasonable. From the analysis above, different criteria should be applied to different strands: 3 m for strands 1 and 2, 5 m for strand 3, 8 m for strand 4, 22 m for strand 5, and 50 m for strand 6. Additionally, adjustments should be made according to different low-flow operation cases.

The transition billet lengths for different strands in each case (ω = 20–80%) are shown in Figure 19, and the specific data are provided in Table 5. The average transition billet lengths for each case are shown in Figure 20. In the normal symmetric Case 0, the transition billet length for strand 1 is 52.13 m, and the lengths for strands 2, 3, and 4 are between 66–72 m, while the lengths for strands 5 and 6 are close to 88 m, with an average transition billet length of 72.41 m. Overall, the transition billets for strands 5 and 6 are too long. In the asymmetric Case 1, the transition billet lengths for strands 2, 3, and 4 range from 63 to 65 m, while the lengths for strands 5 and 6 are 74.16 m and 84.41 m, respectively, with an average transition billet length of 70.16 m. Compared to the normal symmetric Case 0, the transition billet length for strand 5 is reduced by 13.01 m. In the asymmetric Case 2, the transition billet length for strand 1 is 54.60 m, which is an increase of 2.47 m compared to the normal symmetric Case 0. The lengths for strands 3 and 4 are reduced by 2.96 m and 4.0 m, respectively. The length for strand 5 is lower than in the normal symmetric Case 0 but increased by 7.47 m compared to the asymmetric Case 1. The average transition billet length between the strands in the asymmetric Case 2 is 70.30 m, slightly higher than in the asymmetric Case 1. In the asymmetric Case 3, the transition billet length for strand 1 is 55.89 m, which is an increase of 3.76 m compared to the normal symmetric Case 0. In addition, the transition billet lengths for strands 2 to 6 are all smaller than in the normal symmetric Case 0, with reductions ranging from 0.56 m to 5.21 m. The average transition billet length for all strands in the asymmetric Case 3 is 71.68 m.

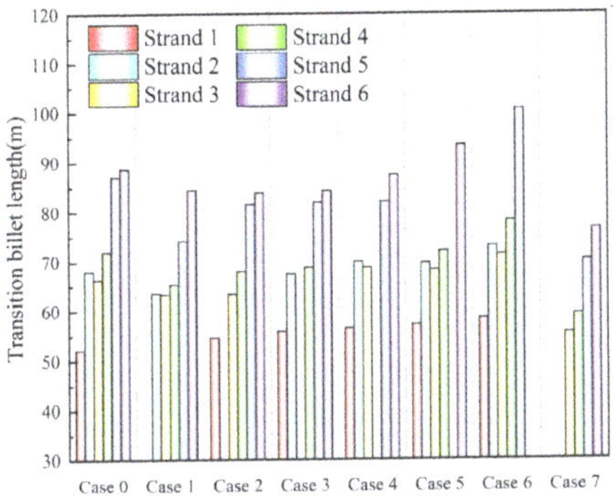

Figure 19. Transition billet lengths (m) for each case.

Table 5. Transition billet lengths and average lengths (m) for each case.

Case	Strand 1	Strand 2	Strand 3	Strand 4	Strand 5	Strand 6
Case 0	52.13	68.04	66.41	72.01	87.17	88.70
Case 1		63.48	63.23	65.50	74.16	84.41
Case 2	54.60		63.45	68.01	81.63	83.83
Case 3	55.89	67.48		68.82	81.96	84.23
Case 4	56.45	70.03	68.69		82.13	87.47
Case 5	57.25	69.64	68.23	72.09		93.37
Case 6	58.48	73.02	71.42	78.14	100.69	
Case 7			55.55	59.38	70.29	76.59

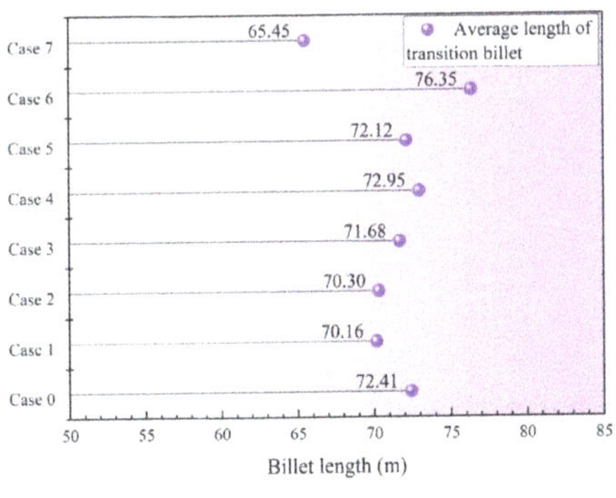

Figure 20. Average length of transition billet for each case.

The calculation results show that fewer strands by closing strands 1 to 3 help to some extent in reducing the transition billet length. This is because by closing the upstream strands (strands 1 to 3), the rapid outflow of new heat from these strands is avoided, allowing for the concentration of new heat to be transferred more effectively to the downstream strands (strands 4 to 6), thus increasing the growth of molten steel concentration at these strands.

In the asymmetric Case 4, the transition billet lengths for strands 1 to 3 are 56.45 m, 70.03 m, and 68.69 m, respectively, all of which are longer than in the normal symmetric Case 0, with increases ranging from 2 to 4 m. The transition billet lengths for strands 5 and 6 are 82.13 m and 87.47 m, respectively, which are shorter than in the normal symmetric Case 0, with reductions of 5.04 m and 1.23 m. The average transition billet length for all strands in the asymmetric Case 4 is 72.95 m. In the asymmetric Case 5, compared to the normal symmetric Case 0, the transition billet lengths for all strands are longer. Specifically, the lengths for strands 1 and 6 are 57.25 m and 93.37 m, respectively, which represent increases of 5.12 m and 4.67 m compared to the normal symmetric Case 0. The increase for strands 2 to 4 is smaller. The average transition billet length for all strands in the asymmetric Case 5 is 72.12 m. In the asymmetric Case 6, compared to the normal symmetric Case 0, the average transition billet lengths for all strands have significantly increased. The lengths for strands 1 to 4 are 58.48 m, 73.02 m, 71.42 m, and 78.14 m, respectively, with increases from 5 to 6 m compared to strand 1. Notably, the transition billet length for strand 5 is 100.69 m, which is an increase of 13.52 m compared to strand 1. The average transition billet length for all strands in the asymmetric Case 6 is 76.35 m. The fewer strands operation that closes strands 4 to 6 increases the transition billet lengths for other strands. This is because, compared to the inlet area, the downstream strands (strands 4 to 6) are farther away, and closing these strands reduces the overall molten steel flow velocity, increasing the slow-flow volume, as shown in Figure 19. Therefore, the case that closes strands 4 to 6 is not beneficial for reducing transition billet lengths, with the asymmetric Case 6 showing a significant increase in transition billet length.

In the asymmetric Case 7, the transition billet lengths for strands 2 to 6 are 55.55 m, 59.38 m, 70.29 m, and 76.59 m, respectively, all of which are reduced by 10–16 m compared to the normal symmetric Case 0. The average transition billet length for the asymmetric Case 7 is 64.45 m, which is significantly lower than that of the normal symmetric Case 0 and the single-strand closing the asymmetric Cases 1 to 6.

Regarding the endpoint of the transition billet, the company determines the endpoint of the transition billet when 100 t of new heat is casting, as shown in Figure 21, which depicts the concentration difference ratio at the tundish exit for different strands when casting 100 t in each case. Specific values are provided in Table 6. Overall, when casting 100 t, the concentration difference ratio for strands 1–6 shows a gradual decrease across all cases, and none of them reach the determination standard of 80%. In the normal symmetric Case 0, the concentration difference ratio for strand 1 is 79.16%, which is the closest to 80%; the concentration difference ratio for strands 2 and 3 is around 70%; while the concentration difference ratios for strands 5 and 6 are significantly lower, at 52.81% and 31.69%, respectively, clearly not meeting the 80% standard. In the asymmetric Case 1, the concentration difference ratios for strands 2–4 are all above 70%, and the ratio for strand 5 increases compared to the normal symmetric Case 0 to 61.3%, while strand 6 is at 31.7%. In the normal symmetric Case 2, the concentration difference ratios for all strands are basically the same as in the normal symmetric Case 0, with the ratio for strand 6 increasing to 35.44%, but still far from the 80% standard. In the asymmetric Case 3, the concentration difference ratio for strand 1 is 77.56%, slightly lower than in the normal symmetric Case 0, but the ratios for the other strands have increased, with the ratio for strand 6 reaching 36.77%. In the asymmetric Case 4, the concentration difference ratios for strands 1–3 are 77.43%, 71.35%, and 71.28%, slightly lower than in the normal symmetric Case 0, while the ratios for strands 5 and 6 are 57.81% and 35.87%, respectively, showing increases of 5% and 4.18% compared to the normal symmetric Case 0. The operation of closing strands 1–4 in the asymmetric Cases 1–4 can promote the concentration difference ratios for the downstream strands 5 and 6, increasing their concentration difference ratios. In the asymmetric Cases 5 and 6, compared to the normal symmetric Case 0, the concentration difference ratios for all strands have decreased, especially for strand 6 in the asymmetric Case 5, which is at 23.17%, and for strand 6 in the asymmetric Case 6, which is at 46.17%. This significant decrease indicates that closing the downstream strands 5 and 6 has a negative impact on the molten steel flow.

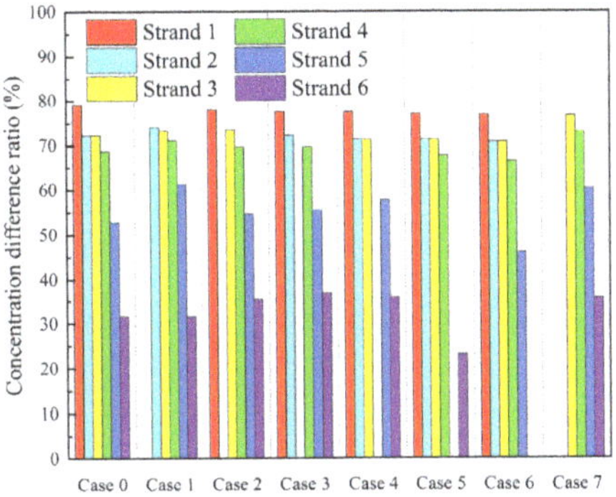

Figure 21. Concentration difference ratio for casting 100 t.

Table 6. Concentration for each strand when casting 100 t.

Case	Strand 1	Strand 2	Strand 3	Strand 4	Strand 5	Strand 6
Case 0	79.16%	72.25%	72.29%	68.72%	52.81%	31.69%
Case 1		74.13%	73.35%	71.10%	61.30%	31.70%
Case 2	78.03%		73.46%	69.75%	54.65%	35.44%
Case 3	77.56%	72.28%		69.69%	55.54%	36.77%
Case 4	77.43%	71.35%	71.28%		57.81%	35.87%
Case 5	77.04%	71.29%	71.26%	67.66%		23.17%
Case 6	76.79%	70.73%	70.75%	66.42%	46.17%	
Case 7			76.59%	73.11%	60.31%	35.68%

In the asymmetric Case 7, the concentration difference ratios for strands 4–6 are 76.59%, 73.11%, 60.31%, and 35.68%, respectively, which are all higher compared to other cases, but still fall short of 80%.

When the concentration difference ratio for each strand reaches 80%, the amount of steel casting is shown in Figure 22, with specific values provided in Table 7. In the normal symmetric Case 0, strand 1 reaches an 80% concentration difference ratio when casting 104.152 t of new heat, which marks the endpoint of mixed pouring for grade transition casting. Strands 2–4 require casting 134.244 t, 133.925 t, and 147.715 t, respectively, to reach the 80% concentration difference ratio. Given that the steel ladle capacity is 150 t, strands 2–4 can only meet the standard for the new steel grade after casting one full ladle of steel. For strands 5 and 6, 200.698 t and 247.135 t of new heat are required, far exceeding the capacity of a single ladle.

In the asymmetric Case 1, strands 2–4 reach the next steel grade standard when casting 125.454 t, 128.477 t, and 137.101 t, respectively, while strands 5 and 6 need 168.648 t and 239.042 t of new heat. Compared to the normal symmetric Case 0, the required casting weight for each strand in the asymmetric Case 1 is reduced, with strand 5 showing a significant reduction of 32.05 t. In the asymmetric Case 2, strand 1 requires casting 108.951 t of new heat, which is an increase of 4.499 t compared to strand 1 in the normal symmetric Case 0. Strands 3–6 require casting 127.887 t, 141.671 t, 189.508 t, and 232.408 t, respectively. Compared to the normal symmetric Case 0, the casting weight for strands 3–6 is reduced by 6–14 t. In the asymmetric Case 3, compared to the normal symmetric Case 0, the casting weight for all strands except strand 1 is reduced, with strands 5 and 6 showing significant reductions of 15.516 t and 16.275 t, respectively. In the asymmetric Case 4, the required casting weights for strands 1–3 are 112.468 t, 137.987 t, and 138.282 t, which are increases of 3–8 t compared to the normal symmetric Case 0. The required casting weights for strands 5 and 6 are 185.014 t and 237.274 t, which are reductions of 15.684 t and 9.861 t, respectively, compared to the normal symmetric Case 0. In the asymmetric Cases 5 and 6, compared to the normal symmetric Case 0, the required casting weight for all strands increases. In the asymmetric Case 5, strands 1–4 require 2–10 t more, and strand 6 requires 270.739 t, which is an increase of 23.6 t. In Case 6, strands 1–4 require 9–14 t more, and strand 5 requires 232.334 t, an increase of approximately 31.6 t.

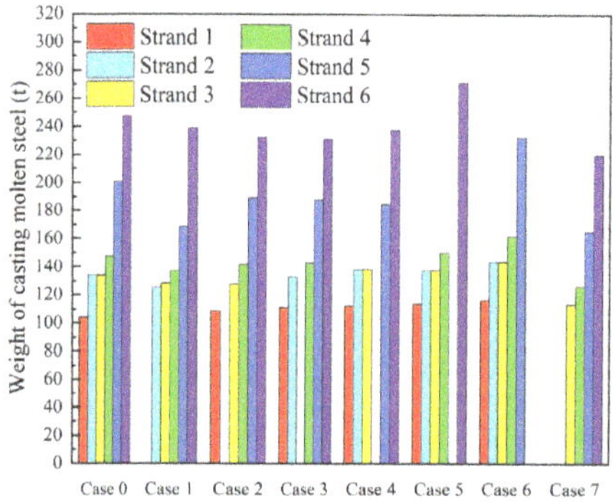

Figure 22. Weight of casting molten steel when the concentration difference ratio reaches 80%.

Table 7. Weight of casting molten steel when the concentration difference ratio reaches 80%.

Case	Strand 1	Strand 2	Strand 3	Strand 4	Strand 5	Strand 6
Normal	104.152	134.244	133.925	147.715	200.698	247.135
Closed strand 1		125.454	128.477	137.101	168.648	239.042
Closed strand 2	108.951		127.887	141.671	189.508	232.408
Closed strand 3	111.405	133.120		142.703	188.182	230.860
Closed strand 4	112.468	137.987	138.282		185.014	237.274
Closed strand 5	113.994	137.250	137.397	150.001		270.739
Closed strand 6	116.314	143.661	143.440	161.499	232.334	
Closed strand 1, 2			113.690	126.368	165.110	220.010

In Case 7, compared to the normal symmetric Case 0 and the single-strand closing cases (Cases 1–6), the required casting weight for all strands significantly decreases. For strands 3–6, the required casting weights are 113.690 t, 126.368 t, 165.110 t, and 220.010 t, respectively. Compared to the normal symmetric Case 0, the required casting weights for strands 3–6 are reduced by 20.235 t, 21.347 t, 35.588 t, and 27.125 t, respectively.

Overall, the weight of casting molten steel required to reach a concentration difference ratio of 80% for different strands in each case exceeds 100 t, and even surpasses 200 t (for strands 5 and 6). Therefore, using 100 t of casting as the endpoint criterion for the grade transition casting process is insufficient. A new determination basis should be established for each strand. According to calculations, in the asymmetric Cases 1–3, strand 1 can meet the new steel grade requirements with around 110 t of casting, strand 2 and strand 3 can meet the new steel grade requirements with 125 t–135 t, and strand 4 can meet the new steel grade requirements with 137 t–147 t. However, strands 5 and 6 cannot meet the new steel grade determination standard within the 150 t capacity of a single ladle. Strands 5 and 6 will need 168 t–200 t and 230 t–250 t of casting molten steel, respectively, to meet the new steel grade requirements. At the same time, in the low-flow operation process, it is advisable to avoid using the asymmetric Cases 5 and 6, which close strands 5 and 6.

For low-flow operations, a single-strand closure can involve strand 1 or strand 2. In the asymmetric Case 7, closing two strands (strands 1 and 2) can significantly reduce the length of the mixed billet during the grade transition casting process and reduce the amount of new steel required for the new steel grade conversion in the next steel pour.

4. Conclusions

(1) In Cases 1–6, the volume of the slow-flow region in the tundish gradually increases from 1.135 m^3 to 1.731 m^3. In Case 7, the slow-flow region volume is 1.398 m^3. The fewer strand casting operations disrupt the symmetric flow pattern observed in the baseline Case 0, thereby enlarging the slow-flow region in the tundish.

(2) Cases 5 and 6 have minimal impact on the minimum response time, peak concentration time, and peak concentration at strands 1–4. However, they exert significantly negative effects on strands 5 and 6, indicating impaired consistency in steel flow.

(3) At 2 min, both symmetric and asymmetric cases show a sharp surface temperature drop near strands 5 and 6. By 30 min, the tundish temperature rises from the far end to the impact zone (1773–1844 K), aligning with the new steel temperature at the impact zone. In asymmetric Case 6, low-temperature zones near strands 5 and 6 remain. Fewer strand operations disrupt the temperature seen in symmetric Case 0, with strand 6 in Case 5 showing a notably slower temperature rise.

(4) In Cases 1–3, the closure of near strands (1–3) prevents the premature outflow of new steel, allowing its concentration to be more effectively transferred to the far strands (4–6), thereby enhancing the new steel concentration at those locations. As a result, these cases contribute to reducing the transition billet length to some extent. However, they also significantly enlarge the slow-flow region, which in turn prolongs the transition billet.

(5) Considering flow behavior, temperature distribution, and transition billet length comprehensively, Case 1 is recommended under fewer strand casting conditions. In contrast, Cases 5 and 6 should be avoided due to their adverse impacts.

Author Contributions: Conceptualization, C.C. and W.L.; methodology, C.C., Z.R. and J.S.; software, Z.R., J.S. and T.W.; validation, H.W. and M.G.; formal analysis, Z.R., J.S., Z.G. and H.W.; investigation, Z.R., J.S. and C.C.; resources, J.W. and Y.S.; data curation, Z.R., J.S., Z.G. and M.G.; writing—original draft preparation, Z.R., J.S. and T.W.; writing—review and editing, C.C., J.W. and Y.S.; visualization, M.G., T.W., H.W. and Z.G.; supervision, W.L. and J.W.; project administration, W.L. and Y.S.; funding acquisition, C.C. All authors have read and agreed to the published version of the manuscript.

Funding: The authors are grateful to the financial support of the Applied Fundamental Research Programs of Shanxi Province (202403021222046), and Graduate Innovation Project of Shanxi Province (2024KY277).

Data Availability Statement: Data are contained within the article.

Conflicts of Interest: The authors declare no conflicts of interest.

References

1. Heaslip, L.J.; Mclean, A. Tundish metallurgy-considerations pertaining to tundish performance and metallurgical treatment during continuous casting. In Proceedings of the 39th Electric Furnace Conference, Houston, TX, USA, 8–11 December 1981.
2. Szekely, J.; Ilegbusi, O.J. *The Physical and Mathematical Modeling of Tundish Operations*; Springer: New York, NY, USA, 1989; pp. 34–52.
3. Sahai, Y. *Tundish Technology for Clean Steel Production*; World Scientific Publisher: Singapore, 2007; pp. 8–12.
4. Geng, M.J.; Wang, T.Y.; Chen, C. Assessment of the volume effect and application of an improved tracer in physical model of a single-strand bare tundish. *Metall. Mater. Trans. B* **2024**, *55*, 4121–4131. [CrossRef]

5. Ding, C.Y.; Lei, H.; Zhang, H.; Xiao, Y.Y.; Zhao, Y.; Zou, Z.S. New understanding on relationship between rtd curve and inclusion behavior in the tundish. *Metall. Mater. Trans. B* **2024**, *55*, 2224–2239. [CrossRef]
6. Lv, A.; Ding, G.Q.; Luo, X.D. Influence of elevation angle of tundish filter on removal rate of impurity in molten steel. *Sci. Rep.* **2025**, *15*, 782. [CrossRef] [PubMed]
7. Chen, C.; Cheng, G.G.; Yang, H.K.; Hou, Z.B. Physical modeling of fluid flow characteristics in a delta shaped, four-strand continuous casting tundish with different flow control devices. *Adv. Mater. Res.* **2011**, *284*, 1071–1079. [CrossRef]
8. Li, Q.H.; Qin, B.M.; Zhang, J.S.; Dong, H.B.; Li, M.; Tao, B.; Mao, X.P.; Liu, Q. Design improvement of four-strand continuous-casting tundish using physical and numerical simulation. *Materials* **2023**, *16*, 849. [CrossRef] [PubMed]
9. Wang, J.H.; Fang, Q.; Huang, L.; Zhao, P.; Xie, X.Q.; Zhang, H. Three-phase flow in a single-strand tundish during start-up operation using top-swirling turbulence inhibitor. *Metall. Mater. Trans. B* **2023**, *54*, 635–649. [CrossRef]
10. Chen, X.Q.; Bai, Y.; Feng, W.W.; Xiao, H.; Tang, H.Y.; Wang, P.; Zhang, J.Q. Structural optimisation of three-strand asymmetric tundish for super large round bloom continuous casting. *Ironmak. Steelmak.* **2024**, *51*, 441–459. [CrossRef]
11. Qin, X.F.; Cheng, C.G.; Li, Y.; Wu, W.L.; Chen, H.; Zhao, C.F.; Jin, Y. Numerical simulation on gas–liquid multiphase flow behavior under coupling effects of annular gas curtain and swirling flow at tundish upper nozzle. *J. Iron Steel Res. Int.* **2024**, *31*, 2693–2709. [CrossRef]
12. Li, Y.; Li, Y.; Cheng, C.G.; Liu, J.G.; Wu, W.L.; Chen, D.L. Theory of movement behavior of argon bubbles in upper nozzle of tundish during argon blowing process. *Metall. Mater. Trans. B* **2023**, *54*, 3568–3578. [CrossRef]
13. Qin, B.M.; Zhang, J.S.; Yang, C.H.; Yang, S.F.; Liu, Q. Study on influencing factors and their combined effects on multiphase behavior in tundish pouring zone. *Metall. Mater. Trans. B* **2025**, *56*, 1176–1191. [CrossRef]
14. Wang, T.Y.; Wang, J.; Chen, C.; Chen, L.; Geng, M.J.; Song, J.T.; Fan, J.P.; Lin, W.M. Physical and numerical study on right side and front side gas blowing at walls in a single-strand tundish. *Steel Res. Int.* **2024**, *95*, 2400037. [CrossRef]
15. Chang, S.; Song, H.; Liu, Z.; Zou, Z.S.; Shao, L.; Li, B.K. Experimental study on bubble entrainment by a free fall jet in an argon-filled continuous casting tundish. *Steel Res. Int.* **2022**, *94*, 2200287. [CrossRef]
16. Zhang, J.S.; Qin, B.M.; Liu, Y.H.; Li, Q.H.; Zuo, X.T.; Wang, C.; Yang, S.F.; Liu, Q. Multiphase flow inside a four-strand continuous casting tundish using three types of ladle shrouds. *J. Iron Steel Res. Int.* **2023**, *30*, 1171–1181. [CrossRef]
17. Wu, D.; Liu, J.D.; Wu, D.L.; Zhou, P. A computational fluid dynamics and support vector regression combined method for predicting the flow performance of the five-strand tundish. *Steel Res. Int.* **2024**, *95*, 2400231. [CrossRef]
18. Merten, D.C.; Hütt, M.T.; Uygun, Y. Effect of slab width on choice of appropriate casting speed in steel production. *J. Iron Steel Res. Int.* **2022**, *29*, 71–79. [CrossRef]
19. Yue, Q.; Li, Y.; Wang, Z.M.; Sun, B.C.; Wang, X.Z. Analysis model of internal residence time distribution for fluid flow in a multi-strand continuous casting tundish. *J. Iron Steel Res. Int.* **2024**, *31*, 2186–2195. [CrossRef]
20. Li, Y.H.; Hu, X.; Wang, L.Y.; Zhang, H.X.; He, Y.B.; Zhang, W.J.; Hua, C.J. Numerical and experimental studies on the effects of molten steel viscosity on fluid flow, inclusion motion, and temperature distribution in a tundish. *Phys. Fluids* **2024**, *36*, 073102. [CrossRef]
21. Zhang, Z.X.; Qu, T.P.; Wang, D.Y.; Li, X.L.; Fan, L.; Zhou, X.Z. Non-uniform thermal transfer of molten steel and its effect on inclusion particles removal behavior in continuous casting tundish. *Metals* **2025**, *15*, 170. [CrossRef]
22. Yao, C.; Wang, M.; Zhu, H.H.; Xing, L.D.; Bao, Y.P. Mathematical study of realistic removal rates of non-metallic inclusions in continuous casting tundish using optimized criterion. *Metall. Mater. Trans. B* **2023**, *54*, 1144–1158. [CrossRef]
23. Quan, Q.; Zhang, Z.X.; Qu, T.P.; Li, X.L.; Tian, J.; Wang, D.Y. Physical and numerical investigation on fluid flow and inclusion removal behavior in a single-strand tundish. *J. Iron Steel Res. Int.* **2023**, *30*, 1182–1198. [CrossRef]
24. Sun, Y.P.; Duan, H.J.; Zhang, L.F. A boundary layer model for capture of inclusions by steel–slag interface in a turbulent flow. *J. Iron Steel Res. Int.* **2023**, *30*, 1101–1108. [CrossRef]
25. Xie, Q.H.; Ni, P.Y.; Li, Y. A model study on collision-coalescence, transport, and removal behavior of non-metallic inclusions in a single-strand tundish. *Metall. Mater. Trans. B* **2024**, *55*, 2783–2804. [CrossRef]
26. Liu, Z.Q.; Yao, Y.C.; Wang, N.; Yang, J.; Xu, G.D.; Li, B.K. Physical and numerical simulations on the unsteady three-phase flow in a continuous casting tundish during ladle changeover process. *Metall. Mater. Trans. B* **2025**, *56*, 1982–1993. [CrossRef]
27. Song, J.T.; Chen, C.; Guo, Z.J.; Geng, M.J.; Rong, Z.R.; Wang, T.Y.; Wang, J.; E, D.Y.; Fan, J.P.; Sun, Y.H. Analysis of flow field, temperature, and inclusion evolution in a 12-strand tundish during ladle changeover process. *Metall. Mater. Trans. B* **2025**, *56*, 2700–2714. [CrossRef]
28. Chen, H.L.; Liu, Z.T.; Li, F.C.; Lyu, B.; Chen, W.; Zhang, L.F. Numerical simulation on multiphase flow and slag entrainment during casting start of a slab continuous casting tundish. *Metall. Mater. Trans. B* **2023**, *54*, 2048–2065. [CrossRef]
29. Wang, J.C.; Liu, Z.T.; Chen, W.; Chen, H.L.; Zhang, L.F. Numerical simulation on the multiphase flow and reoxidation of the molten steel in a two-strand tundish during ladle change. *Int. J. Miner. Metall. Mater.* **2024**, *31*, 1540–1553. [CrossRef]
30. Zhao, S.; Lu, Z.Y.; Zhang, Z.; Ma, Z.Y.; Cao, Y.B.; Tang, H.Y. Formation factors of confluence vortex in double nozzle ladle at end of teeming. *Iron Steel* **2024**, *59*, 71–79. [CrossRef]

31. Zhang, J.S.; Yang, S.F.; Chen, Y.F.; Chen, Z.X.; Zhao, J.W.; Li, J.S.; Jiang, Z.Y. Comparison of multiphase flow in a continuous casting tundish using two types of industrialized ladle shrouds. *J. Miner. Met. Mater. Soc.* **2018**, *70*, 2886–2892. [CrossRef]
32. Liu, J.; Zhou, P.; Zuo, X.; Wu, D.; Wu, D. Optimization of the liquid steel flow behavior in the tundish through water model experiment, numerical simulation and industrial trial. *Metals* **2022**, *12*, 1480. [CrossRef]
33. Wu, D.; Wu, D.L.; Liu, B.; Zhou, P.; Liu, J.D. A method of reducing trace deviation of the flow field in the model experiment: An example of a tundish. *Metall. Mater. Trans. B* **2024**, *55*, 2515–2524. [CrossRef]
34. Wang, J.; Liu, W.; Yang, S.; Zuo, X.T.; Zhao, L.; Li, J.S. Effect of tundish impact zone optimization on inclusion removal in steel: Industrial and simulation studies. *Metall. Mater. Trans. B* **2024**, *55*, 808–820. [CrossRef]
35. Wang, X.; Wang, S.; Hu, H.; Xie, X.; Wu, C.; Chen, D.; Long, M. Flow behavior of liquid steel in fewer strands casting of six-strand bloom tundish. *Metals* **2023**, *13*, 706. [CrossRef]
36. Diener, A.; Görl, E.; Pluschkell, W.; Sardemann, K. Vermischungsvorgänge beim stranggießen. *Steel Res.* **1990**, *61*, 449–454. [CrossRef]
37. Bruch, C.; Valentin, P. Mathematical modelling of fluid flow in a continuous casting tundish with regard to extraordinary casting conditions. *Steel Res. Int.* **2004**, *75*, 659–665. [CrossRef]
38. Merder, T. Numerical investigation of the hydrodynamic conditions in a multi-strand CC tundish with closed outlets. *Arch. Metall. Mater.* **2014**, *59*, 887–892. [CrossRef]
39. Braun, A.; Warzecha, M.; Pfeifer, H. Numerical and physical modeling of steel flow in a two-strand tundish for different casting conditions. *Metall. Mater. Trans. B* **2010**, *41*, 549–559. [CrossRef]
40. Mishra, S.K.; Jha, P.K.; Sharma, S.C.; Ajmani, S.K. Effect of blockage of outlet nozzle on fluid flow and heat transfer in continuously cast multistrand billet caster tundish. *Can. Metall. Q.* **2012**, *51*, 170–183. [CrossRef]
41. Zhang, L.F. Fluid flow, heat transfer and inclusion motion in a four-strand billet continuous casting tundish. *Steel Res. Int.* **2005**, *76*, 784–796. [CrossRef]
42. Zhao, P.; Zhang, H.; Fang, Q.; Wang, J.H.; Wu, G.L.; Ni, H.W. Numerical study on strand-blocking operation of a six-strand square billet tundish. *J. Iron Steel Res. Int.* **2022**, *34*, 438–450. [CrossRef]
43. Zhong, L.C.; Wang, M.A.; Chen, B.Y.; Wang, C.R.; Zhu, Y.X. Flow control in six-strand billet continuous casting tundish with different configurations. *J. Iron Steel Res. Int.* **2010**, *17*, 7–12. [CrossRef]
44. Yao, C.; Wang, M.; Pan, M.; Bao, Y.P. Optimization of large capacity six-strand tundish with flow channel for adapting situation of fewer strands casting. *J. Iron Steel Res. Int.* **2021**, *28*, 1114–1124. [CrossRef]
45. Song, J.T.; Luo, Y.Z.; Li, Y.Q.; Guo, Z.J.; Wang, T.Y.; Geng, M.J.; Lin, W.M.; Fan, J.P.; Chen, C. Comparison of fluid flow and tracer dispersion in four-strand tundish under fewer strand casting and sudden blockage of strand conditions. *Metals* **2024**, *14*, 571. [CrossRef]
46. Fan, J.P.; Li, Y.Q.; Chen, C.; Ouyang, X.; Wang, T.Y.; Lin, W.M. Effect of uniform and non-uniform increasing casting flow rate on dispersion and outflow percentage of tracers in four strand tundishes under strand blockage conditions. *Metals* **2022**, *12*, 1016. [CrossRef]
47. Tsai, M.C.; Green, M.J. A three-dimensional concurrent numerical simulation of molten steel behavior and chemical transition at inland steel's no. 2 caster tundish. In Proceedings of the Steelmaking Conference Proceedings, Paris, French, 4–5 December 1991.
48. Burns, M.; Schade, J.; Brown, W.A.; Minor, K. Transition model for armco steel's ashland slab caster. *Iron Steelmak.* **1992**, *19*, 35–39. Available online: https://api.semanticscholar.org/CorpusID:115002623 (accessed on 10 April 2025).
49. Chen, H.S.; Pehlke, R.D. Mathematical modeling of tundish operation and flow control to reduce transition slabs. *Metall. Mater. Trans. B* **1996**, *27*, 745–756. [CrossRef]
50. Piccone, T.J.; Sinha, A.K.; Madden, M. Intermix trials at Gary No. 2 Caster using different tundish configurations. In Proceedings of the 84th Steelmaking Conference, Baltimore, MD, USA, 25–28 March 2001.
51. Kant, S.; Jha, P.K.; Kumar, P. Investigation of effect of dam on intermixing during ladle changeover in six strand billet caster tundish. *Ironmak. Steelmak.* **2011**, *38*, 391–397. [CrossRef]
52. Guarneros, J.D.; Morales, R.; Gutierrez, E. Optimized fluid flow control system for a tundish used in frequent steel grade change operations. *Steel Res. Int.* **2023**, *94*, 2200809. [CrossRef]
53. Chakraborty, S.; Hirose, T.; Jones, B. Transition reduction and simultaneous maintenance of steel quality in Granite City tundishes. In Proceedings of the 82nd Steelmaking Conference, Chicago, IL, USA, 21–24 March 1999.
54. Xu, D.; Rogler, J.P.; Heaslip, L.J.; Dorricott, J.D.; Foss, R. Dynamic flow behavior in the tundish: Optimization of grade transitions. In Proceedings of the 60th Electric Furnace Conference, San Antonio, TX, USA, 10–13 November 2002.
55. Amorim, L.C.; Silva, C.A.; Resende, A.D.; Silva, I.A.; Oliveira, M.J.M. A study of intermix in a six-strand billet caster. *Metall. Mater. Trans. A* **2018**, *49*, 6308–6324. [CrossRef]
56. Cwudziński, A.; Pieprzyca, J.; Merder, T. Numerical and physical modeling of liquid steel asymmetric behavior during non-isothermal conditions in a two-strand slab tundish-"Butterfly Effect". *Materials* **2023**, *16*, 6920. [CrossRef]

57. Song, S.; Sun, Y.; Chen, C. Numerical simulation of macro-segregation phenomena in transition blooms with various carbon contents. *Metals* **2024**, *14*, 263. [CrossRef]
58. Song, S.C.; Sun, Y.H.; Zhou, W.H.; Yang, J.; Yang, W.Z. The effect of steel grade casting sequence on the length of transition bloom. *Metall. Mater. Trans. B* **2024**, *55*, 1795–1811. [CrossRef]
59. Patankar, S.V. *Numerical Heat Transfer and Fluid Flow*; Hemispere Publishing Corporation: New York, NY, USA, 1980; p. 16.
60. Wang, J.H.; Ni, P.Y.; Chen, C.; Ersson, M.; Li, Y. Effect of gas blowing nozzle angle on multiphase flow and mass transfer during RH refining process. *Int. J. Min. Met. Mater.* **2023**, *30*, 844–856. [CrossRef]
61. Ouyang, X.; Lin, W.M.; Luo, Y.Z.; Zhang, Y.X.; Fan, J.P.; Chen, C.; Cheng, G.G. Effect of salt tracer dosages on the mixing process in water model of single snorkel refining furnace. *Metals* **2022**, *12*, 1948. [CrossRef]
62. Zhang, J.S.; Liu, Y.H.; Qin, B.M.; Li, D.Y.; Yang, S.F.; Liu, Q. A structured mesh generation tool for cfd simulations in process metallurgy. *Metall. Mater. Trans. B* **2023**, *54*, 481–486. [CrossRef]
63. Siemens, P.L.M. *STAR-CCM + User Guide*, version 16.06; Siemens PLM Software Inc.: Munich, Germany, 2021.
64. He, F.; Wang, H.J.; Zhu, Z.H. Numerical investigation of effect of casting speed on flow characteristics of molten steel in multistrand tundish. *Iron Steel Inst. Jpn. Int.* **2019**, *59*, 1250–1258. [CrossRef]
65. Zhang, J.S.; Li, J.S.; Yang, S.F. Optimization of fluid flow in a twelve-strand continuous casting tundish with two strands closed. *Metal. Int.* **2014**, *19*, 10–13.

symmetry

MDPI

Article

Assessment of the Measured Mixing Time in a Water Model of Asymmetrical Gas-Stirred Ladle with a Low Gas Flowrate Part II: Effect of the Salt Solution Tracer Volume and Concentration

Yansong Zhao [1], Xin Tao [1], Linbo Li [1], Zhijie Guo [2], Hongyu Qi [1], Jia Wang [1,3], Kun Yang [4], Wanming Lin [1], Jinping Fan [1,*] and Chao Chen [1,*]

1 College of Materials Science and Engineering, Taiyuan University of Technology, Taiyuan 030024, China; 2023520441@link.tyut.edu.cn (Y.Z.); taoxin0290@link.tyut.edu.cn (X.T.); lilinbo0316@link.tyut.edu.cn (L.L.); hyqi0414@163.com (H.Q.); wangjia01@tyut.edu.cn (J.W.); linwanming@tyut.edu.cn (W.L.)

2 Collaborative Innovation Center of Steel Technology, University of Science and Technology Beijing, Beijing 100083, China; d202210640@xs.ustb.edu.cn

3 College of Architecture and Arts, Taiyuan University of Technology, Taiyuan 030024, China

4 College of Mechanics, Taiyuan University of Technology, Taiyuan 030024, China; yangkun@tyut.edu.cn

* Correspondence: fanjinping@tyut.edu.cn (J.F.); chenchao@tyut.edu.cn (C.C.)

Abstract: Mixing time, as a key parameter for evaluating ladle refining efficiency, has long attracted extensive attention from researchers. In typical experimental studies, salt solution tracers are introduced into ladle water models to assess the degree of mixing within the ladle. Previous studies have demonstrated that the volume of tracer can significantly influence the measured mixing time. However, the gas flow rates employed in these studies are generally relatively high, whereas, in industrial operations, especially during final composition adjustments, lower gas flow rates are often applied. To systematically investigate the effect of the salt solution tracer volume on the mixing efficiency in a ladle water model under asymmetrical gas stirring with a low gas flow rate, a 1:3-scaled water model was developed based on a 130-ton industrial ladle. The mixing behaviors corresponding to different tracer volumes were comprehensively analyzed. The results indicate that the relationship between tracer volume and mixing time is non-monotonic. As the tracer volume increases, the mixing time first decreases and then increases, reaching a minimum at 185 mL. When the tracer volume was small, the dimensionless concentration curves at Monitoring Point 4 exhibited two distinct patterns: A parabolic profile, which was when the tracer initially moved through the left and central regions and then slowly crossed the gas plume to reach the monitoring point. A sinusoidal profile, which was when the tracer predominantly circulated along the right side of the ladle. When the tracer volume exceeded 277 mL, the concentration curves at Monitoring Point 4 consistently exhibited a sinusoidal pattern. Compared with moderate gas flow conditions (8.3 L/min), the peak concentration at Monitoring Point 3 was significantly lower under a low gas flow (2.3 L/min), and the overall mixing time was longer, indicating reduced mixing efficiency. Based on the findings, a recommended tracer volume range of 185–277 mL is proposed for low gas flow conditions (2.3 L/min) to achieve accurate and efficient mixing time measurements with minimal disturbance to the flow field. It was also observed that when the tracer concentration was relatively low, the mixing behavior throughout the ladle became more uniform.

Keywords: ladle; water model; transport path; mixing time; salt solution tracer

1. Introduction

The ladle plays a crucial role in the secondary refining process of steelmaking [1–5]. Gas is injected into the molten pool through porous plugs installed at the ladle bottom, forming bubbles that rise through the melt. The circulating flow induced by bubble ascent provides several key metallurgical functions: (1) promoting homogenization of temperature and composition fields in the molten steel [6–13]; and (2) assisting in the removal of non-metallic inclusions through flotation and enhancing their removal efficiency [14–16]. Due to the difficulty of directly observing the internal flow characteristics of the ladle under industrial conditions, most scholars have employed physical modeling [17–24] and mathematical modeling [25–34] to investigate the flow homogenization [35–38] and inclusion behavior [39–42] inside the ladle. Due to the low stirring efficiency associated with central gas injection [43], asymmetric stirring is commonly employed in small-capacity ladles [44].

Among various performance indicators, the mixing time is a key parameter for evaluating ladle refining efficiency. Understanding the mixing time is essential for optimizing refining operations. It helps avoid insufficient stirring, which may lead to poor refining quality, and it also prevents over-stirring, which can accelerate refractory erosion. However, the measurement of mixing time is influenced by multiple factors. Classic studies have indicated that the injection location of the tracer and the position of the monitoring point significantly affect the measured mixing time [2,45,46]. Contrarily, Krishna et al. [47] demonstrated through optimized experimental design that, when properly measured, the mixing time is independent of tracer injection and sampling locations. Several researchers have examined the effects of the gas flow rate [48–50], slag layer properties [51–54], and the position and number of porous plugs [55–60] on the mixing and flow behavior in ladles. Notably, Renato et al. [61] found that lower gas flow rates can lead to shorter mixing times due to a more inclined primary recirculation loop, which enhances turbulent diffusion. In addition, the effect of slag on mixing time is dual in nature. Kim et al. [62] observed that the presence of a slag layer increases the mixing time, which becomes more pronounced with increasing slag thickness and viscosity. However, Han et al. [53] pointed out that, under extremely low gas flow rates, a low-density slag layer may actually reduce the mixing time.

Current methods for measuring the mixing time can be broadly categorized into two types: (1) invasive techniques, and (2) non-invasive techniques. Non-invasive methods include colorimetry [63,64] and the Planar Laser-Induced Fluorescence (PLIF) technique [65–67]. Although PLIF involves higher costs and greater operational complexity, it is increasingly favored by researchers due to its non-intrusive nature and its ability to capture instantaneous concentration fields without disturbing the flow. In contrast, invasive measurement techniques, such as conductivity probes [46] and pH probes [68], may introduce minor disturbances to the flow field. However, their lower cost and ability to provide high measurement accuracy at specific locations make them the most widely used tools in ladle water model experiments.

Existing studies show limitations in the selection of tracers. In most ladle water models, saturated NaCl or KCl solutions are used. These solutions have densities that are 16.6% and 19.9% higher than water, respectively, at room temperature. This density difference may influence the accuracy of mixing time measurements. Furthermore, the volume of tracer introduced could also impact the results. Table 1 summarizes the ratio of the tracer-to-total-water volume used in various studies [69–82]. A wide variation has been observed in this ratio, ranging from 0.038×10^{-3} to 4.42×10^{-3}, which has been shown to significantly affect experimental results [83–88]. Chen et al. [86] were among the first to identify that increasing the volume of salt solution tracer significantly reduces the measured mixing time. Gómez et al. [87] attributed this to the greater buoyancy of the salt solution tracer, noting that this effect becomes more pronounced under low gas flow conditions. Furthermore, Zhang et al. [88], using numerical simulations, confirmed that the mixing process of saturated KCl solution is faster than that of pure water tracers in water models. It is important to note that the gas flow rates used in the above experiments, when converted to industrial scale, correspond to high gas flow conditions. In practice, however, excessively high gas flow rates can lead to emulsification between slag and steel [89] and may erode refractory linings or increase operating costs. Therefore, lower gas flow rates are typically employed during the later stages of industrial ladle refining.

Table 1. Ratio of the tracer-to-total-water water model.

Researchers	Tracer			Tracer to Water Volume Ratio	Flow Rate (L/min)	Year
	Kind	Concentration	Dosages (mL)			
Pan et al. [69]	NaCl	-	15	0.038×10^{-3}	5/10/20/30/40	1997
Wen et al. [70]	KCl	Saturated	200	0.56×10^{-3}	2.67/7.33/8	2007
Ek et al. [71]	NaCl	-	400	4.42×10^{-3}	2.5/5/7	2010
Liu et al. [72]	NaCl	Saturated	50	1.09×10^{-3}	250	2019
Tan et al. [73]	NaCl	Saturated	150	0.96×10^{-3}	0.7/2.26/3.01/3.76	2020
Aguilar et al. [74]	KCl	3.35 mol/L	20	0.41×10^{-3}	4.9	2021
Conejo et al. [75]	KCl	Saturated	100/20/10	0.3×10^{-3}	5.15/8.83/18.4	2021
Shi et al. [76]	NaCl	Saturated	500	0.96×10^{-3}	500–700	2021
Ortega et al. [77]	KCl	Saturated	35	0.41×10^{-3}	800	2021
Cheng et al. [78]	KCl	Saturated	500	1.17×10^{-3}	5/10/20/30/40	2021
Wang et al. [79]	NaCl	Saturated	24	0.11×10^{-3}	73	2021
Wu et al. [80]	KCl	Saturated	100	0.267×10^{-3}	65/130/195	2022
Zhou et al. [32]	NaCl	Saturated	200	0.215×10^{-3}	5	2022
Li et al. [81]	KCl	Saturated	200	0.52×10^{-3}	0.6/1.1/1.5/2.5	2023
Shan et al. [82]	KCl	Saturated	100	0.27×10^{-3}	4.2–21	2023

In previous research conducted by our group, it was found that, under moderate gas flow conditions (Q = 8.3 L/min), the tracer volume mainly influences the mixing behavior in the lower region of the ladle [90]. However, under low gas flow conditions (Q = 2.3 L/min), tracer transport exhibits multiple possible paths, resulting in diverse concentration–time curves [91]. Accordingly, a 1:3-scaled water model was established in this study based on a 130-ton industrial ladle from a steel plant. Multiple conductivity probes were installed at various positions within the ladle to monitor spatial variations in conductivity. The objective was to investigate the influence of the salt solution tracer volume on mixing time under low bottom-blowing gas flow conditions, as well as to analyze the transport paths of tracers within the ladle corresponding to different injection volumes.

2. Method of Experiment

2.1. Principle of Experiment

To ensure the reliability of physical simulation results, the experimental water model must satisfy both geometric and dynamic similarity criteria.

2.1.1. Geometric Similarity

Based on a 130-ton industrial ladle prototype, a water model was constructed with a geometric similarity ratio of λ = 1:3. The model strictly maintained dimensionless geometric characteristics of the prototype, including the height-to-diameter ratio (H/D) and bottom-blowing porous plug position (0.2 R offset from center). The comparative geometric parameters between the prototype and model are presented in Table 2.

Table 2. Size parameters of the industrial prototype and water model.

Parameters	Prototype	Model
Diameter of Ladle Top (mm)	2925	975
Diameter of Ladle Bottom (mm)	2690	897
Ladle Height (mm)	3150	1050
Liquid Level (mm)	3000	1000
Gas Flow Rate (L/min)	200	2.3

2.1.2. Dynamic Similarity

Dynamic similarity was achieved through the dimensionless number G proposed by K. Krishnapisharody and G.A. Irons [92], which is expressed as follows:

$$G = \frac{\rho_{m1}^2 Q_m^2}{\rho_{m2}^2 H_m d_m^4} = \frac{\rho_{p1}^2 Q_p^2}{\rho_{p2}^2 H_p d_p^4}. \tag{1}$$

In the equation, ρ_{m1} and ρ_{m2} are the densities of gas and liquid used in the model, kg/m^3; ρ_{p1} and ρ_{p2} are the densities of Ar and steel used in the plant, kg/m^3; Q_p is the Ar flow rate used in the plant, m^3/h; and Q_m is the air flow rate used in the model, m^3/h.

During the actual steelmaking process, when gas enters the ladle, the gas flow rate will change due to variations in temperature and pressure. Therefore, it is necessary to correct Q_p as follows:

$$Q_p' = \frac{T_m}{T_P}\frac{P_p}{P_m} Q_p. \tag{2}$$

In the equation, T_p and T_m are the temperatures of the steel and room temperature, K; P_m and P_p are the standard atmospheric pressure and the gas pressure at the ladle outlet, Pa; and Q_p' is the Ar flow rate under standard conditions, m^3/h.

Numerical calculation demonstrates that the prototype's soft blowing process with 200 L/min argon flow corresponds to 2.3 L/min gas injection in the water model.

2.2. *Experimental Facility and Scheme*

The schematic diagram of the water model apparatus is shown in Figure 1. During experiments, the liquid level was maintained at 1000 mm with single-hole bottom gas injection through the right side of the ladle bottom (0.2 R offset from the center). To minimize the disturbance to the ladle flow field during tracer injection, a funnel was employed to buffer the addition of the tracer. Gas was continuously injected for 5 min to achieve stable flow field conditions. The stimulus-response method was employed to evaluate fluid mixing efficiency by measuring the mixing time at monitoring points. To minimize the disturbance of the flow field caused by the probes, the conductivity probes were installed along the sidewalls of the ladle model during the experiments [85]. Tracer was introduced vertically above the ladle center, with real-time monitoring of liquid conductivity variations using a DDSJ308A conductivity meter (±5% accuracy) and DJS—1D Platinum Black Electrodes. Electrical conductivity data were recorded and analyzed across multiple measurement points to determine the complete mixing status. Mixing time was defined as the duration required for dimensionless conductivity values to stabilize within ±5% of the equilibrium values. The experiment was repeated 10 times, and the final results were calculated as the average of repeated measurements.

To investigate the tracer dosage effects on mixing time under low gas flow conditions, six tracer volumes (92 mL, 185 mL, 277 mL, 370 mL, 463 mL, and 695 mL) were systematically tested. The volumetric ratios between tracer and water in the ladle are detailed in

Table 3. Comparative analysis of the mixing time across different tracer dosages provides insights into their quantitative relationship with mixing efficiency.

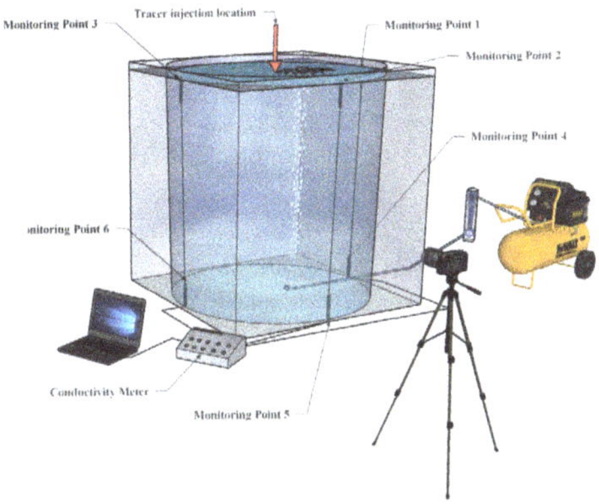

Figure 1. Schematic diagram of the experimental installation.

Table 3. The tracer-to-water-volume ratio.

Tracer Volume (mm)	92	185	277	370	463	695
Tracer-to-Water-Volume Ratio	0.13×10^{-3}	0.26×10^{-3}	0.39×10^{-3}	0.52×10^{-3}	0.65×10^{-3}	0.97×10^{-3}

3. Results and Discussion

3.1. Experimental Error Analysis

To ensure the reliability and repeatability of the experimental results, a preliminary error analysis was conducted prior to formal experimentation, focusing on three key factors that may affect measurement accuracy: (1) the impact of the probe volume on fluid flow behavior; (2) the influence of tracer addition on the overall fluid properties; and (3) the measurement errors associated with dimensionless concentration curves and mixing time.

In the experiments, Platinum Black electrodes were employed as conductivity probes. Each probe had a volume of 0.118 cm^3, and a total of six probes were installed, resulting in a combined volume of 0.708 cm^3. Given that the total volume of the ladle model was 0.53 m^3, the volume ratio of probes to the model was only 1.336×10^{-6}. Therefore, the effect of probe volume on the flow behavior in the ladle water model can be considered negligible.

A saturated NaCl solution was used as the tracer, with a maximum injection volume of 695 mL, which accounted for 0.131% of the total water volume (530 L), corresponding to a mass fraction of 0.047%. At 20 °C, a 7% NaCl solution has a dynamic viscosity of 1.079 mPa·s compared to 1.005 mPa·s for pure water. Additionally, a 1% NaCl solution has a density of 1005 kg/m^3, which is slightly higher than the density of water (998.2 kg/m^3) [93]. Therefore, in the context of this study, the introduction of the tracer had a minimal effect on the viscosity and density of the fluid, and its impact on flow behavior was deemed insignificant. The water in the ladle was replaced after each experiment to maintain consistency.

To evaluate the repeatability of the experiment, a representative case involving the addition of 185 mL of saturated NaCl solution was selected for ten repeated trials. The resulting dimensionless concentration curves were compared and analyzed. As shown in Figure 2, the maximum and minimum concentration curves were used to generate a shaded region over time [90], visually representing the range of variation. These figures indicate that the variation range of the repeated experimental data at these monitoring points was relatively small compared to the average curve. Table 4 summarizes the mixing times obtained from ten experimental repetitions. The average mixing time from the ten trials was adopted as the representative value for each monitoring point. The standard deviation between the individual mixing times and their average was also calculated. Results showed that the standard deviation at all monitoring points remained within 12%, indicating good repeatability and high reliability of the measured data.

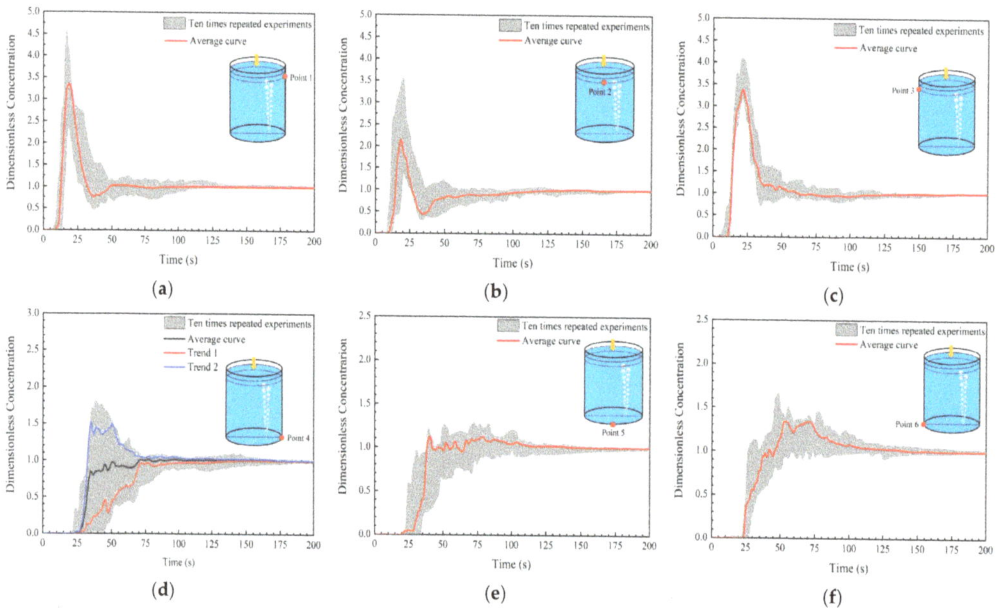

Figure 2. Dimensionless concentration curve summary chart: (**a**) Monitoring Point 1, (**b**) Monitoring Point 2, (**c**) Monitoring Point 3, (**d**) Monitoring Point 4, (**e**) Monitoring Point 5, and (**f**) Monitoring Point 6.

Table 4. Compare the mixing times of 10 trials and the average mixing time.

	1	2	3	4	5	6	7	8	9	10	AVG	S.D.
Monitoring Point 1	96.9	93.3	90.6	87.6	90.4	66.2	70.0	72.5	75.7	72.8	81.6	10.61
Monitoring Point 2	132.7	129.6	126.3	123.8	121.0	100.3	103.4	106.7	109.1	112.6	116.5	10.99
Monitoring Point 3	117.6	115.1	111.4	109.2	105.8	83.7	86.8	89.8	93.3	95.3	100.8	11.82
Monitoring Point 4	117.3	114.2	111.4	108.5	105.7	89.1	92.3	95.5	98.4	101.6	103.2	9.05
Monitoring Point 5	140.5	136.2	131.8	127.3	124.9	107.1	111.6	116.0	120.4	123.2	124	11.7
Monitoring Point 6	138.7	135.4	132.1	128.8	125.5	109.3	112.6	115.9	119.2	122.5	115.4	10.4

AVG: the average mixing time. S.D.: the standard deviation of the mixing time of ten trials.

In conclusion, the disturbance caused by the probe volume and tracer properties can be considered negligible. Furthermore, the dimensionless concentration curves and mixing time data demonstrated excellent repeatability. Thus, the experimental results were considered reliable and suitable for subsequent analysis.

3.2. Flow in Ladle Unit

The computational fluid dynamics (CFD) model results for the streamlines in the ladle are shown in Figure 3. In this study, the Euler–Euler multiphase flow model combined with the Realizable k-ε turbulence model was employed to solve the continuity and momentum equations, as well as to compute the steady-state flow field. The modeling strategy, mesh testing, simulation software (STAR-CCM+ 13.04), and validation followed the methodology used in our previous work [91]. Since these CFD results were not the primary focus of this study, a detailed simulation setup is not described here. From the simulation results, it can be observed that, after the gas entered through the bottom-blowing nozzle, a gas plume formed and rose toward the free surface. The flow field in the ladle can be divided into three regions: a large recirculating flow that forms on the left side of the ladle; a smaller recirculating flow that develops in the upper right region of the gas plume; and the left-side recirculating flow and the smaller right-side recirculating flow that meets in the lower-right region, merging into the gas plume, completing the circulation, and forming a fully developed flow field.

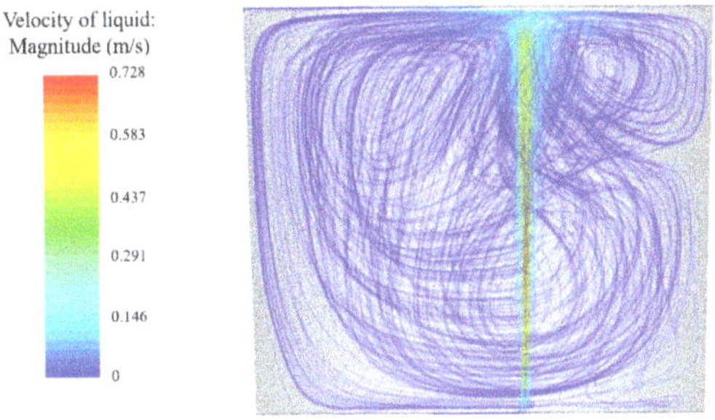

Figure 3. Prediction results of the streamline in a ladle water model.

3.3. Transport Path of Tracer

Figure 4 illustrates the movement of a 100 mL ink–salt solution mixed tracer in the ladle under a bottom-blowing gas flow rate of 2.3 L/min after being injected at the center of the ladle's liquid surface. At 1 s, the tracer penetrated the liquid surface and began convection transport downward. At 5 s, the tracer reached approximately one-third of the liquid depth (330 mm below the surface). By this time, the ink component had already spread across the surface, forming a black layer covering the entire top region. At 10 s, the ink started descending along the left-side wall, with significantly more ink observed on the left than on the right. At 15 s, the downward convection transport of the ink on both sides became nearly symmetrical. However, due to the gas injection port being located on the right side, the main recirculating flow occurred on the left, resulting in a higher ink concentration on the left side. Between 20 s and 30 s, the ink continued to convection transport downward at a synchronized and slow rate on both sides, indicating that, at 2.3 L/min, the gas flow had a significantly weaker influence on tracer movement compared to a higher gas flow rate of 8.3 L/min. At 35 s, the tracer began spreading from the bottom right toward the left, eventually dispersing throughout the ladle.

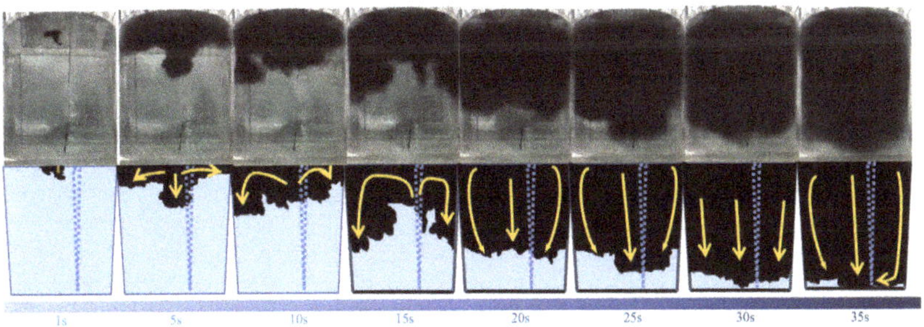

Figure 4. The tracer transport pathway diagram under soft stirring with a flow rate of 2.3 L/min is shown.

3.4. Tracer Transport Path and Analysis

Figure 5 presents the concentration–time curves at Monitoring Point 1 after adding tracer agents of different volumes into the water model of the ladle. Table 5 summarizes the key time points extracted from the dimensionless concentration curves. The tracer agent volumes were 92 mL, 185 mL, 277 mL, 370 mL, 463 mL, and 695 mL. All dimensionless concentration curves were obtained by averaging the results of ten experimental runs. After introducing 92 mL of saturated NaCl solution into the ladle, the dimensionless concentration at Monitoring Point 1 began to respond at 8 s, which was followed by a rapid increase. At 15 s, the concentration reached a peak value of 2.4, after which it rapidly decreased to 0.69 at 32 s, which represents the lowest concentration observed after the peak. Subsequently, the dimensionless concentration gradually rose and stabilized around 1. For the 185 mL tracer volume, the dimensionless concentration started to rise at 8 s and reached a maximum value of 3.37 at 18 s. Between 18 s and 36 s, the concentration decreased, reaching its lowest value of 0.75 at 36 s, after which it gradually increased and fluctuated around 1. When the tracer volume was 277 mL, the dimensionless concentration began to increase at 5 s, reaching its peak of 3.2 at 18 s. Subsequently, the concentration declined to 0.92, followed by a gradual rise after 35 s. A minor fluctuation was observed at 52 s, after which the curve stabilized around 1. For the 370 mL tracer volume, the dimensionless concentration started to rise after 8 s, reaching a peak value of 2.6 at 19 s. It then declined to 0.81, followed by a gradual increase after 43 s, and then eventually stabilizing within a $\pm 5\%$ standard range. After adding 463 mL of saturated NaCl solution, the concentration variation began at 8 s. The concentration continued to increase between 8 s and 19 s, reaching a peak of 2.76. From 19 s to 38 s, the concentration decreased to its lowest value of 0.75, then slowly rose and stabilized around 1. For the 695 mL tracer volume, the concentration started increasing at 9 s and reached a peak of 1.9 at 21 s. The concentration then declined to its lowest value of 0.64 at 39 s before gradually rising and stabilizing within the $\pm 5\%$ standard range. The initiation time of the concentration variation at Monitoring Point 1 remained around 8 s for all tracer volumes, indicating that changes in tracer volume do not significantly affect the response time at this location. However, the time required to reach the concentration peak varies with volume, recorded as 15 s, 18 s, 18 s, 19 s, 19 s, and 21 s for 92 mL, 185 mL, 277 mL, 370 mL, 463 mL, and 695 mL, respectively. This suggests that increasing the tracer volume delays the peak concentration occurrence. The peak concentration values for increasing tracer volumes were 2.4, 3.37, 3.2, 2.6, 2.76, and 1.9, respectively, indicating an initial increase followed by a decrease. The lower peak concentration at 92 mL was attributed to the small tracer volume, which, upon introduction, was largely carried by strong flow fields to the left side of the ladle, resulting in a lower tracer concentration on the right side. Consequently, the peak concentration at Monitoring Point 1 remained relatively low at 18 s. When the tracer volume exceeded 185 mL, a noticeable decline in peak concentration was observed. The duration from the concentration increase to peak occurrence for each tracer volume was recorded as 7 s, 10 s, 13 s, 11 s, 11 s, and 12 s, respectively, demonstrating that a larger tracer volume extended the time required to reach peak concentration. However, the duration of concentration declined following the peak, which was 17 s, 18 s, 17 s, 24 s, 19 s, and 18 s, respectively. Except for the 370 mL tracer volume, the time interval from peak-to-minimum concentration remained approximately 18 s for all cases, indicating that the tracer volume had minimal influence on the time span from peak-to-minimum concentration.

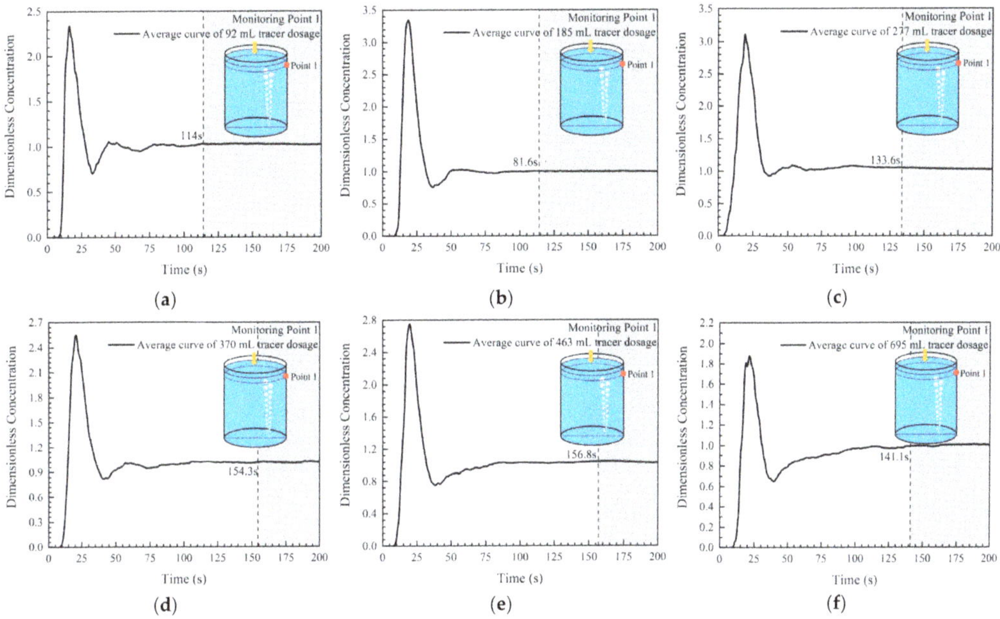

Figure 5. Dimensionless concentration curves of different volume tracers at Monitoring Point 1: (**a**) 92 mL, (**b**) 185 mL, (**c**) 277 mL, (**d**) 370 mL, (**e**) 463 mL, and (**f**) 695 mL.

Table 5. The key points in the dimensionless concentration curves of Monitoring Point 1.

Tracer Volume (mL)	92	185	277	370	463	695
Response time (s)	8	8	5	8	8	9
Mixing time (s)	114	81.6	133.6	154.3	156.8	141.1

Figure 6 presents the concentration–time curves at Monitoring Point 2 after adding saturated NaCl solutions of different volumes into the ladle water model. Table 6 summarizes the key time points extracted from the dimensionless concentration curves. For a tracer volume of 92 mL, the dimensionless concentration began to change at 8 s and reached its peak value of 3.75 at 16 s. Between 16 s and 32 s, the concentration gradually decreased, reaching a minimum value of 0.67 at 32 s. Subsequently, a slight increase was observed, leading to a secondary peak at 51 s with a value of 1.4. When the tracer volume was 185 mL, the concentration started increasing at 8 s and reached a maximum of 2.2 at 18 s. Between 18 s and 33 s, the concentration rapidly decreased to its post-peak minimum of 0.44, after which it gradually stabilized around 1. For the 277 mL tracer volume, the dimensionless concentration began to rise at 9 s, reaching a peak of 2.2 at 19 s. The concentration then gradually decreased, reaching its lowest value of 0.56 at 34 s. After 34 s, the concentration gradually increased and stabilized within a $\pm5\%$ standard range. When the tracer volume was 370 mL, the concentration response began at 10 s, reaching a peak value of 1.79 at 20 s. Subsequently, the concentration decreased, following a downward trend until reaching its minimum value of 0.45 at 37 s. Afterward, the concentration slowly rose and gradually stabilized around 1. For the 463 mL tracer volume, the conductivity sensor detected a response at 9 s, which was followed by a rapid increase in concentration between 9 s and 18 s. After reaching a maximum value of 2.1, the concentration began to decrease and reached its minimum of 0.51 at 36 s, subsequently stabilizing around 1. When the tracer volume reached 695 mL, the dimensionless concentration started rising at 9 s, reaching its peak of 1.35 at 20 s. It then underwent a sharp decline, reaching its minimum of 0.37 at 38 s, after which it gradually increased and stabilized around 1. For all six tracer volumes, the response time at Monitoring Point 2 was approximately 9 s, indicating that the tracer volume did not significantly affect the electrical conductivity response time at this location. As the tracer volume increased from 92 mL to 695 mL, the times required to reach peak concentration were 16 s, 18 s, 19 s, 20 s, 18 s, and 20 s, respectively. This suggests that a larger tracer volume delays the time to reach the maximum concentration, indicating an influence on the transport velocity of dissolved salts. The peak concentration values for increasing the tracer volumes were recorded as 3.75, 2.2, 2.2, 1.8, 2.1, and 1.35, respectively, demonstrating a general decreasing trend. This trend was attributed to the increased volume reducing the salt concentration at this point, as well as the excessive tracer volumes leading to salt settling at the bottom. By comparing the six dimensionless concentration curves under two different flow rates, it was observed that, at a flow rate of 2.3 L/min, secondary peaks were rarely present, and this phenomenon became even less frequent as the tracer volume increased. This was due to the lower flow rate limiting the movement of dissolved salts along the ladle's flow field, preventing a substantial amount of salts from re-accumulating and reaching the upper monitoring point, which would otherwise result in a secondary peak. Additionally, the minimum concentration values at 2.3 L/min were consistently lower than those observed at 8.3 L/min. This is because the lower flow rate failed to keep salts suspended in the upper region, which caused them to settle due to buoyancy force, leading to lower minimum values at the upper monitoring point. Furthermore, the recovery from the minimum concentration to the ±5% standard level predominantly exhibited a slow growth trend.

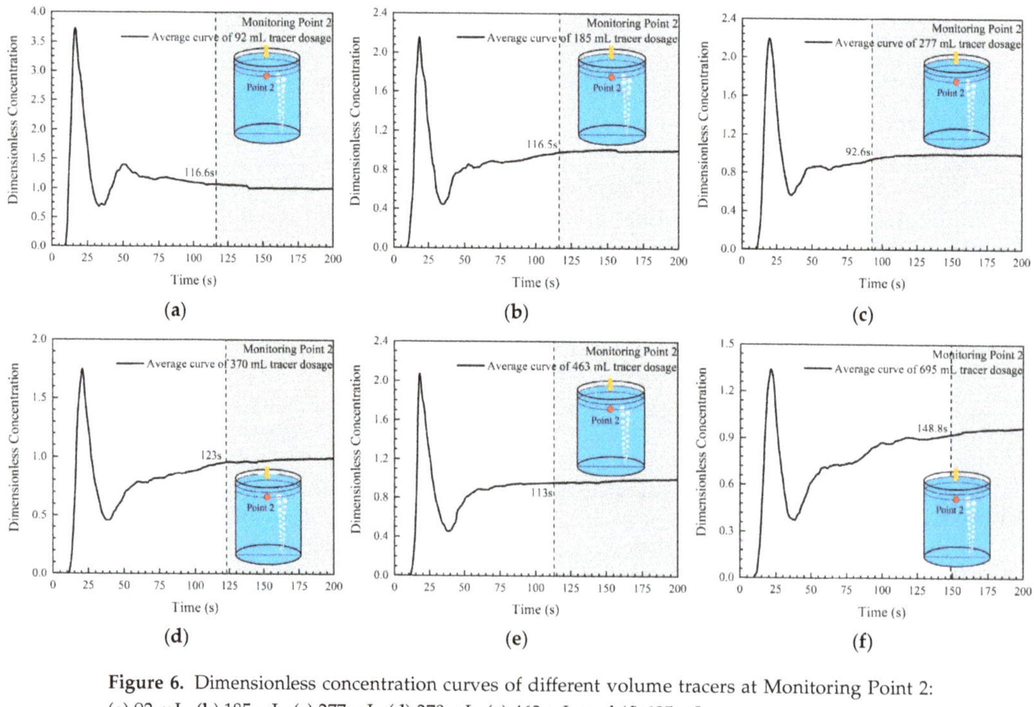

Figure 6. Dimensionless concentration curves of different volume tracers at Monitoring Point 2: (**a**) 92 mL, (**b**) 185 mL, (**c**) 277 mL, (**d**) 370 mL, (**e**) 463 mL, and (**f**) 695 mL.

Table 6. The key points in the dimensionless concentration curves of Monitoring Point 2.

Tracer Volume (mL)	92	185	277	370	463	695
Response time (s)	8	8	9	10	9	9
Mixing time (s)	116.6	116.5	92.6	123	113	148.8

Figure 7 presents the concentration–time curves at Monitoring Point 3 after adding tracer agents of six different volumes into the ladle water model. Table 7 summarizes the key time points extracted from the dimensionless concentration curves. For a tracer volume of 92 mL, the dimensionless concentration began to rise at 8 s and reached its maximum value of 8 at 15 s. The concentration increased rapidly, with a steep slope observed in the concentration curve between 8 s and 15 s. In the interval from 15 s to 40 s, the concentration exhibited a rapid downward trend, indicating that a large amount of dissolved salt quickly passed through this location. At 50 s, a secondary peak appeared with a value of 1.52, which was followed by a gradual decrease until the concentration stabilized at 1. When the tracer volume was 185 mL, the concentration started increasing at 10 s and reached a maximum value of 3.4 at 21 s. Subsequently, the concentration declined until it stabilized at 1. For the 277 mL tracer volume, the electrical conductivity response was still detected at 8 s, with the concentration reaching a peak value of 2.7 at 19 s. The concentration then continued to decrease until 50 s, reaching a value of 0.76, after which it gradually rose and stabilized at 1. When 370 mL of saturated NaCl solution was introduced, the concentration began to increase slowly at 10 s, which was followed by a rapid rise after 15 s, reaching a peak value of 1.6 at 23 s. Thereafter, the concentration followed a general downward trend, reaching its lowest value of 0.66 at 55 s. After 55 s, the concentration gradually increased until complete mixing was achieved. For the 463 mL tracer volume, the concentration started rising at 8 s, reaching a peak of 2.6 at 20 s. The overall trend of the concentration curve was similar to that observed for the 370 mL tracer volume. When the tracer volume was 695 mL, the concentration increased from 9 s, reaching its peak value of 1.9 at 20 s. From 20 s to 46 s, the concentration followed a decreasing trend, reaching a minimum value of 0.59 at 46 s, after which it slowly increased and eventually stabilized within a ±5% standard range. The response time at Monitoring Point 3 remained around 8 s, indicating that the upper monitoring point's response time was not significantly affected by the tracer volume. However, at a flow rate of 8.3 L/min, the response time at the upper monitoring point varied noticeably with tracer volume, suggesting a clear relationship between the gas injection rate and the time required for the tracer to reach the upper monitoring point. Since the response times at the three upper monitoring points were similar (all around 8 s), it can be concluded that the gas injection nozzle had little effect on the liquid velocity near the free surface. By comparing the six dimensionless concentration curves in Figure 7, it was observed that the first peak value generally decreased with increasing tracer volume. This phenomenon occurred because, at smaller tracer volumes, nearly all of the tracer followed the strong recirculating flow to Monitoring Point 3. As the tracer volume increased, the added tracer initially moved downward, temporarily disrupting the flow field in the ladle water model, leading to partial salt deposition. Once the flow field stabilized, only a small amount of salt was transported to the upper Monitoring Point 3 by recirculating flow, resulting in a lower first peak as the tracer volume increased. Furthermore, at Monitoring Point 3, the first peak values under a lower flow rate (2.3 L/min) were consistently lower than those observed at a higher flow rate (8.3 L/min), further confirming that variations in gas flow rate significantly influence the transport behavior of the tracer within the ladle.

Figure 8 presents the concentration–time curves at Monitoring Point 4 after introducing tracer agents of different volumes into the ladle. Table 8 summarizes the key time points extracted from the dimensionless concentration curves. When 92 mL of saturated NaCl solution was added, two distinct concentration trends, Trend 1 and Trend 2, were observed in the dimensionless concentration curve at Monitoring Point 4. The black curve in the figure represents the average of both trends. The primary difference between these trends lay in their initial concentration behavior: Trend 1 exhibited a gradual increase from zero without pronounced peaks before reaching a dimensionless concentration of 1, whereas Trend 2 showed a rapid increase in concentration with noticeable but relatively small peaks immediately after the response began. A similar phenomenon was observed when the tracer volume increased to 185 mL, with the concentration curve displaying both trends similar to those observed for the 92 mL tracer. This behavior was mainly attributed to the small tracer volume being more significantly affected by the recirculating flow within the ladle, making it highly sensitive to minor variations in the flow field. Consequently, subtle changes in the ladle's flow conditions altered the tracer transport direction, leading to the emergence of two distinct trends [90,91]. Trend 1 closely resembled the previously studied Path 2 curve, indicating that, after injection, the tracer was primarily transported through the left and central regions of the ladle before gradually crossing the gas plume and reaching Monitoring Point 4. In contrast, Trend 2 was similar to the Path 5 curve, suggesting that the tracer predominantly circulated on the right side, leading to the appearance of peak values in the concentration curve. When the tracer volume increased to 277 mL, the dimensionless concentration began to change at 18 s and reached its peak of 1.92 at 40 s. A minor trough appeared at 52 s, with the concentration dropping to 1.36. Between 52 s and 57 s, the concentration increased again, reaching 1.7, before gradually declining after 57 s, until then stabilizing within the range of 0.95–1.05. For a tracer volume of 370 mL, the concentration response began at 13 s, which was followed by a rapid increase, reaching a maximum of 5.4 at 32 s. Subsequently, the concentration declined sharply, reaching its lowest value of 1.39 at 86 s. A temporary increase was observed between 86 s and 103 s, leading to a second peak at 103 s with a value of 1.82. Afterward, the concentration continuously decreased until it stabilized around 1. When 463 mL of saturated NaCl solution was introduced, the concentration started increasing at 14 s, showing a rapid growth phase until 22 s, after which the growth rate slowed. The concentration reached a maximum of 2.27 at 47 s, which was followed by a continuous decline. By 72 s, the concentration decreased to 1.2. Between 72 s and 104 s, fluctuations were observed, with a secondary peak appearing at 104 s with a value of 1.5. Afterward, the concentration gradually decreased with minor oscillations until stabilizing within the required range. For the 695 mL tracer volume, the concentration began to respond at 13 s, which was followed by a rapid increase, reaching a peak of 3.7 at 37 s. After 37 s, the concentration continuously declined with significant fluctuations until complete mixing was achieved. The initial response time of the dimensionless concentration curve for tracer volumes of 92 mL and 185 mL was 24 s and 23 s. For tracer volumes of 277 mL, 370 mL, 463 mL, and 695 mL, the initial response times were 18 s, 13 s, 14 s, and 13 s, respectively. This suggests that, when the tracer volume was small, its transport path was significantly influenced by the flow field, leading to a delayed response at Monitoring Point 4. However, as the tracer volume increased, the influence of the flow field on transport pathways decreased, resulting in a more consistent response time at Monitoring Point 4. Furthermore, with increasing tracer volume, Trend 2 became dominant in the concentration curves at Monitoring Point 4.

Table 7. The key points in the dimensionless concentration curves of Monitoring Point 3.

Tracer Volume (mL)	92	185	277	370	463	695
Response time (s)	8	10	8	10	8	9
Mixing time (s)	129.6	100.8	102.9	125.5	126.1	143

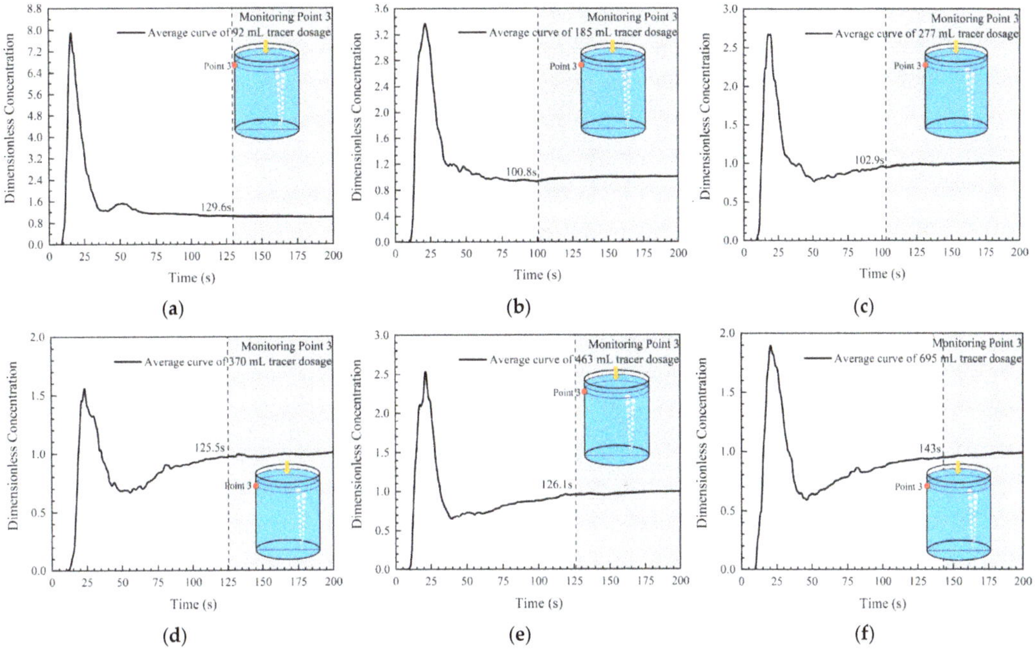

Figure 7. Dimensionless concentration curves of different volume tracers at Monitoring Point 3: (**a**) 92 mL, (**b**) 185 mL, (**c**) 277 mL, (**d**) 370 mL, (**e**) 463 mL, and (**f**) 695 mL.

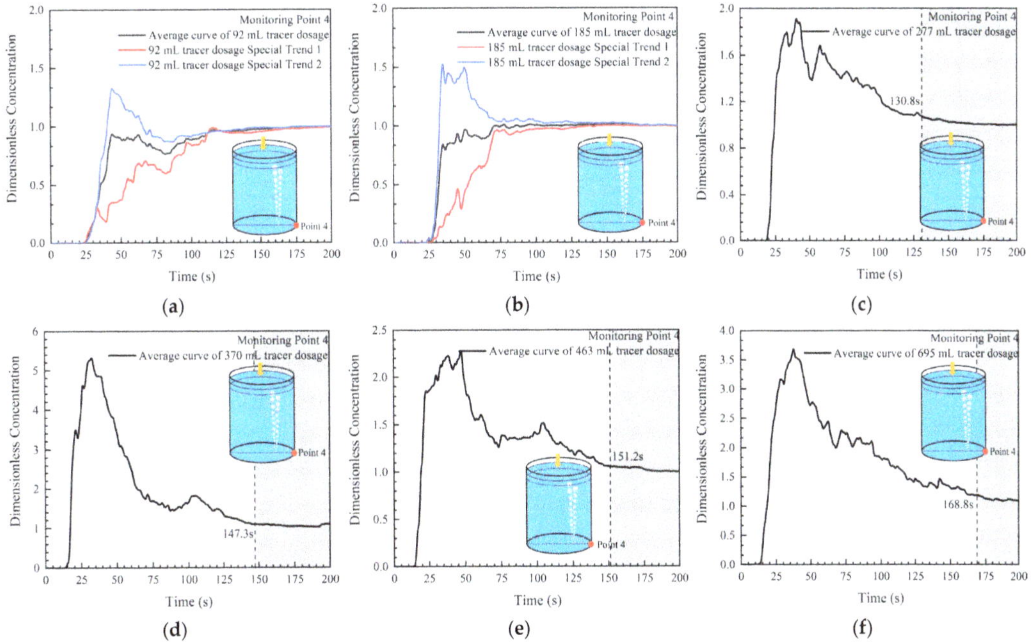

Figure 8. Dimensionless concentration curves of different volume tracers at Monitoring Point 4: (**a**) 92 mL, (**b**) 185 mL, (**c**) 277 mL, (**d**) 370 mL, (**e**) 463 mL, and (**f**) 695 mL.

Table 8. The key points in the dimensionless concentration curves of Monitoring Point 4.

Tracer Volume (mL)	92	185	277	370	463	695
Response time (s)	24	23	18	13	14	13
Mixing time (s)	123.6	103.2	130.8	147.3	151.2	168.8

Figure 9 presents the concentration–time curves at Monitoring Point 5 after adding tracer agents of different volumes into the ladle. Table 9 summarizes the key time points extracted from the dimensionless concentration curves. For a tracer volume of 92 mL, the concentration curve began to change at 25 s. Before complete mixing was achieved, the concentration remained below 1, indicating that the dimensionless concentration increased gradually from zero to 1 with minor fluctuations. A relatively faster increase was observed only between 15 s and 25 s. When the tracer volume was 185 mL, the dimensionless concentration started changing at 20 s. However, between 20 s and 27 s, the concentration increased very slowly, indicating that the tracer reached this location at approximately 27 s. Between 27 s and 39 s, the concentration increased rapidly. After 39 s, noticeable fluctuations appear in the concentration curve, which was followed by a gradual decrease until mixing was complete. For the 277 mL tracer volume, the concentration began to change at 20 s. Similar to the 185 mL case, the concentration exhibited minimal variation between 20 s and 28 s, suggesting that only a small amount of salt reached this location initially. Between 28 s and 43 s, the concentration increased rapidly, reaching a maximum value of 1.38 at 49 s. The concentration then declined from 49 s to 61 s, briefly reaching a local minimum of 1.1, which was followed by a slight increase to 1.3. After 69 s, the concentration decreased with minor fluctuations until stabilizing within the range of 0.95–1.05. When the tracer volume increased to 370 mL, the concentration response began at 15 s, which was followed by a rapid increase, reaching 4 at 30 s. Subsequently, the concentration gradually declined, exhibiting significant fluctuations that resulted in a jagged pattern of peaks. Eventually, the concentration stabilized around 1. For the 463 mL tracer volume, the concentration starts increasing at 22 s, with multiple peaks appearing in the curve. The highest peak, reaching 1.9, occurred at 59 s. However, due to significant fluctuations, a clear trend was not observed. When the tracer volume reached 695 mL, the concentration began changing at 15 s, followed by a rapid increase, reaching 3.2 at 26 s. After a brief fluctuation, the concentration continued to rise, reaching a maximum value of 3.7 at 39 s. Beyond 39 s, the concentration exhibited an overall decreasing trend but was accompanied by significant fluctuations, resulting in a jagged concentration curve. These oscillations persisted until approximately 150 s before stabilizing. The response times for different tracer volumes at Monitoring Point 5 were recorded as 25 s, 20–27 s, 20–28 s, 15 s, 22 s, and 15 s, respectively. Most of the response times fell within the 20–25 s range, with the remaining cases occurring at 15 s. This suggests that the first arrival time of the tracer at Monitoring Point 5 was not uniform, indicating the possibility of multiple transport pathways leading to that location. Additionally, the maximum concentrations recorded for different tracer volumes were 1, 1.13, 1.38, 4, 1.9, and 3.7, respectively. The significant variation in both peak concentration values and their occurrence times further confirmed that the tracer reached this point via multiple, possibly non-unique transport pathways.

Table 9. The key points in the dimensionless concentration curves of Monitoring Point 5.

Tracer Volume (mL)	92	185	277	370	463	695
Response time (s)	25	20	20	15	22	15
Mixing time (s)	141	124	134.4	136.1	142.7	191.2

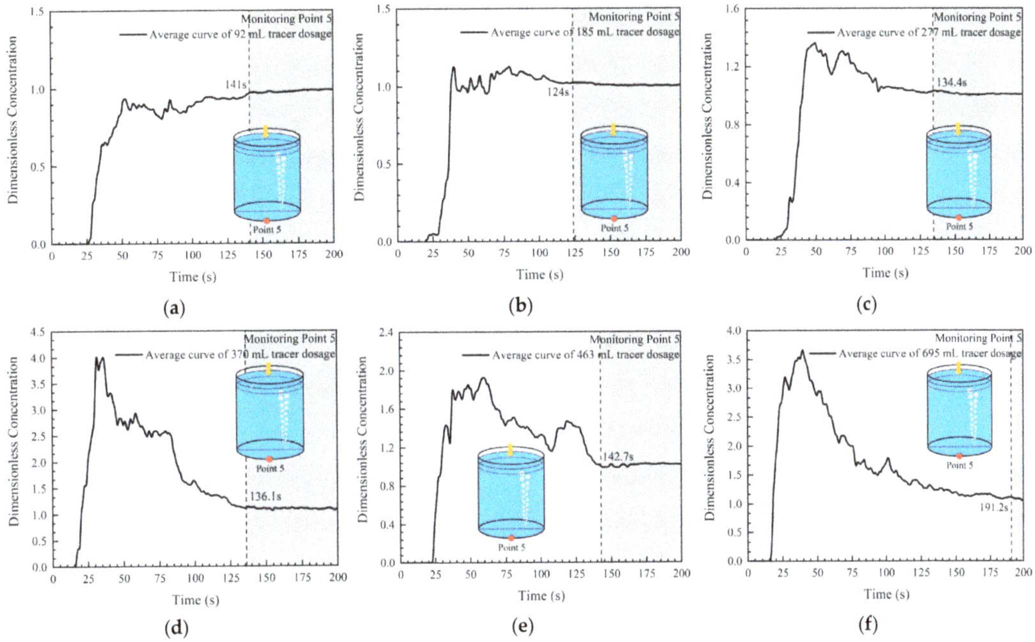

Figure 9. Dimensionless concentration curves of different volume tracers at Monitoring Point 5: (**a**) 92 mL, (**b**) 185 mL, (**c**) 277 mL, (**d**) 370 mL, (**e**) 463 mL, and (**f**) 695 mL.

Figure 10 presents the concentration–time curves at Monitoring Point 6 after introducing saturated NaCl solutions of different volumes (92 mL, 185 mL, 277 mL, 370 mL, 463 mL, and 695 mL) into the ladle water model. Table 10 summarizes the key time points extracted from the dimensionless concentration curves. For a 92 mL tracer volume, the dimensionless concentration curve began to change at 25 s. However, between 25 s and 35 s, the concentration remained relatively stable, showing no significant increase. After 35 s, a rapid rise was observed, reaching 1.0 at 58 s. From 58 s to 68 s, the concentration decreased to 0.83, which was followed by a gradual increase until it fluctuated around 1.0. When the 185 mL tracer was added, the concentration began to rise at 23 s, showing rapid growth and reaching a maximum of 1.35 at 53 s. After a slight decrease, the concentration increased again to 1.34 at 71 s, which was followed by a gradual decline until complete mixing was achieved. For the 277 mL tracer volume, the concentration started changing at 25 s. Between 25 s and 35 s, the concentration increased rapidly. After 35 s, the concentration curve exhibited a sawtooth pattern, which is characterized by fluctuations with relatively small variations in overall concentration. The peak concentration of 1.37 was recorded at 85 s. When the 370 mL tracer volume was introduced, the concentration began increasing at 18 s, showing a very rapid growth phase. At 35 s, the concentration momentarily stabilized before experiencing a slight decrease, reaching 3.3 at 46 s. The concentration then increased again to 4.5 at 50 s, which was followed by a general downward trend. Throughout the decline, significant fluctuations were observed until the concentration eventually stabilized near 1.0. For the 463 mL tracer volume, the conductivity sensor detected a response at 20 s, which was followed by a rapid increase between 20 s and 40 s. The concentration reached a peak value of 2.6 at 40 s, after which it gradually decreased until complete mixing was achieved. Minor fluctuations were observed throughout the decline. When the 695 mL tracer was added, the concentration started changing at 17 s and exhibited a very rapid increase, reaching 3.7 at 31 s. A short-lived decrease followed, with the concentration dropping to 2.5 at 55 s. After 55 s, the concentration increased slightly again, peaking at 2.9 at 61 s. Subsequently, the concentration followed a general downward trend, which was accompanied by fluctuations. The dimensionless concentration curve at Monitoring Point 6 showed significant fluctuations, with multiple secondary peaks and sawtooth-like oscillations during the concentration decline. This suggests that the tracer, after reaching this location, accumulated and settled before being gradually transported away in batches under the influence of the gas plume, rather than being immediately and entirely dispersed. This stepwise removal process resulted in the observed oscillatory behavior. The sensor response times for different tracer volumes were recorded as 25 s, 23 s, 25 s, 18 s, 20 s, and 17 s, respectively. As the tracer volume increased, the response time of the conductivity sensor gradually shortened, indicating that larger tracer volumes were transported to Monitoring Point 6 more quickly under the influence of the gas plume. Additionally, as the tracer volume increased, sedimentation at Monitoring Point 6 became more pronounced, leading to a prolonged mixing time at this location.

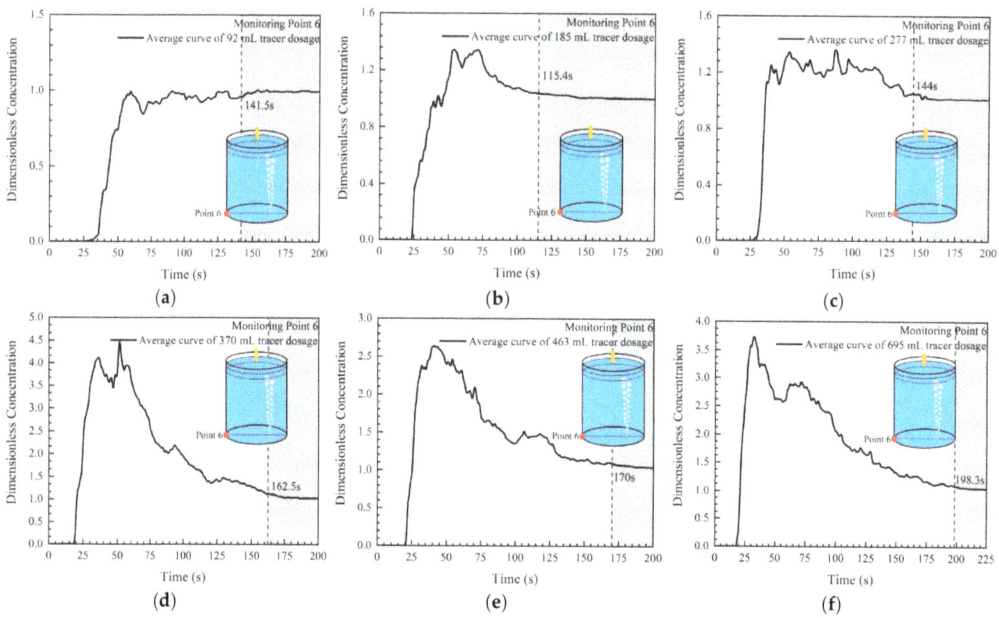

Figure 10. Dimensionless concentration curves of different volume tracers at Monitoring Point 6: (**a**) 92 mL, (**b**) 185 mL, (**c**) 277 mL, (**d**) 370 mL, (**e**) 463 mL, and (**f**) 695 mL.

Table 10. The key points in the dimensionless concentration curves of Monitoring Point 6.

Tracer Volume (mL)	92	185	277	370	463	695
Response time (s)	25	23	25	18	20	17
Mixing time (s)	141.5	115.4	144	162.5	170	198.3

3.5. Analysis of Mixing Time

Figure 11 illustrates the mixing times at various monitoring points after introducing tracer agents of different volumes into the ladle water model. At Monitoring Point 1, the mixing times corresponding to tracer volumes of 92 mL, 185 mL, 277 mL, 370 mL, 463 mL, and 695 mL were 114 s, 81.6 s, 133.6 s, 154.3 s, 156.8 s, and 141.1 s, respectively. The mixing time initially decreased, which was followed by an increase and then another decrease. Two transition points were identified at 185 mL and 463 mL, where the mixing time significantly decreased when the tracer volume increased from 92 mL to 185 mL, which was followed by a continuous increase from 185 mL to 463 mL. However, when the volume increased to 695 mL, the mixing time shortened again. At Monitoring Point 2, the mixing times for increasing tracer volumes were 116.6 s, 116.5 s, 92.6 s, 123 s, 113 s, and 148.8 s, respectively. The mixing time exhibited two increasing phases, occurring between 277 mL to 370 mL and 463 mL to 695 mL. For Monitoring Point 3, the mixing times corresponding to 92 mL, 185 mL, 277 mL, 370 mL, 463 mL, and 695 mL were 129.6 s, 100.8 s, 102.9 s, 125.5 s, 126.1 s, and 143 s, respectively. The mixing time first decreased, then increased, and then remained in an upward trend when the tracer volume exceeded 185 mL. At Monitoring Point 4, the mixing times for increasing tracer volumes were 123.6 s, 103.2 s, 130.8 s, 147.3 s, 151.2 s, and 168.8 s. Notably, when the tracer volume was 185 mL, the mixing time was significantly shorter than for other tracer volumes. For Monitoring Point 5, the mixing times corresponding to different tracer volumes were 141 s, 124 s, 134.4 s, 136.1 s, 142.7 s, and 191.2 s, respectively. A trend of initial reduction followed by an increase was observed. At Monitoring Point 6, the mixing times for increasing tracer volumes were 141.5 s, 115.4 s, 144 s, 162.5 s, 170 s, and 198.3 s. Similar to other locations, the mixing time first decreased and then increased, with the longest mixing time observed at 695 mL, making it the longest among all six monitoring points. As the tracer volume changed, the mixing time at all six monitoring points also varied accordingly, generally exhibiting a trend of initial shortening followed by an increase. Figure 11 also shows that the mixing times at the bottom monitoring points were relatively concentrated, following similar trends, with a high degree of overlap among different tracer volumes. In contrast, the mixing time variations at the upper monitoring points were less consistent, suggesting that the upper regions were more susceptible to changes in tracer volume.

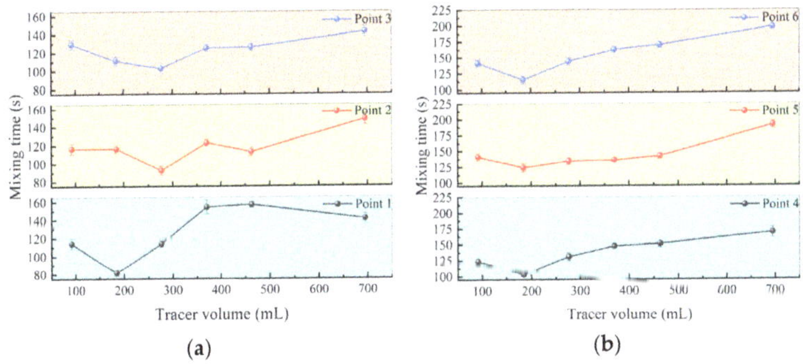

Figure 11. Mixing time at different monitoring points: the (**a**) top monitoring points and (**b**) bottom monitoring points.

Figure 12 shows that the mixing times at the upper monitoring points were generally shorter than those at the lower monitoring points. This difference was particularly pronounced when larger tracer volumes were introduced as the mixing times at the lower monitoring points became significantly longer than those at the upper monitoring points. This observation suggests that the gas flow rate was insufficient to fully disperse larger volumes of tracer within the ladle. The three-dimensional visualization further confirmed that when the tracer volume was 185 mL, the mixing time was notably shorter compared to other tracer volumes. Additionally, the mixing time initially decreased and then increased as the tracer volume increased.

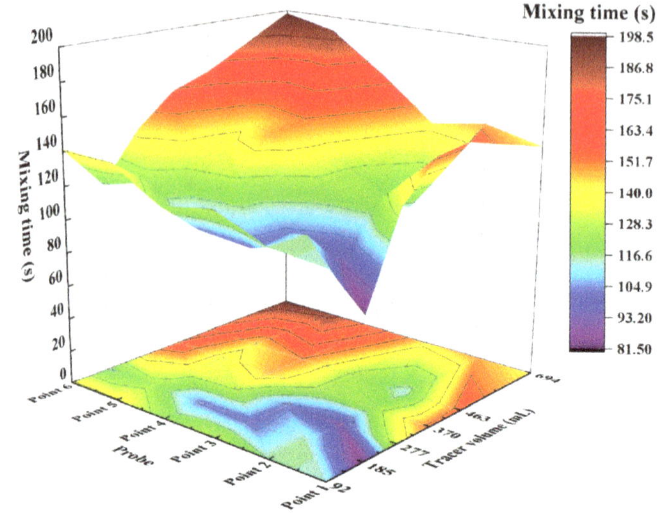

Figure 12. The mixing time of different volume tracers at monitoring points.

3.6. Analysis of Tracer Concentration Experiments

To further clarify the relative influence of the tracer volume and concentration on mixing behavior, a set of controlled experiments was designed in which the tracer volume was held constant at 185 mL, thereby keeping the total liquid volume in the system unchanged. The effect of different NaCl mass concentrations on the mixing process was evaluated by varying the NaCl concentration while maintaining a constant volume. The selected concentrations were 6.63%, 13.25%, 19.68%, and 26.5%.

As shown in Figure 13a–c, changes in NaCl concentration had a limited impact on the dimensionless concentration curves at the upper monitoring points. The tracer arrival time at the top was nearly identical across all concentrations, and the differences in the time to initial response were minimal. This indicates that the tracer response in the upper region was relatively stable. However, the peak concentration decreased progressively as the NaCl concentration decreased, suggesting that, although the diffusion rate was not significantly affected, the signal intensity detected by the probes was influenced by the tracer concentration. In contrast, as shown in Figure 13d–f, the effect of the NaCl concentration was more pronounced at the bottom monitoring points. Notably, when the NaCl concentration was 19.68%, the bottom monitoring points exhibited significantly higher peak values than those observed under the other three conditions. This indicates that, at this concentration, the transport and accumulation of the tracer in the lower region was more effective. Moreover, at higher concentrations (19.68% and 26.5%), larger differences between the concentration curves were observed, whereas, at lower concentrations (6.63% and 13.25%), the curves tended to converge, indicating reduced sensitivity to concentration changes. Further analysis from Figure 14 reveals that the influence of NaCl concentration on the mixing time varied among different monitoring points. The most significant effects were observed at Monitoring Points 3, 4, and 6. Nevertheless, the overall mixing trends remained consistent, indicating that, while concentration changes can affect local signal intensity and mixing time, they do not significantly alter the global mixing behavior. In summary, the results of the controlled concentration experiments suggest that, under a constant tracer volume, the effect of the NaCl concentration on mixing behavior is region-specific, with the bottom of the ladle being more sensitive to concentration variations.

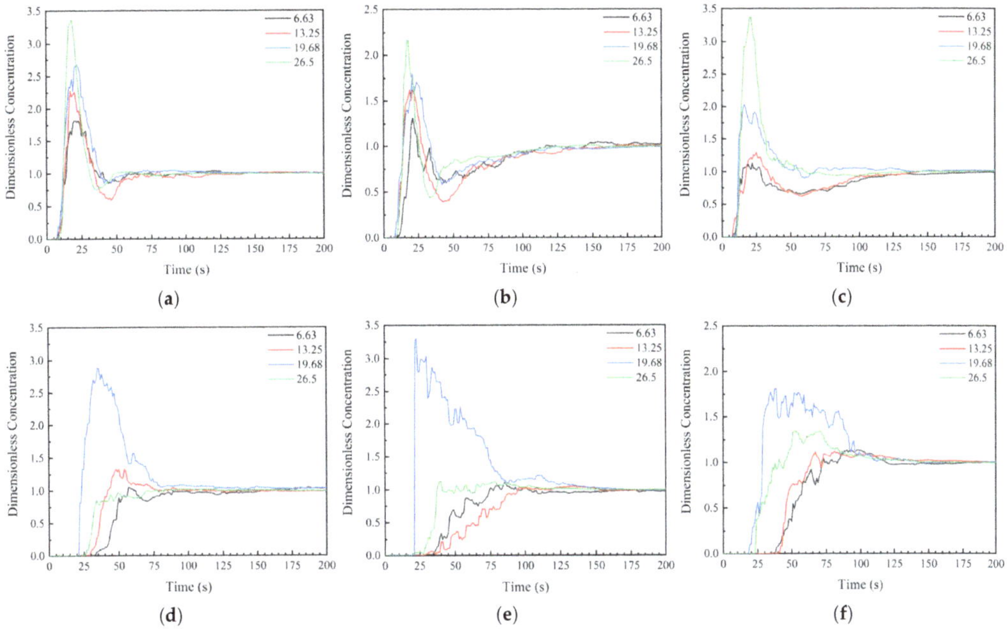

Figure 13. Dimensionless concentration curves of different tracer concentrations: (**a**) Monitoring Point 1, (**b**) Monitoring Point 2, (**c**) Monitoring Point 3, (**d**) Monitoring Point 4, (**e**) Monitoring Point 5, and (**f**) Monitoring Point 6.

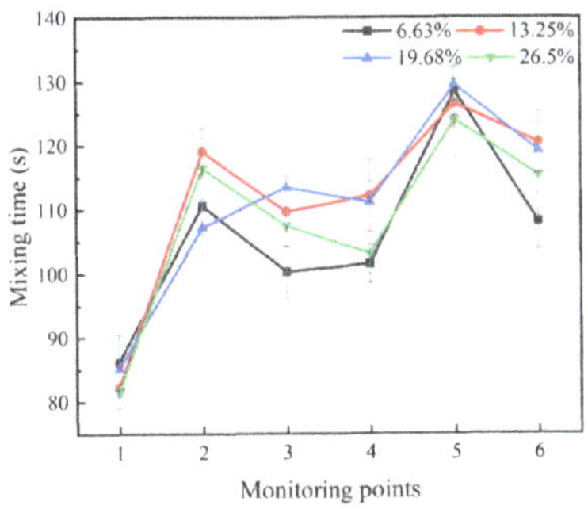

Figure 14. Mixing time at different monitoring points.

4. Conclusions

A water model with a geometric similarity ratio of 1:3 based on the 130 t steel ladle of the steel plant was established to study the transmission path and mixing time of the tracer under low flow conditions, and the following conclusions were made:

1. At a bottom-blowing gas flow rate of 2.3 L/min, two distinct trends appeared in the dimensionless concentration curve at Monitoring Point 4 when the tracer volume was 92 mL or 185 mL. This indicates that, at low tracer volumes, the tracer transport path to Monitoring Point 4 was unstable due to the significant influence of the gas plume. As the tracer volume increased, only one dominant transport path was observed. When the bottom-blowing gas flow rate was 8.3 L/min, only two trends were observed at a tracer volume of 92 mL.

2. When the tracer volume was too small, the dimensionless concentration curve at Monitoring Point 4 exhibited either a sinusoidal or parabolic shape. If the tracer first moved through the left-side and central regions before slowly crossing the gas plume to reach Monitoring Point 4, the curve followed a parabolic shape. If the tracer predominantly circulated on the right side, the curve exhibited a sinusoidal pattern. When the tracer volume exceeded 277 mL, the concentration curve at Monitoring Point 4 was always sinusoidal.
3. Regardless of the gas flow rate, Monitoring Point 3 exhibited the highest peak concentration among the six monitoring points. At a bottom-blowing gas flow rate of 8.3 L/min, the peak concentration at Monitoring Point 3 was significantly higher than that observed at 2.3 L/min.
4. Different bottom-blowing gas flow rates led to distinct trends in the mixing time variation with increasing tracer volume. At 2.3 L/min, the mixing time at the bottom first decreased and then increased, reaching the shortest duration at 185 mL. At 8.3 L/min, the mixing time at the bottom monitoring points decreased continuously with increasing tracer volume, whereas, at the top monitoring points, the mixing time first decreased and then increased.
5. At different bottom-blowing gas flow rates, the mixing time at the upper monitoring points was consistently shorter than at the lower monitoring points.
6. At a bottom-blowing gas flow rate of 2.3 L/min, the optimal tracer volume range was 185 mL–277 mL. This ensured a stable transport path while minimizing the flow field disturbances caused by tracer injection.
7. The tracer concentration experiments indicate that variations in concentration have a considerable impact on the mixing behavior at the bottom of the ladle, while their influence on the overall mixing trend remains limited. When the tracer concentration is relatively low, the mixing behavior throughout the ladle becomes more uniform.

Author Contributions: Conceptualization, J.F. and C.C.; methodology, J.F., C.C., K.Y., W.L. and Y.Z.; software, Y.Z., X.T., L.L., Z.G., H.Q., J.W. and K.Y.; validation, Y.Z., X.T. and J.F.; formal analysis, Y.Z., X.T. and L.L.; investigation, Y.Z., L.L., Z.G., H.Q. and J.W.; resources, J.F., C.C. and W.L.; data curation, Z.G., H.Q., J.W. and K.Y; writing—original draft preparation, Y.Z., X.T. and L.L.; writing—review and editing, Y.Z., X.T., L.L., J.F. and C.C.; visualization, Y.Z., X.T., Z.G. and H.Q.; supervision, J.F. and C.C.; project administration, C.C., J.F. and W.L.; funding acquisition, C.C., J.F. and J.W. All authors have read and agreed to the published version of the manuscript.

Funding: The authors are grateful to the financial support of the Applied Fundamental Research Programs of Shanxi Province (202403021222046) and to the Graduate Innovation Project of Shanxi Province (2024KY277).

Data Availability Statement: Data are contained within the article.

Acknowledgments: The comments from the reviewers are appreciated as they helped to improve this manuscript.

Conflicts of Interest: The authors declare no conflicts of interest.

References

1. Szekely, J.; Carlsson, G.; Helle, L. *Ladle Metallurgy*; Springer: Berlin/Heidelberg, Germany, 1989; pp. 27–71.
2. Mazumdar, D.; Guthrie, R.L. The Physical and mathematical modelling of gas stirred ladle systems. *ISIJ Int.* **1995**, *35*, 1–20. [CrossRef]
3. Jönsson, P.; Jonsson, L.I. The Use of Fundamental Process Models in Studying Ladle Refining Operations. *ISIJ Int.* **2001**, *41*, 1289–1302. [CrossRef]
4. Mazumdar, D.; Evans, J.W. Macroscopic Models for Gas Stirred Ladles. *ISIJ Int.* **2004**, *44*, 447–461. [CrossRef]

5. Xin, Z.; Zhang, J.; Peng, K.; Zhang, J.; Zhang, C.; Liu, Q. Modeling of LF refining process: A review. *J. Iron Steel Res. Int.* **2024**, *31*, 289–317. [CrossRef]
6. Niu, K.J.; Feng, W.H.; Conejo, A.N.; Ramírez-Argáez, M.A.; Yan, H. 3D CFD Model of Ladle Heat Transfer with Gas Injection. *Metall. Mater. Trans. B* **2023**, *54*, 2066–2079. [CrossRef]
7. Conejo, A.N. Physical and Mathematical Modelling of Mass Transfer in Ladles due to Bottom Gas Stirring: A Review. *Processes* **2020**, *8*, 750. [CrossRef]
8. Niu, K.J.; Feng, W.H.; Conejo, A.N. Effect of the Nozzle Radial Position and Gas Flow Rate on Mass Transfer during Bottom Gas Injection in Ladles with One Nozzle. *Metall. Mater. Trans. B* **2022**, *53*, 1344–1350. [CrossRef]
9. Yang, R.; Chen, C.; Lin, Y.; Zhao, Y.; Zhao, J.; Zhu, J.; Yang, S. Water model experiment on motion and melting of scarp in gas stirred reactors. *Chin. J. Process Eng.* **2022**, *22*, 954–962. [CrossRef]
10. Ji, S.; Niu, K.J.; Conejo, A.N. Multiphase modeling of steel-slag mass transfer through distorted interface in bottom-stirred ladle. *ISIJ Int.* **2024**, *64*, 52–58. [CrossRef]
11. Podder, A.; Coley, K.S.; Phillion, A.B. Modeling Study of Steel–Slag–Inclusion Reactions During the Refining of Si–Mn Killed Steel. *Steel Res. Int.* **2022**, *93*, 2100831. [CrossRef]
12. Podder, A.; Coley, K.S.; Phillion, A.B. Simulation of Ladle Refining Reactions in Si–Mn-Killed Steel. *Steel Res. Int.* **2024**, *95*, 2300330. [CrossRef]
13. Duan, H.J.; Zhang, L.F.; Thomas, B.G.; Conejo, A.N. Fluid Flow, Dissolution, and Mixing Phenomena in Argon-Stirred Steel Ladles. *Metall. Mater. Trans. B* **2018**, *49*, 2722–2743. [CrossRef]
14. Li, X.L.; Wang, H.H.; Tian, J.; Wang, D.Y.; Qu, T.P.; Hou, D.; Hu, S.Y.; Wu, G.J. Investigation on the Alloy Mixing and Inclusion Removement through Using a New Slot-Porous Matched Tuyeres. *Metals* **2023**, *13*, 667. [CrossRef]
15. Wen, X.; Ren, Y.; Zhang, L.F. Effect of CaF_2 Contents in Slag on Inclusion Absorption in a Bearing Steel. *Steel Res. Int.* **2022**, *94*, 2200218. [CrossRef]
16. Kim, T.S.; Yang, J.; Park, J.H. Effect of Physicochemical Properties of Slag on the Removal Rate of Alumina Inclusions in the Ruhrstahl–Heraeus (RH) Refining Conditions. *Metall. Mater. Trans. B* **2022**, *53*, 2523–2533. [CrossRef]
17. Conejo, A.N.; Kitamura, S.; Maruoka, N.; Kim, S.J. Effects of Top Layer, Nozzle Arrangement, and Gas Flow Rate on Mixing Time in Agitated Ladles by Bottom Gas Injection. *Metall. Mater. Trans. B* **2013**, *44*, 914–923. [CrossRef]
18. Terrazas, M.S.C.; Conejo, A.N. Effect of Nozzle Diameter on Mixing Time During Bottom-Gas Injection in Metallurgical Ladles. *Metall. Mater. Trans. B* **2014**, *45*, 711–718. [CrossRef]
19. Feng, W.H.; Conejo, A.N. Physical and Numerical Simulation of the Optimum Nozzle Radial Position in Ladles with One Nozzle and Bottom Gas Injection. *ISIJ Int.* **2022**, *62*, 1211–1221. [CrossRef]
20. Conejo, A.N.; Feng, W.H. Ladle Eye Formation Due to Bottom Gas Injection: A Reassessment of Experimental Data. *Metall. Mater. Trans. B* **2022**, *53*, 999–1017. [CrossRef]
21. Wondrak, T.; Timmel, K.; Bruch, C.; Gardin, P.; Hackl, G.; Lachmund, H.; Lüngen, H.B.; Odenthal, H.J.; Eckert, S. Large-Scale Test Facility for Modeling Bubble Behavior and Liquid Metal Two-Phase Flows in a Steel Ladle. *Metall. Mater. Trans. B* **2022**, *53*, 1703–1720. [CrossRef]
22. Wang, R.D.; Jin, Y.; Cui, H. The Flow Behavior of Molten Steel in an RH Degasser Under Different Ladle Bottom Stirring Processes. *Metall. Mater. Trans. B* **2022**, *53*, 342–351. [CrossRef]
23. Jojo-Cunningham, Y.; Guo, X.P.; Zhou, C.; Liu, Y. Volumetric Flow Field inside a Gas Stirred Cylindrical Water Tank. *Fluids* **2024**, *9*, 11. [CrossRef]
24. Chen, C.; Cheng, G.G.; Yang, H.K.; Hou, Z.B. Physical Modeling of Fluid Flow Characteristics in a Delta Shaped, Four-Strand Continuous Casting Tundish with Different Flow Control Devices. *Adv. Mater. Res.* **2011**, *284*, 1071–1079. [CrossRef]
25. Schwarz, M.P.; Koh, P.T.L. Numerical Modelling of Bath Mixing by Swirled Gas Injection. In Proceedings of the SCANINJECT IV-4th International Conference on Injection Metallurgy, Luleå, Sweden, 11–13 July 1986.
26. Schwarz, M.P.; Turner, W.J. Applicability of the standard k-ϵ turbulence model to gas-stirred baths. *Appl. Math. Model.* **1988**, *12*, 273–279. [CrossRef]
27. Zhao, J.X.; Zhu, H.Y.; Chen, J.; Wang, L.Q.; Yan, X.B.; Sun, J. Numerical Simulation on the Motion Behavior of Micro-Inclusions at the Steel-Slag Interface. *Metall. Mater. Trans. B* **2024**, *55*, 1700–1711. [CrossRef]
28. Zhong, J.; Xu, J.; Luo, D.Q.; Liu, S.D.; Zhou, Z.Y.; Lai, C.B. Numerical Simulation on Dehydrogenation Behavior During Argon Blowing Enhanced Vacuum Refining Process of High-Quality Steel. *J. Mater. Res. Technol.* **2024**, *33*, 6481–6493. [CrossRef]
29. Nick, R.S.; Teng, L.; Yang, H.L.; Tilliander, A.; Glaser, B.; Sheng, D.Y.; Jönsson, P.G.; Björkvall, J. Mathematical modelling of novel combined stirring method during the final stage of ladle refining. *Ironmak. Steelmak.* **2023**, *50*, 721–733. [CrossRef]
30. Wang, J.H.; Ni, P.Y.; Chen, C.; Ersson, M.; Li, Y. Effect of gas blowing nozzle angle on multiphase flow and mass transfer during RH refining process. *Int. J. Miner. Metall. Mater.* **2023**, *30*, 844–856. [CrossRef]

31. Li, X.L.; Wang, D.Y.; Tian, J.; Wang, H.H.; Qu, T.P.; Hou, D.; Hu, S.Y.; Zhang, Z.X.; Zhou, X.Z.; Wu, G.J. Modeling of Bubble Transportation, Expansion, as Well as Adhesion of Inclusions in a Ladle with Different Tuyeres. *Metall. Mater. Trans. B* **2024**, *55*, 14–31. [CrossRef]
32. Zhou, X.B.; Zhang, Y.; He, Q.L.; Ni, P.Y.; Yue, Q.; Ersson, M. Novel Evaluation Method to Determine the Mixing Time in a Ladle Refining Process. *Metall. Mater. Trans. B* **2022**, *53*, 4114–4123. [CrossRef]
33. Hua, C.J.; Bao, Y.P.; Wang, M. Multiphysics numerical simulation model and hydraulic model experiments in the argon-stirred ladle. *Processes* **2022**, *10*, 1563. [CrossRef]
34. Tiwari, R.; Girard, B.; Labrecque, C.; Isac, M.M.; Guthrie, R.I.L. CFD Predictions for Mixing Times in an Elliptical Ladle Using Single- and Dual-Plug Configurations. *Processes* **2023**, *11*, 1665. [CrossRef]
35. Schwarz, M.P. Buoyancy and expansion power in gas-agitated baths. *ISIJ Int.* **1991**, *31*, 947–951. [CrossRef]
36. Schwarz, M.P.; Dang, P. Simulation of Blowthrough in Smelting Baths with Bottom Gas Injection. In Proceedings of the 13th Process Technology Conference Proceedings, Nashville, TN, USA, 2–5 April 1995; pp. 415–421.
37. Ramírez-López, A. Analysis of Mixing Efficiency in a Stirred Reactor Using Computational Fluid Dynamics. *Symmetry* **2024**, *16*, 237. [CrossRef]
38. Ramírez-López, A. Analysis of the Hydrodynamics Behavior Inside a Stirred Reactor for Lead Recycling. *Fluids* **2023**, *8*, 268. [CrossRef]
39. Li, X.L.; Hu, S.Y.; Wang, D.Y.; Qu, T.P.; Wu, G.J.; Zhang, P.; Quan, Q.; Zhou, X.Z.; Zhang, Z.X. Inclusion Removements in a Bottom-Stirring Ladle with Novel Slot-Porous Matched Dual Plugs. *Metals* **2022**, *12*, 162. [CrossRef]
40. Sun, Y.P.; Duan, H.J.; Zhang, L.F. A Boundary Layer Model for Capture of Inclusions by Steel-Slag Interface in a Turbulent Flow. *J. Iron Steel Res. Int.* **2023**, *30*, 1101–1108. [CrossRef]
41. Huang, C.; Duan, H.J.; Zhang, L.F. Modeling of Motion of Inclusions in Argon-Stirred Steel Ladles. *Steel Res. Int.* **2024**, *95*, 2300537. [CrossRef]
42. Cao, J.Q.; Li, Y.; Lin, W.M.; Che, J.L.; Zhou, F.; Tan, Y.F.; Li, D.L.; Dang, J.; Chen, C. Assessment of Inclusion Removal Ability in Refining Slags Containing Ce_2O_3. *Crystals* **2023**, *13*, 202. [CrossRef]
43. Wei, T.X.; Oeters, F. A model test for emulsion in gas-stirred ladles. *Steel Res.* **1992**, *63*, 60–68. [CrossRef]
44. Lou, W.; Zhu, M. Numerical simulations of inclusion behavior and mixing phenomena in gas-stirred ladles with different arrangment of tuyeres. *ISIJ Int.* **2014**, *54*, 9–18. [CrossRef]
45. Villarreal-Medina, R.; Jardón–pérez, L.E.; Amaro–Villeda, A.M.; Trápaga–Martínez, G.; Ramírez–Argáez, M.A. Effect of the Injection Position on Mixing Time in a Centric Gas-stirred Ladle Water Model Assisted by a Systematic CFD Study. *ISIJ Int.* **2024**, *64*, 2079–2083. [CrossRef]
46. Xue, L.Q.; Zhao, Y.S.; Wang, J.; Wang, H.; Miao, Z.; Xiong, R.X.; Xu, Z.B.; Lin, W.M.; Niu, X.F.; Chen, C. The Flow Performance of Combined Gas Blowing at the Vacuum Chamber and Up-Snorkel in the Ruhrstahl–Heraeus (RH) Degasser for Electrical Steel Production. *Processes* **2025**, *13*, 448. [CrossRef]
47. Krishna-Murthy, G.G.; Elliott, J.F. Definition and Determination of Mixing Time in Gas Agitated Liquid Baths. *ISIJ Int.* **1992**, *32*, 190–195. [CrossRef]
48. Joo, S.; Guthrie, R.I.L. Modeling flows and mixing in steelmaking ladles designed for single- and dual-plug bubbling operations. *Metall. Mater. Trans. B* **1992**, *23*, 765–778. [CrossRef]
49. Zhu, M.Y.; Inomoto, T.; Sawada, I.; Hsiao, T.C. Fluid flow and mixing phenomena in the ladle stirred by argon through multi-tuyere. *ISIJ Int.* **1995**, *35*, 472–479. [CrossRef]
50. Llanos, C.A.; Garcia, S.; Ramos-Banderas, J.A.; Barreto, J.D.J.; Solorio, G. Multiphase Modeling of the Fluidynamics of Bottom Argon Bubbling during Ladle Operations. *ISIJ Int.* **2010**, *50*, 396–402. [CrossRef]
51. Yamashita, S.; Miyamoto, K.I.; Iguchi, M.; Zeze, M. Model Experiments on the Mixing Time in a Bottom Blown Bath Covered with Top Slag. *ISIJ Int.* **2003**, *43*, 1858–1860. [CrossRef]
52. Patil, S.P.; Satish, D.; Peranandhanathan, M.; Mazumdar, D. Mixing Models for Slag Covered, Argon Stirred Ladles. *ISIJ Int.* **2010**, *50*, 1117–1124. [CrossRef]
53. Han, J.W.; Heo, S.H.; Kam, D.H.; You, B.D.; Park, J.J.; Song, H.S. Transient Fluid Flow Phenomena in a Gas Stirred Liquid Bath with Top Oil Layer-Approach by Numerical Simulation and Water Model Experiments. *ISIJ Int.* **2001**, *41*, 1165–1172. [CrossRef]
54. Liu, H.P.; Qi, Z.Y.; Xu, M.G. Numerical Simulation of Fluid Flow and Interfacial Behavior in Three-phase Argon-Stirred Ladles with One Plug and Dual Plugs. *Steel Res. Int.* **2011**, *82*, 440–458. [CrossRef]
55. Conejo, A.N. It is, either refining impurities or liquid steel mixing in the ladle furnace, but not both. In Proceedings of the 2023 International Conference on Secondary Refining and Inclusion Controlling (SRIC), Kunming, China, 23–25 August 2023; The Chinese Society for Metals: Kunming, China, 2023; Volume 25.
56. Xie, Y.K.; Oeters, F. Experimental studies on the flow velocity of molten metals in a ladle model at centric gas blowing. *Steel Res.* **1992**, *63*, 93–104. [CrossRef]

57. Xie, Y.K.; Oeters, F. Experimental studies on the bath oscillation during gas blowing into liquids part 1 measurements using a single nozzle. *Steel Res.* **1992**, *63*, 227–233. [CrossRef]
58. Xie, Y.K.; Oeters, F. Measurements of bubble plume behaviour and flow velocity in gas stirred liquid woods metal with an eccentric nozzle position. *Steel Res.* **1994**, *65*, 315–319. [CrossRef]
59. Jauhiainen, A.; Jonsson, L.; Sheng, D. Modelling of alloy mixing into steel: The influence of porous plug placement in the ladle bottom on the mixing of alloys into steel in a gas-stirred ladle. A comparison made by numerical simulation. *Scand. J. Metall.* **2001**, *30*, 242–253. [CrossRef]
60. Ganguly, S.; Chakraborty, S. Numerical modelling studies of flow and mixing phenomena in gas stirred steel ladles. *Ironmak. Steelmak.* **2008**, *35*, 524–530. [CrossRef]
61. González-Bernal, R.; Solorio-Diaz, G.; Ramos-Banderas, A.; Torres-Alonso, E.; Hernández-Bocanegra, C.A.; Zenit, R. Effect of the Fluid-Dynamic Structure on the Mixing Time of a Ladle Furnace. *Steel Res. Int.* **2018**, *89*, 1700281. [CrossRef]
62. Kim, S.H.; Fruehan, R.J. Physical modeling of liquid/liquid mass transfer in gas stirred ladles. *Metall. Mater. Trans. B* **1987**, *18*, 381–390. [CrossRef]
63. Nunes, R.P.; Pereira, J.A.M.; Vilela, A.C.F.; Derlaan, F.T.V. Visualisation and analysis of the fluid flow structure inside an elliptical steelmaking ladle through image processing techniques. *J. Eng. Sci. Technol.* **2007**, *2*, 139–150.
64. Michalek, K.; Gryc, K.; Morávka, J. Physical modelling of bath homogenisation in argon stirred ladle. *Metalurgija* **2009**, *48*, 215–218.
65. Jardón-Pérez, L.E.; Amacro-Villeda, A.M.; Conejo, A.N.; Gonzalez-Rivera, C. Optimizing Gas Stirred Ladles by Physical Modeling and PIV Measurements. *Mater. Manuf. Process.* **2018**, *33*, 882–890. [CrossRef]
66. Jardón-Pérez, L.E.; Amacro-Villeda, A.; González-Rivera, C.; González-Rivera, C.; Trápaga, G.; Conejo, C.A.; Ramírez-Argáez, M.A. Introducing the Planar Laser-Induced FluorescenceTechnique (PLIF) to Measure Mixing Time in Gas-Stirred Ladles. *Metall. Mater. Trans. B* **2019**, *50*, 2121–2133. [CrossRef]
67. Jardón-Pérez, L.E.; Amacro-Villeda, A.M.; Trápaga-Martínez, G.; González-Rivera, C.; Ramírez-Argáez, M.A. Utilization of the Planar Laser-Induced Fluorescence Technique (PLIF) to Measure Temperature Fields in a Gas-Stirred Ladle. *Metall. Mater. Trans. B* **2020**, *51*, 2510–2521. [CrossRef]
68. Xie, J.X.; Liu, L.; Lin, T.C.; Yang, L.B. Physical Simulation of 210 t RH Furnace. *J. Iron Steel Res.* **2020**, *32*, 780–787. [CrossRef]
69. Pan, S.; Chiang, J.D.; Hwang, W.S. Effects of gas injection condition on mixing efficiency in the ladle refining process. *J. Mater. Eng. Perform.* **1997**, *6*, 113–117. [CrossRef]
70. Wen, D.S.; Li, J.S.; Xie, C.H.; Tang, H.Y.; Zhang, L.; Wang, Z.X. Physical model of 150 t ladle refining process. *J. Univ. Sci. Technol. Beijing* **2007**, *29*, 101–104. [CrossRef]
71. Ek, M.; Wu, L.; Valentin, P.; Sichen, D. Effect of Inert Gas Flow Rate on Homogenization and Inclusion Removal in a Gas Stirred Ladle. *ISIJ Int.* **2010**, *81*, 1056–1063. [CrossRef]
72. Liu, Y.; Bai, H.T.; Liu, H.P.; Ersson, M.; Jönsson, P.G.; Gan, Y. Physical and Numerical Modelling on the Mixing Condition in a 50 t Ladle. *Metals* **2019**, *9*, 1136. [CrossRef]
73. Tan, F.G.; He, Z.; Jin, S.L.; Pan, L.P.; Li, Y.W.; Li, B.K. Physical Modeling Evaluation on Refining Effects of Ladle with Different Purging Plug Designs. *Steel Res. Int.* **2020**, *91*, 1900606. [CrossRef]
74. Aguilar, G.; Solorio-Diaz, G.; Aguilar-Corona, A.; Ramos-Banderas, J.A.; Hernández, C.A.; Saldaña, F. Study of the Effect of a Plug with Torsion Channels on the Mixing Time in a Continuous Casting Ladle Water Model. *Metals* **2021**, *11*, 1942. [CrossRef]
75. Conejo, A.N.; Zhao, G.; Zhang, L. On the Limits of the Geometric Scale Ratio Using Water Modeling in Ladles. *Metall. Mater. Trans. B* **2021**, *52*, 2263–2274. [CrossRef]
76. Shi, P.J.; Tian, Y.S.; Xu, L.J.; Guo, W.B.; Qiu, S.T. Physical simulation of 120 t ladle bottom blown argon gas. *China Metall.* **2021**, *31*, 30–36. [CrossRef]
77. Johansen, M.; Ramos-Banderas, J.Á.; Hernández-Bocanegra, C.A.; Montes-Rodríguez, J.J. Effect of the location of tracer addition in a ladle on the mixing time through physical and numerical modeling. *ISIJ Int.* **2021**, *61*, 2185–2192. [CrossRef]
78. Cheng, R.; Zhang, L.J.; Yin, Y.B.; Zhang, J.M. Effect of side blowing on fluid flow and mixing phenomenon in gas-stirred ladle. *Metals* **2021**, *11*, 369. [CrossRef]
79. Wang, X.; Zheng, S.G.; Zhu, M.Y. Optimization of the Structure and Injection Position of Top Submerged Lance in Hot Metal Ladle. *ISIJ Int.* **2021**, *61*, 792–801. [CrossRef]
80. Wu, W.Q.; Dong, J.F.; Wei, G.S. Physical Simulation of Bottom Blowing Argon in 150 t Ladle. *Contin. Cast.* **2022**, *2*, 45–48. [CrossRef]
81. Li, Z.W.; Ouyang, W.; Wang, Z.L.; Zheng, R.X.; Bao, Y.P.; Gu, C. Physical Simulation Study on Flow Field Characteristics of Molten Steel in 70t Ladle Bottom Argon Blowing Process. *Metals* **2023**, *13*, 639. [CrossRef]
82. Shan, Q.L.; Sun, Y.; Duan, H.J.; Li, Z.H.; Jia, N.; Chen, W. Water modeling on the transport phenomenon during bottom gas injection of a 210 t ladle. *Steelmaking* **2023**, *39*, 35–43.

83. Zhang, Y.X.; Chen, C.; Lin, W.M.; Yu, Y.C.; Dianyu, E.; Wang, S.B. Numerical Simulation of Tracers Transport Process in Water Model of a Vacuum Refining Unit: Single Snorkel Refining Furnace. *Steel Res. Int.* **2020**, *91*, 202000022. [CrossRef]
84. Ouyang, X.; Lin, W.M.; Luo, Y.Z.; Zhang, Y.X.; Fan, J.P.; Chen, C.; Cheng, G.G. Effect of Salt Tracer Dosages on the Mixing Process in the Water Model of a Single Snorkel Refining Furnace. *Metals* **2022**, *12*, 1948. [CrossRef]
85. Xu, Z.B.; Ouyang, X.; Chen, C.; Li, Y.H.; Wang, T.Y.; Ren, R.J.; Yang, M.M.; Zhao, Y.S.; Xue, L.Q.; Wang, J. Numerical Simulation of the Density Effect on the Macroscopic Transport Process of Tracer in the Ruhrstahl–Heraeus (RH) Vacuum Degasser. *Sustainability* **2024**, *16*, 3923. [CrossRef]
86. Chen, C.; Rui, Q.X.; Cheng, G.G. Effect of Salt Tracer Amount on the Mixing Time Measurement in a Hydrodynamic Model of Gas-Stirred Ladle System. *Steel Res. Int.* **2013**, *84*, 900–907. [CrossRef]
87. Gómez, A.S.; Conejo, A.N.; Zenit, V.R. Effect of Separation Angle and Nozzle Radial Position on Mixing Time in Ladles with Two Nozzles. *J. Appl. Fluid Mec.* **2018**, *11*, 11–20. [CrossRef]
88. Zhang, D.; Chen, C.; Zhang, Y.X.; Wang, S.B. Numerical Simulation of Tracer Transport Process in Water Model of Gas-stirred Ladle. *J. Taiyuan Univ. Technol.* **2020**, *51*, 50–58. [CrossRef]
89. Valentin, P.; Bruch, C.; Kyrylenko, Y.; Kochner, H.; Dannert, C. Influence of the Stirring Gas in a 170-t Ladle on Mixing Phenomena-Formation and On-line Control of Open-Eye at an Industrial LD Steel Plant. *Steel Res. Int.* **2010**, *80*, 552–558. [CrossRef]
90. Li, L.B.; Chen, C.; Tao, X.; Qi, H.Y.; Liu, T.; Yan, Q.J.; Deng, F.; Allayev, A.; Lin, W.M.; Wang, J. Effect of Salt Solution Tracer Dosage on the Transport and Mixing of Tracer in a Water Model of Asymmetrial Gas-Stirred Ladle with a Moderate Gas Flowrate. *Symmetry* **2024**, *16*, 619. [CrossRef]
91. Tao, X.; Qi, H.Y.; Guo, Z.J.; Wang, J.; Wang, X.G.; Yang, J.D.; Zhao, Q.; Lin, W.M.; Yang, K.; Chen, C. Assessment of Measured Mixing Time in a Water Model of Eccentric Gas-Stirred Ladle with a Low Gas Flow Rate: Tendency of Salt Solution Tracer Dispersions. *Symmetry* **2024**, *16*, 1241. [CrossRef]
92. Krishnapisharody, K.; Irons, G.A. A Critical Review of the Modified Froude Number in Ladle Metallurgy. *Metall. Mater. Trans. B.* **2013**, *44*, 1486–1498. [CrossRef]
93. Liu, G.Q.; Ma, L.X.; Xiang, S.G. Chlorides and Chlorates. In *Chemical and Chemical Engineering Physical Properties Data Manual*, 2nd ed.; Chemical Industry Press: Beijing, China, 2012; pp. 575–660. (In Chinese)

symmetry

MDPI

Article

Porous and Magnetic Effects on Axial Couette Flows of Second Grade Fluids in Cylindrical Domains

Constantin Fetecau [1] and Dumitru Vieru [2,3,*]

1 Academy of Romanian Scientists, 3 Ilfov, 050044 Bucharest, Romania; c_fetecau@yahoo.com
2 Department of Mathematics, Saveetha School of Engineering, Saveetha Institute of Medical and Technical Sciences, Saveetha University, Chennai 602105, Tamil Nadu, India
3 Department of Theoretical Mechanics, Technical University of Iasi, 700050 Iasi, Romania
* Correspondence: dumitru_vieru@yahoo.com

Abstract: Axial Couette flows of electrically conducting incompressible second grade fluids are analytically and numerically investigated through a porous medium in the presence of a constant magnetic field. General exact analytical expressions are derived for the dimensionless velocities corresponding to unidirectional unsteady motions in an infinite circular cylinder and between two infinite coaxial circular cylinders. They can be immediately particularized to give similar results for Newtonian fluids in same flows. Exact expressions for steady velocities of a large class of flows were provided. Due to the generality of boundary conditions the problems in discussion are completely solved. For illustration, some case studies with engineering applications are considered and the corresponding velocity fields are provided. Their correctness is graphically proved. It was also proved that the fluid flows slower and the steady state is rather touched in the presence of a magnetic field or porous medium. Moreover, the steady state is rather touched in the case of the motions between circular coaxial cylinders as compared with same motions in an infinite circular cylinder.

Keywords: axial couette flows; second grade fluids; porous and magnetic effects; cylindrical domains

MSC: 76A05

1. Introduction

Flows of incompressible second grade fluids have been extensively investigated in the existing literature. Ting [1] was among the first authors who derived exact solutions for motions of such fluids both in rectangular and cylindrical domains. Other important studies regarding isothermal flows of these fluids in cylindrical domains have been offered by Rajagopal [2], Bandelli and Rajagopal [3], Fetecacau and Corina Fetecau [4], Erdogan and Imrak [5] and Bano et al. [6]. Exact general solutions for unsteady unidirectional motions of same fluids induced by an infinite circular cylinder that applies longitudinal or rotational shear stresses to the fluid have been established by Corina Fetecau et al. [7]. The natural convection of these fluids in an oscillating circular cylinder has been recently investigated by Maria Javaid et al. [8]. Interesting results concerning unsteady flows of incompressible second grade fluids in rectangular domains were obtained, for instance, by Erdogan [9], Erdogan and Imrak [10], Yao and Liu [11], Sultan et al. [12], Imran et al. [13], Baranovskii and Artemov [14], Veera Krishna and Subba Reddy [15], Fetecau and Vieru [16], Baranovskii [17] and Kanuri et al. [18].

It is also known that the motion of a fluid through a porous medium or in the presence of a magnetic field has many engineering applications. Motions through porous media have practical applications in geophysics, astrophysics, composite manufacturing processes, oil reservoir technology, the petroleum industry and agriculture. Interaction between a moving electrically conducting fluid and a magnetic field generates effects that have multiple applications in physics, chemistry and engineering. However, in the existing literature, are few results regarding flows of electrically conducting incompressible second grade fluids (ECISGFs) in cylindrical domains under the influence of a magnetic field or porous medium. Jamil and Zafarullah [19] provided interesting exact solutions for some magnetohydrodynamic (MHD) motions of such fluids between circular cylinders when porous effects are taken into consideration. Unfortunately, their work contains some writing mistakes (see Eqs. (9) and (19), for instance) and it does not contain any proof of the results' correctness. General solutions for some MHD rotational motions of ECISGFs through a porous medium in an infinite circular cylinder have been recently established by Fetecau et al. [20].

The main purpose of this work is to study axial Couette flows of ECISGFs through a porous medium in cylindrical domains under the influence of a magnetic field. Exact general expressions are derived for the dimensionless velocity fields of the fluid motion induced by an infinite circular cylinder that moves along its symmetry axis with a time-dependent velocity. Actually, they represent the first general solutions for such flows of ECISGFs in cylindrical domains and are obtained in the simplest way using the finite Hankel transform only. For the results validation, as well as to bring to light the fluid behavior in some concrete situations, three particular cases are considered and the corresponding velocity fields are provided. They are written as sum of steady state and transient components. The steady state components of these velocities are presented in two different forms whose equivalency is graphically proved. Graphical representations clearly show that the fluid flows faster and the steady state comes later in the absence of a magnetic field or porous medium.

2. Problem Presentation

Let us consider an ECISGF at rest in a porous medium in an infinite circular cylinder of radius R or between two infinite coaxial circular cylinders of radii R_1 and $R_2(> R_1)$. Its constitutive equation is given by the relation

$$T = -p\boldsymbol{I} + \mu \boldsymbol{A}_1 + \alpha_1 \boldsymbol{A}_2 + \alpha_2 \boldsymbol{A}_1^2, \tag{1}$$

in which $\boldsymbol{T}$ is the stress tensor, $\boldsymbol{A}_1$ and $\boldsymbol{A}_2$ are the first two Rivlin-Ericksen tensors, $\boldsymbol{I}$ is the unit tensor, p is the hydrostatic pressure, μ is the dynamic viscosity of the fluid, while α_1 and α_2 are material constants. Taking $\alpha_1 = \alpha_2 = 0$ in Equation (1), the governing equation of incompressible Newtonian fluids is obtained.

At the moment $t = 0^+$ the cylinder in the first case or the outer cylinder in the second case begins to slide along its symmetry axis with the velocity $Vf(t)$ and a circular magnetic field of constant strength B acts in the azimuthally direction on the fluid motion [19,21]. The function $f(\cdot)$ is piecewise continuous and has the zero value in $t = 0$ while V is a constant velocity. Owing to the shear the fluid begins to move and, bearing in mind the previous results in the field, we are looking for a velocity field whose velocity vector $\boldsymbol{v}$ is given by the relation

$$\boldsymbol{v} = \boldsymbol{v}(r,t) = v(r,t)\boldsymbol{k}, \tag{2}$$

reported to a convenient cylindrical coordinate system r, θ and z in which $\mathbf{k}$ is the unit vector along z-direction. Introducing $\boldsymbol{v}$ from Equation (2) in (1) one obtains the relation

$$\tau(r,t) = \left(\mu + \alpha_1 \frac{\partial}{\partial t}\right) \frac{\partial v(r,t)}{\partial r}. \tag{3}$$

for the non-trivial shear stress $\tau(r,t) = T_{rz}(r,t)$.

In the next we assume that the fluid is finitely conducting, its magnetic permeability is constant, the induced magnetic field is negligible and there is no surplus electric charge distribution. In these conditions, in absence of a pressure gradient in the flow direction, the balance of linear momentum reduces to the differential equation (see [19])

$$\rho \frac{\partial v(r,t)}{\partial t} = \frac{\partial \tau(r,t)}{\partial r} + \frac{1}{r}\tau(r,t) - \sigma B^2 v(r,t) - \frac{\varphi}{k}\left(\mu + \alpha_1 \frac{\partial}{\partial t}\right) v(r,t), \tag{4}$$

where ρ is the fluid density, σ is the electrical conductivity while φ and k are the porosity and the permeability of the porous medium. Replacing $\tau(r,t)$ from Equation (3) in (4) one finds the governing equation [19,21]

$$\rho \frac{\partial v(r,t)}{\partial t} = \left(\mu + \alpha_1 \frac{\partial}{\partial t}\right)\left[\frac{\partial^2 v(r,t)}{\partial r^2} + \frac{1}{r}\frac{\partial v(r,t)}{\partial r}\right] - \sigma B^2 v(r,t) - \frac{\varphi}{k}\left(\mu + \alpha_1 \frac{\partial}{\partial t}\right) v(r,t), \tag{5}$$

for the velocity field $v(r,t)$. The corresponding initial and boundary conditions are

$$v(r,0) = 0,\ \ 0 \le r \le R;\ \ v(R,t) = Vf(t),\ \ t > 0, \tag{6}$$

for motions in an infinite circular cylinder.

3. Fluid Velocity for Flows Through a Circular Cylinder

In order to determine the velocity field corresponding to this flow, the governing Equation (5) together with the initial and boundary conditions (6) has to be solved. Using the next non-dimensional functions and variables

$$r^* = \frac{1}{R}r,\ \ t^* = \frac{\nu}{R^2}t,\ \ v^* = \frac{1}{V}v,\ \ \tau^* = \frac{R}{\mu V}\tau,\ \ f^*(t^*) = f\left(\frac{R^2}{\nu}t^*\right), \tag{7}$$

one attains to the next initial-boundary value problem

$$\frac{\partial v(r,t)}{\partial t} = \left(1 + \alpha \frac{\partial}{\partial t}\right)\left[\frac{\partial^2 v(r,t)}{\partial r^2} + \frac{1}{r}\frac{\partial v(r,t)}{\partial r}\right] - Mv(r,t) - K\left(1 + \alpha \frac{\partial}{\partial t}\right) v(r,t), \tag{8}$$

$$v(r,0) = 0,\ \ 0 \le r \le 1;\ v(1,t) = f(t),\ \ t > 0. \tag{9}$$

The constant α and the magnetic and porous parameters M and K, respectively, from Equation (8) are defined by the relations

$$\alpha = \frac{\alpha_1}{\rho R^2},\ \ M = \frac{R^2}{\mu}\sigma B^2 = \frac{R^2}{\nu}\frac{\sigma B^2}{\rho},\ \ K = \frac{\varphi}{k}R^2. \tag{10}$$

The magnetic parameter M is an important indicator of the influence of the magnetic field on the fluid flow. It represents the ratio between the electromagnetic forces and the inertial forces in the fluid. The non-dimensional form of Equation (3) is

$$\tau(r,t) = \left(1 + \alpha \frac{\partial}{\partial t}\right) \frac{\partial v(r,t)}{\partial r};\ \ 0 < r < 1,\ \ t > 0. \tag{11}$$

To determine the dimensionless velocity field $v(r,t)$ corresponding to the motions in discussion, the finite Hankel transform and its inverse defined by the relations (A1) from Appendix A and the identity (A2) are used. In these relations $J_0(\cdot)$ and $J_1(\cdot)$ are Bessel functions of the first kind of zero and one order, respectively, and r_n are the positive roots of the transcendental equation $J_0(r) = 0$. Consequently, multiplying Equation (8) by $rJ_0(rr_n)$, integrating the result between 0 and 1 one finds

$$\frac{\partial v_H(r_n,t)}{\partial t} + \frac{r_n^2 + K_{eff}}{1+\alpha(r_n^2+K)} v_H(r_n,t) - \frac{f(t)+\alpha f'(t)}{1+\alpha(r_n^2+K)} r_n J_1(r_n) = 0, \; v_H(r_n,0) = 0. \quad (12)$$

In above relation $K_{eff} = M + K$ is the effective permeability. The solution of the problem with initial value (12) is given by the relation

$$\begin{aligned} v_H(r_n,t) &= \alpha f(t) \frac{r_n J_1(r_n)}{1+\alpha(r_n^2+K)} \\ &+ (1-\alpha M) \frac{r_n J_1(r_n)}{[1+\alpha(r_n^2+K)]^2} \int_0^t f(s) \exp\left[-\frac{r_n^2+K_{eff}}{1+\alpha(r_n^2+K)}(t-s)\right] ds. \end{aligned} \quad (13)$$

Applying the inverse finite Hankel transform to Equality (13) and using the first entry of Table X from Appendix C of the reference [22] one finds that the dimensionless starting velocity $v(r,t)$ of the fluid can be written in the suitable form

$$\begin{aligned} v(r,t) &= f(t) - 2(1+\alpha K) f(t) \sum_{n=1}^{\infty} \frac{J_0(rr_n)}{[1+\alpha(r_n^2+K)] r_n J_1(r_n)} \\ &+ 2(1-\alpha M) \sum_{n=1}^{\infty} \frac{r_n J_0(rr_n)}{[1+\alpha(r_n^2+K)]^2 J_1(r_n)} \int_0^t f(s) \exp\left[-\frac{r_n^2+K_{eff}}{1+\alpha(r_n^2+K)}(t-s)\right] ds, \end{aligned} \quad (14)$$

which satisfies the initial and boundary conditions (9). In above relations $J_0(\cdot)$ and $J_1(\cdot)$ are standard Bessel functions of the first kind of zero and one order, respectively.

Making $\alpha = 0$ in Equation (14) one finds the dimensionless starting velocity field

$$v_N(r,t) = f(t) - 2f(t) \sum_{n=1}^{\infty} \frac{J_0(rr_n)}{r_n J_1(r_n)} + 2\sum_{n=1}^{\infty} \frac{r_n J_0(rr_n)}{J_1(r_n)} \int_0^t f(s) e^{-(r_n^2+K_{eff})(t-s)} ds, \quad (15)$$

corresponding to MHD axial Couette flow of electrically conducting incompressible Newtonian fluids through a porous medium in an infinite circular cylinder that moves along its symmetry axis with the velocity $Vf(t)$. This velocity does not depend of the parameters M and K independently and a two parameter approach in this case is superfluous. The dimensionless starting shear stresses $\tau(r,t)$ and $\tau_N(r,t)$ corresponding to the same motion of incompressible second grade and Newtonian fluids can be immediately obtained substituting $v(r,t)$ and $v_N(r,t)$ from Equations (14) and (15) in (11).

The relation (14) gives the general expression of the dimensionless velocity field $v(r,t)$ corresponding to MHD axial Couette flow of ECISGFs through a porous medium in an infinite circular cylinder that moves along its symmetry axis with the time dependent velocity $Vf(t)$. Using this expression it is possible to derive the velocity field for any motion of this kind of the respective fluids. Consequently, the problem in discussion is completely solved. For illustration, some particular study cases are considered in the following and the corresponding velocity fields are provided. They will be later used to bring to light some characteristics of the fluid motion and to find the necessary time to reach the steady state. This time is very important for experimental researchers who want to know the transition moment of the motion to the steady or permanent sate.

3.1. Case $f(t) = H(t)\cos(\omega t)$ or $f(t) = H(t)\sin(\omega t)$ (Motion in a Cylinder)

Replacing $f(t)$ by $H(t)\cos(\omega t)$ or $H(t)\sin(\omega t)$ in Equation (14) (where $H(t)$ is the Heaviside unit step function and ω is the non-dimensional frequency of the oscillations) one finds the dimensionless starting velocity fields $v_c(r,t)$ and $v_s(r,t)$ corresponding to the axial Couette flows induced by cosine or sine oscillations, respectively, of the cylinder along its symmetry axis. Since these flows become steady in time, the corresponding starting velocities can be written as sum of their steady state (long time) and transient components, i.e.,

$$v_c(r,t) = [v_{cs}(r,t) + v_{ct}(r,t)]H(t),\ v_s(r,t) = [v_{ss}(r,t) + v_{st}(r,t)]H(t), \tag{16}$$

where

$$\begin{aligned} v_{cs}(r,t) &= \cos(\omega t) - 2(1+\alpha K)\cos(\omega t)\sum_{n=1}^{\infty}\frac{J_0(rr_n)}{[1+\alpha(r_n^2+K)]r_nJ_1(r_n)} \\ &+2(1-\alpha M)\cos(\omega t)\sum_{n=1}^{\infty}\frac{r_n(r_n^2+K_{eff})J_0(rr_n)}{[1+\alpha(r_n^2+K)]a_nJ_1(r_n)} + 2\omega(1-\alpha M)\sin(\omega t)\sum_{n=1}^{\infty}\frac{r_nJ_0(rr_n)}{a_nJ_1(r_n)}, \end{aligned} \tag{17}$$

$$v_{ct}(r,t) = -2(1-\alpha M)\sum_{n=1}^{\infty}\frac{r_n(r_n^2+K_{eff})J_0(rr_n)}{[1+\alpha(r_n^2+K)]a_nJ_1(r_n)}\exp\left[-\frac{r_n^2+K_{eff}}{1+\alpha(r_n^2+K)}t\right], \tag{18}$$

$$\begin{aligned} v_{ss}(r,t) &= \sin(\omega t) - 2(1+\alpha K)\sin(\omega t)\sum_{n=1}^{\infty}\frac{J_0(rr_n)}{[1+\alpha(r_n^2+K)]r_nJ_1(r_n)} \\ &+2(1-\alpha M)\sin(\omega t)\sum_{n=1}^{\infty}\frac{r_n(r_n^2+K_{eff})J_0(rr_n)}{[1+\alpha(r_n^2+K)]a_nJ_1(r_n)} - 2\omega(1-\alpha M)\cos(\omega t)\sum_{n=1}^{\infty}\frac{r_nJ_0(rr_n)}{a_nJ_1(r_n)}, \end{aligned} \tag{19}$$

$$v_{st}(r,t) = 2\omega(1-\alpha M)\sum_{n=1}^{\infty}\frac{r_nJ_0(rr_n)}{a_nJ_1(r_n)}\exp\left[-\frac{r_n^2+K_{eff}}{1+\alpha(r_n^2+K)}t\right]. \tag{20}$$

The constants a_n from the above relations are given by the relation

$$a_n = (r_n^2+K_{eff})^2 + \omega^2[1+\alpha(r_n^2+K)]^2,\quad n = 1,2,3\ldots \tag{21}$$

For a check of the obtained results, let us provide equivalent expressions for the steady state velocities $v_{cs}(r,t)$ and $v_{ss}(r,t)$, namely

$$v_{cc}(r,t) = \mathrm{Re}\left\{\frac{I_0(r\sqrt{\gamma})}{I_0(\sqrt{\gamma})}e^{i\omega t}\right\}, \tag{22}$$

$$v_{ss}(r,t) = \mathrm{Im}\left\{\frac{I_0(r\sqrt{\gamma})}{I_0(\sqrt{\gamma})}e^{i\omega t}\right\}, \tag{23}$$

where Re and Im mean the real part and the imaginary part of that which follows. The complex constant γ from these expressions, that have been determined using the complex velocity $v_{com}(r,t) = v_{cs}(r,t) + iv_{ss}(r,t)$, is given by the relation

$$\gamma = \frac{K_{eff} + i\omega(1+\alpha K)}{1+i\omega\alpha} \tag{24}$$

and Figure 1 shows the equivalence of the expressions of $v_{cs}(r,t)$ and $v_{ss}(r,t)$ given by the relations (17), (22) and (19), (23), respectively. Of course, in the case of incompressible

Newtonian fluids, all previous relations take simpler forms. The steady state velocities from Equations (22) and (23), for instance, become

$$v_{cs}(r,t) = \text{Re}\left\{\frac{I_0(r\sqrt{K_{eff}+i\omega})}{I_0(\sqrt{K_{eff}+i\omega})}e^{i\omega t}\right\},\quad v_{ss}(r,t) = \text{Im}\left\{\frac{I_0(r\sqrt{K_{eff}+i\omega})}{I_0(\sqrt{K_{eff}+i\omega})}e^{i\omega t}\right\}. \quad (25)$$

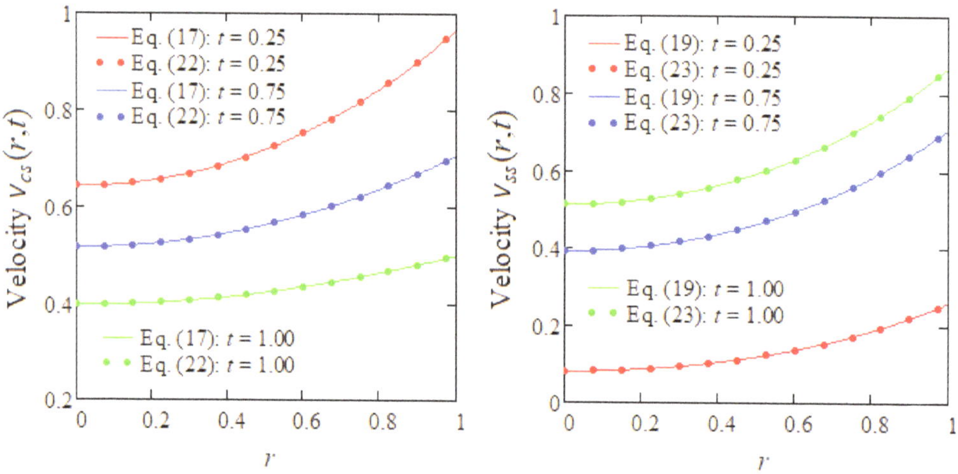

Figure 1. Equivalence of the expressions of $v_{cs}(r,t)$ and $v_{ss}(r,t)$ given by the relations (17), (22) and (19), (23), respectively, for $\alpha = 0.3,\ \omega = \pi/3,\ M = 1,\ K = 0.7$.

In above relations $I_0(\cdot)$ is the modified Bessel function of the first kind of zero order.

3.2. *Case $f(t) = H(t)$ (Motion in a Cylinder)*

In this case the fluid motion is generated by the cylinder that slides along its symmetry axis with the constant velocity V. The expressions of steady and transient components $v_{Cs}(r)$ and $v_{Ct}(r,t)$ of the corresponding starting velocity $v_C(r,t)$, namely

$$v_{Cs}(r) = 1 - 2K_{eff}\sum_{n=1}^{\infty}\frac{J_0(rr_n)}{r_n(r_n^2+K_{eff})J_1(r_n)}, \quad (26)$$

$$v_{Ct}(r,t) = -2(1-\alpha M)\sum_{n=1}^{\infty}\frac{r_nJ_0(rr_n)}{[1+\alpha(r_n^2+K)](r_n^2+K_{eff})J_1(r_n)}\exp\left[-\frac{r_n^2+K_{eff}}{1+\alpha(r_n^2+K)}t\right], \quad (27)$$

have been directly obtained putting $\omega = 0$ in Equations (17) and (18). An equivalent expression for the dimensionless steady velocity $v_{Cs}(r)$, namely

$$v_{Cs}(r) = \frac{I_0(r\sqrt{K_{eff}})}{I_0(\sqrt{K_{eff}})}, \quad (28)$$

has been derived from Equation (22).

3.3. Case When $\lim\limits_{t\to\infty} f(t) = \updownarrow < \infty$

It is interesting to observe that the steady velocity field corresponding to such a motion can be generally determined using the following property

$$\lim_{s\to 0} s\overline{g}(s) = \lim_{t\to\infty} g(t), \tag{29}$$

if $\overline{g}(s)$ is the Laplace transform of $g(t)$. Indeed, applying the Laplace transform to the equality (14) one finds that

$$\begin{aligned}\overline{v}(r,s) &= \overline{f}(s) - 2(1+\alpha K)\overline{f}(s)\sum_{n=1}^{\infty}\frac{J_0(rr_n)}{[1+\alpha(r_n^2+K)]r_nJ_1(r_n)}\\ &+2(1-\alpha M)\overline{f}(s)\sum_{n=1}^{\infty}\frac{r_nJ_0(rr_n)}{[1+\alpha(r_n^2+K)]^2J_1(r_n)}\overline{f}(s)\frac{1}{s+\dfrac{r_n^2+K_{eff}}{1+\alpha(r_n^2+K)}}.\end{aligned} \tag{30}$$

Multiplying the equality (30) with s, taking its limit when $s\to 0$ and bearing in mind the property (29) one finds the steady or permanent velocity field

$$v_p(r) = \updownarrow - 2\updownarrow K_{eff}\sum_{n=1}^{\infty}\frac{J_0(rr_n)}{r_n(r_n^2+K_{eff})J_1(r_n)}. \tag{31}$$

Now, taking $\updownarrow = 1$ in the last equality one recovers $v_{Cs}(r)$ from Equality (26).

4. Fluid Velocity for Flows Between Infinite Coaxial Circular Cylinders

The initial and boundary conditions for this motion are

$$v(r,0) = 0,\ \ R_1 \le r \le R_2;\ \ v(R_1,t) = 0,\ \ v(R_2,t) = Vf(t),\ \ t>0. \tag{32}$$

Introducing the next non-dimensional variables and functions

$$r^* = \frac{1}{R_2}r,\ \ t^* = \frac{\nu}{R_2^2}t,\ \ v^* = \frac{1}{V}v,\ \ \tau^* = \frac{R_2}{\mu V}\tau,\ \ f^*(t^*) = f\left(\frac{R_2^2}{\nu}t^*\right), \tag{33}$$

and again dropping out the star notation one obtains the same dimensionless form (8) for the governing equation of the dimensionless velocity field $v(r,t)$. The corresponding initial and boundary conditions are

$$v(r,0) = 0,\ \ r_0 \le r \le 1;\ v(r_0,t) = 0,\ \ v(1,t) = f(t),\ \ t>0, \tag{34}$$

where the constant $r_0 = R_1/R_2$.

In order to solve the partial differential Equation (8) with the initial and boundary conditions (34), we use the finite Hankel transform and its inverse defined by the relations (A3) from Appendix A [23]. Multiplying Equation (8) by $rA(r,r_n)$ where r_n are the positive roots of the transcendental equation $A(1,r) = 0$ and

$$A(r,r_n) = Y_0(r_0r_n)J_0(rr_n) - J_0(r_0r_n)Y_0(rr_n) \tag{35}$$

and integrating the result between the limits r_0 and 1 and bearing in mind the identity (A4) from Appendix A one finds the following problem with initial value

$$\frac{\partial v_{H0}(r_n,t)}{\partial t} + \frac{r_n^2+K_{eff}}{1+\alpha(r_n^2+K)}v_{H0}(r_n,t) = \frac{2}{\pi}\frac{f(t)+\alpha f'(t)}{1+\alpha(r_n^2+K)}\frac{J_0(r_0r_n)}{J_0(r_n)},\ \ v_{H0}(r_n,0) = 0. \tag{36}$$

The solution of this problem is given by the relation

$$
\begin{aligned}
v_{H0}(r_n,t) &= \frac{2}{\pi}\frac{\alpha f(t) J_0(r_0 r_n)}{[1+\alpha(r_n^2+K)]J_0(r_n)} \\
&+\frac{2(1-\alpha M)J_0(r_0 r_n)}{\pi J_0(r_n)[1+\alpha(r_n^2+K)]^2}\int_0^t f(s)\exp\left[-\frac{r_n^2+K_{eff}}{1+\alpha(r_n^2+K)}(t-s)\right]ds.
\end{aligned}
\tag{37}
$$

Applying the inverse finite Hankel transform to the last relation, one finds the dimensionless velocity field

$$
\begin{aligned}
v_0(r,t) &= \alpha\pi f(t)\sum_{n=1}^{\infty}\frac{r_n^2 J_0(r_0 r_n) J_0(r_n) A(r,r_n)}{[1+\alpha(r_n^2+K)][J_0^2(r_0 r_n)-J_0^2(r_n)]}+\pi(1-\alpha M) \\
&\times\sum_{n=1}^{\infty}\frac{r_n^2 J_0(r_0 r_n) J_0(r_n) A(r,r_n)}{[1+\alpha(r_n^2+K)]^2[J_0^2(r_0 r_n)-J_0^2(r_n)]}\int_0^t f(s)\exp\left[-\frac{r_n^2+K_{eff}}{1+\alpha(r_n^2+K)}(t-s)\right]ds.
\end{aligned}
\tag{38}
$$

In this form, the dimensionless velocity $v_0(r,t)$ satisfies the initial condition and the first boundary condition. The second boundary condition seems to be unsatisfied. An equivalent form for the velocity field $v_0(r,t)$, namely

$$
\begin{aligned}
v_0(r,t) &= \frac{r^2-r_0^2}{1-r_0^2}f(t)-\pi(1+\alpha K)f(t)\sum_{n=1}^{\infty}\frac{J_0(r_0 r_n) J_0(r_n) A(r,r_n)}{[1+\alpha(r_n^2+K)][J_0^2(r_0 r_n)-J_0^2(r_n)]} \\
&+\frac{4\pi f(t)}{1-r_0^2}\sum_{n=1}^{\infty}\frac{J_0(r_n)}{r_n^2[J_0(r_0 r_n)+J_0(r_n)]}A(r,r_n)+\pi(1-\alpha M) \\
&\times\sum_{n=1}^{\infty}\frac{r_n^2 J_0(r_0 r_n) J_0(r_n) A(r,r_n)}{[1+\alpha(r_n^2+K)]^2[J_0^2(r_0 r_n)-J_0^2(r_n)]}\int_0^t f(s)\exp\left[-\frac{r_n^2+K_{eff}}{1+\alpha(r_n^2+K)}(t-s)\right]ds,
\end{aligned}
\tag{39}
$$

has been obtained using the identities (A5) and (A6) from Appendix A. In this form, $v_0(r,t)$ satisfies all imposed initial and boundary conditions. Taking $\alpha=0$ in Equation (39) one finds the dimensionless velocity field corresponding to the MHD axial Couette flows of electrically conducting incompressible Newtonian fluids through a porous medium induced by the outer cylinder that moves along its symmetry axis with the time dependent velocity $Vf(t)$. Its expression is

$$
\begin{aligned}
v_{N0}(r,t) &= \frac{r^2-r_0^2}{1-r_0^2}f(t)+\frac{4\pi f(t)}{1-r_0^2}\sum_{n=1}^{\infty}\frac{J_0(r_n)}{r_n^2[J_0(r_0 r_n)+J_0(r_n)]}A(r,r_n) \\
&-\pi f(t)\sum_{n=1}^{\infty}\frac{J_0(r_0 r_n) J_0(r_n) A(r,r_n)}{J_0^2(r_0 r_n)-J_0^2(r_n)}+\pi\sum_{n=1}^{\infty}\frac{r_n^2 J_0(r_0 r_n) J_0(r_n) A(r,r_n)}{J_0^2(r_0 r_n)-J_0^2(r_n)}\int_0^t f(t-s)\,e^{-(r_n^2+K_{eff})s}ds.
\end{aligned}
$$

4.1. Case $f(t)=H(t)\cos(\omega t)$ or $f(t)=H(t)\sin(\omega t)$ (Motion Between Two Cylinders)

Substituting $f(t)$ by $H(t)\cos(\omega t)$ or $H(t)\sin(\omega t)$ in Equation (39) one finds the dimensionless starting velocities $v_{0c}(r,t)$ and $v_{0s}(r,t)$ corresponding to the MHD axial Couette flows of ECISGFs through a porous medium between coaxial circular cylinders due to cosine or sine oscillations, respectively, of the outer cylinder. The dimensionless steady state and transient components $v_{0cs}(r,t)$, $v_{0ct}(r,t)$ and $v_{0ss}(r,t)$, $v_{0st}(r,t)$ of the these velocities, respectively, are given by the relations

$$
\begin{aligned}
v_{0cs}(r,t) &= \frac{r^2-r_0^2}{1-r_0^2}\cos(\omega t)-\pi(1+\alpha K)\cos(\omega t)\sum_{n=1}^{\infty}\frac{J_0(r_0 r_n) J_0(r_n) A(r,r_n)}{[1+\alpha(r_n^2+K)][J_0^2(r_0 r_n)-J_0^2(r_n)]} \\
&+\frac{4\pi\cos(\omega t)}{1-r_0^2}\sum_{n=1}^{\infty}\frac{J_0(r_n)}{r_n^2[J_0(r_0 r_n)+J_0(r_n)]}A(r,r_n)+\pi(1-\alpha M) \\
&\times\sum_{n=1}^{\infty}\frac{r_n^2\left[(r_n^2+K_{eff})\cos(\omega t)+\omega[1+\alpha(r_n^2+K)]\sin(\omega t)\right]J_0(r_0 r_n) J_0(r_n) A(r,r_n)}{[1+\alpha(r_n^2+K)][J_0^2(r_0 r_n)-J_0^2(r_n)]a_n},
\end{aligned}
\tag{40}
$$

$$v_{0ct}(r,t) = -\pi(1-\alpha M)\sum_{n=1}^{\infty}\frac{r_n^2(r_n^2+K_{eff})J_0(r_0r_n)J_0(r_n)A(r,r_n)}{[1+\alpha(r_n^2+K)][J_0^2(r_0r_n)-J_0^2(r_n)]a_n}\exp\left[-\frac{r_n^2+K_{eff}}{1+\alpha(r_n^2+K)}t\right], \quad (41)$$

$$\begin{aligned} v_{0ss}(r,t) = {} & \frac{r^2-r_0^2}{1-r_0^2}\sin(\omega t) - \pi(1+\alpha K)\sin(\omega t)\sum_{n=1}^{\infty}\frac{J_0(r_0r_n)J_0(r_n)A(r,r_n)}{[1+\alpha(r_n^2+K)][J_0^2(r_0r_n)-J_0^2(r_n)]} \\ & +\frac{4\pi\sin(\omega t)}{1-r_0^2}\sum_{n=1}^{\infty}\frac{J_0(r_n)}{r_n^2[J_0(r_0r_n)+J_0(r_n)]}A(r,r_n) + \pi(1-\alpha M) \\ & \times\sum_{n=1}^{\infty}\frac{r_n^2\left[(r_n^2+K_{eff})\sin(\omega t)] - \omega[1+\alpha(r_n^2+K)]\cos(\omega t)\right]J_0(r_0r_n)J_0(r_n)A(r,r_n)}{[1+\alpha(r_n^2+K)][J_0^2(r_0r_n)-J_0^2(r_n)]a_n}, \end{aligned} \quad (42)$$

$$v_{0st}(r,t) = \omega\pi(1-\alpha M)\sum_{n=1}^{\infty}\frac{r_n^2J_0(r_0r_n)J_0(r_n)A(r,r_n)}{[J_0^2(r_0r_n)-J_0^2(r_n)]a_n}\exp\left[-\frac{r_n^2+K_{eff}}{1+\alpha(r_n^2+K)}t\right]. \quad (43)$$

Newtonian solutions corresponding to same motions are obtained making $\alpha = 0$ in above relations. As before, they do not depend of the magnetic and porous parameters M and K independently, but only by means of the effective permeability $K_{eff} = M + K$.

Equivalent expressions for the steady state velocities $v_{0cs}(r,t)$ and $v_{0ss}(r,t)$ are given by the simple relations

$$v_{0cs}(r,t) = \mathrm{Re}\left\{\frac{K_0(r_0\sqrt{\gamma})I_0(r\sqrt{\gamma}) - I_0(r_0\sqrt{\gamma})K_0(r\sqrt{\gamma})}{K_0(r_0\sqrt{\gamma})I_0(\sqrt{\gamma}) - I_0(r_0\sqrt{\gamma})K_0(\sqrt{\gamma})}e^{i\omega t}\right\}, \quad (44)$$

$$v_{0ss}(r,t) = \mathrm{Im}\left\{\frac{K_0(r_0\sqrt{\gamma})I_0(r\sqrt{\gamma}) - I_0(r_0\sqrt{\gamma})K_0(r\sqrt{\gamma})}{K_0(r_0\sqrt{\gamma})I_0(\sqrt{\gamma}) - I_0(r_0\sqrt{\gamma})K_0(\sqrt{\gamma})}e^{i\omega t}\right\}, \quad (45)$$

where $K_0(\cdot)$ is the modified Bessel function of second kind and zero order. The equivalence of the expressions of the dimensionless steady state velocities fields $v_{0cs}(r,t)$ and $v_{0ss}(r,t)$ given by the relations (40), (44) and (42), (45), respectively, is proved by means of Figure 2.

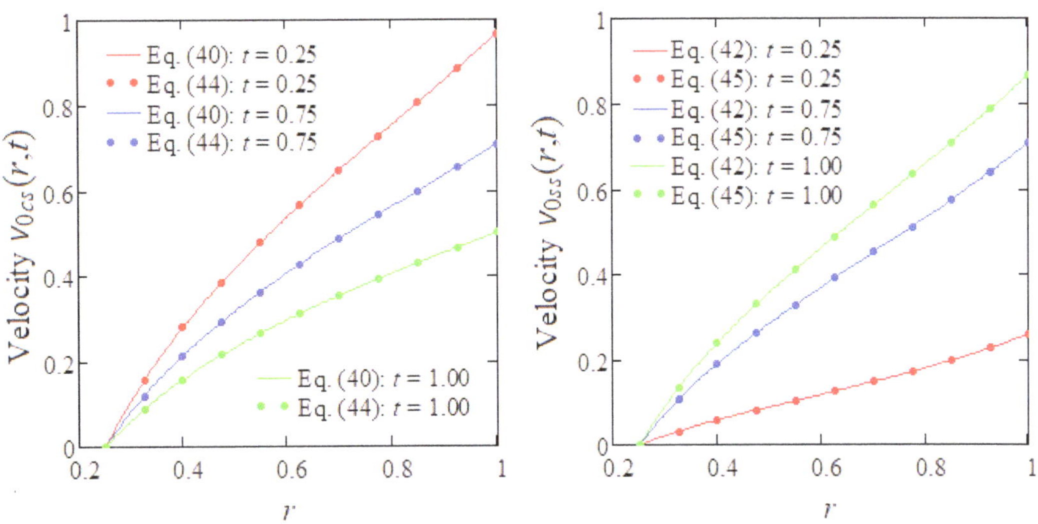

Figure 2. Equivalence of the expressions of $v_{0cs}(r,t)$ and $v_{0ss}(r,t)$ given by the relations (40), (44) and (42), (45), respectively, for $\alpha = 0.3$, $\omega = \pi/3$, $M = 1$, $K = 0.7$ and $r_0 = 0.25$.

As it was to be expected, simple computations show that the limits of the steady state velocities $v_{0cs}(r,t)$ and $v_{0ss}(r,t)$ of the motions between infinite circular cylinders give even the steady state velocities $v_{cs}(r,t)$ and $v_{ss}(r,t)$, respectively, of the fluid motions through an infinite circular cylinder when $r_0 \to 0$, i.e.,

$$\lim_{r_0 \to 0} v_{0cs}(r,t) = \mathrm{Re}\left\{ \frac{I_0(r\sqrt{\gamma})}{I_0(\sqrt{\gamma})} \mathrm{e}^{i\omega t} \right\} = v_{cs}(r,t), \tag{46}$$

$$\lim_{r_0 \to 0} v_{0ss}(r,t) = \mathrm{Im}\left\{ \frac{I_0(r\sqrt{\gamma})}{I_0(\sqrt{\gamma})} \mathrm{e}^{i\omega t} \right\} = v_{ss}(r,t). \tag{47}$$

For completion, it is showed in Figure 3 that the diagrams of the dimensionless starting velocities fields $v_{0ct}(r,t)$ and $v_{0st}(r,t)$ corresponding to flows between circular cylinders overlap over the diagrams of the starting velocities $v_{ct}(r,t)$ and $v_{st}(r,t)$ of the motions through an infinite circular cylinder when $r_0 \to 0$.

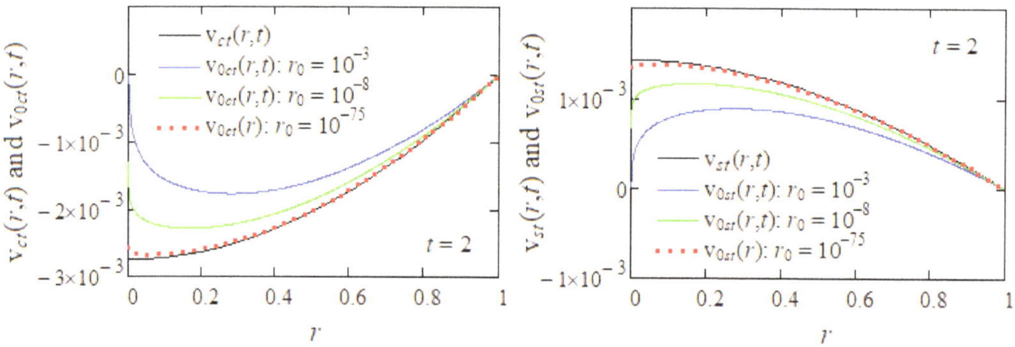

Figure 3. Convergence of the dimensionless starting velocities $v_{0ct}(r,t)$ and $v_{0st}(r,t)$ to the corresponding staring velocities $v_{ct}(r,t)$ and $v_{st}(r,t)$, respectively, of the motion in a circular cylinder when $\alpha = 0.4$, $\omega = \pi/3$, $M = 0.8$, $K = 0.6$ and $r_0 \to 0$.

4.2. Case $f(t) = H(t)$ (Motion Between Two Cylinders)

Substituting $f(t)$ by $H(t)$ in Equation (39) one finds the dimensionless starting velocity $v_{0C}(r,t)$ corresponding to the motion of ECISGFs between two infinite circular coaxial cylinders induced by the outer cylinder that slides along its symmetry axis with the constant velocity V. The expressions of the steady and transient components $v_{0Cs}(r)$ and $v_{0Ct}(r,t)$ of $v_{0C}(r,t)$, which are given by the relations

$$\begin{aligned} v_{0Cs}(r) = {} & \frac{r^2 - r_0^2}{1 - r_0^2} + \frac{4\pi}{1 - r_0^2} \sum_{n=1}^{\infty} \frac{J_0(r_n) A(r, r_n)}{r_n^2 [J_0(r_0 r_n) + J_0(r_n)]} \\ & - \pi K_{eff} \sum_{n=1}^{\infty} \frac{J_0(r_0 r_n) J_0(r_n) A(r, r_n)}{(r_n^2 + K_{eff})][J_0^2(r_0 r_n) - J_0^2(r_n)]}, \end{aligned} \tag{48}$$

$$\begin{aligned} v_{0Ct}(r,t) = {} & -\pi(1 - \alpha M) \sum_{n=1}^{\infty} \frac{r_n^2 J_0(r_0 r_n) J_0(r_n) A(r, r_n)}{(r_n^2 + K_{eff})[1 + \alpha(r_n^2 + K)][J_0^2(r_0 r_n) - J_0^2(r_n)]} \\ & \times \exp\left[-\frac{r_n^2 + K_{eff}}{1 + \alpha(r_n^2 + K)} t \right], \end{aligned} \tag{49}$$

can be directly obtained taking $\omega = 0$ in Equations (40) and (41). An equivalent form for the steady component $v_{0Cs}(r)$, namely

$$v_{0Cs}(r) = \text{Re}\left\{ \frac{K_0(r_0\sqrt{K_{eff}})I_0(r\sqrt{K_{eff}}) - I_0(r_0\sqrt{K_{eff}})K_0(r\sqrt{K_{eff}})}{K_0(r_0\sqrt{K_{eff}})I_0(\sqrt{K_{eff}}) - I_0(r_0\sqrt{K_{eff}})K_0(\sqrt{K_{eff}})} \right\}, \tag{50}$$

has been obtained replacing ω by zero in Equation (44). Taking the limit of the equality (50) when $r_0 \to 0$ one recovers the velocity field $v_{Cs}(r)$ given by Equation (28). Finally, following the same way as in Section 3.3. and using the expression of $v_{0c}(r,t)$ from Equation (39) with $\omega = 0$, it is easily to show that the permanent velocity field $v_{0p}(r)$ corresponding to the case when $\lim_{t\to\infty} f(t) = \updownarrow < \infty$ is given by the simple relation

$$\begin{aligned} v_{0p}(r) = {} & \frac{r^2 - r_0^2}{1 - r_0^2}\updownarrow + \frac{4\pi\updownarrow}{1 - r_0^2}\sum_{n=1}^{\infty} \frac{J_0(r_n)}{r_n^2[J_0(r_0 r_n) + J_0(r_n)]}A(r, r_n) \\ & -\pi\updownarrow K_{eff}\sum_{n=1}^{\infty} \frac{J_0(r_0 r_n)J_0(r_n)}{(r_n^2 + K_{eff})][J_0^2(r_0 r_n) - J_0^2(r_n)]}A(r, r_n). \end{aligned} \tag{51}$$

Making $\updownarrow = 1$ in above relation one finds the expression of $v_{0Cs}(r)$ from Equation (48).

5. Numerical Results

General analytic expressions for the dimensionless velocity fields corresponding to MHD axial Couette flows of ECISGFs through a porous medium in an infinite circular cylinder and between two infinite coaxial circular cylinders have been established. Such flows describe a fairly complex fluid dynamics involving several physical phenomena. Models shearing flow situations which are useful in applications like lubrication, polymer processing and geophysical flows. For validation of the obtained results, some particular flows with technical relevance have been considered. The corresponding motions become steady in time and the starting velocity fields have been presented as sum of their steady state and transient components. The steady state components of these velocities have been presented in different forms whose equivalence was graphically proved in Figures 1 and 2. Interesting expressions for the steady velocity fields corresponding to a large class of flows of these fluids have been also provided. In all cases similar solutions for electrical conducting incompressible Newtonian fluids performing the same flows can be immediately determined.

Now, in order to bring to light some characteristics of the fluid behavior and to determine the dimensionless time that is necessary to get the steady state for both types of unsteady motions, Figures 4–7 have been prepared for increasing values of the time t, different values of the parameters K and M and fixed values for ω and α. This time is very important for the experimental researchers who want to know the moment of transition of the motion to the steady state. It is the time after which the profile of starting velocity superposes over that of its steady component. In all cases the boundary conditions are clearly satisfied and the necessary time to reach the steady state diminishes for increasing values of the parameters K or M. In addition, as it clearly results from these figures and Table 1, the fluid velocity grows up in time but is a decreasing function with regard to the porous and magnetic parameters K and M. Consequently, as expected, the fluid flows slower and the steady state is rather obtained under the influence of a porous medium or magnetic field. Furthermore, it also results from these figures that the steady state is earlier touched in the case of motion between infinite coaxial circular cylinders than in an infinite circular cylinder in the same conditions. This is due to the stationary inner cylinder that slows down the fluid motion.

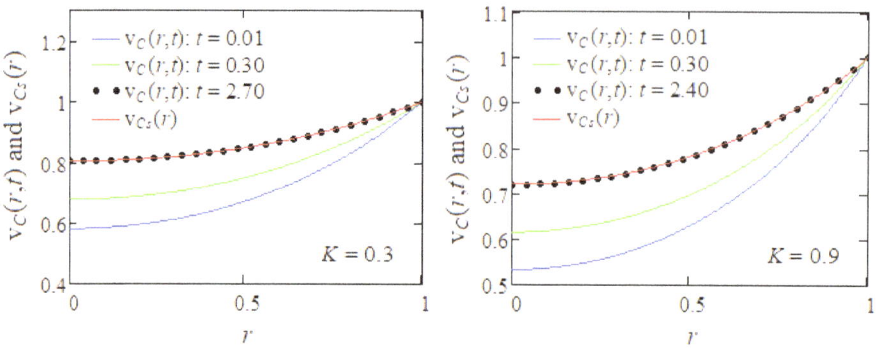

Figure 4. Convergence of the starting velocity $v_C(r,t)$ to its steady component $v_{Cs}(r)$ given by Equation (26) for $\alpha = 0.4,\ M = 0.8,\ K = 0.3$ and $K = 0.9$ and increasing values of t.

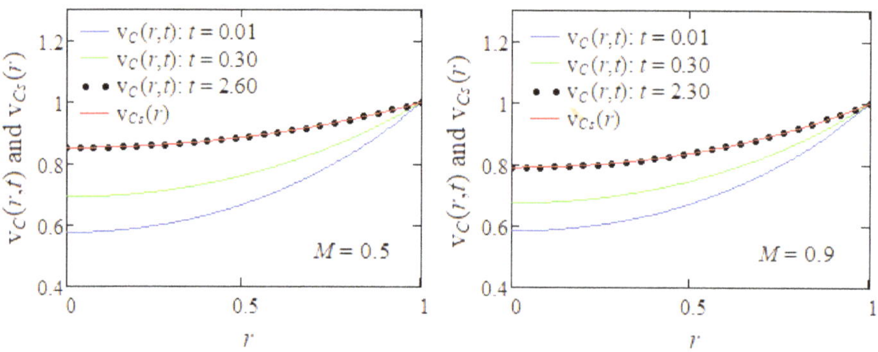

Figure 5. Convergence of the starting velocity $v_C(r,t)$ to its steady component $v_{Cs}(r)$ given by Equation (26) for $\alpha = 0.4,\ K = 0.3,\ M = 0.5$ and $M = 0.9$ and increasing values of t.

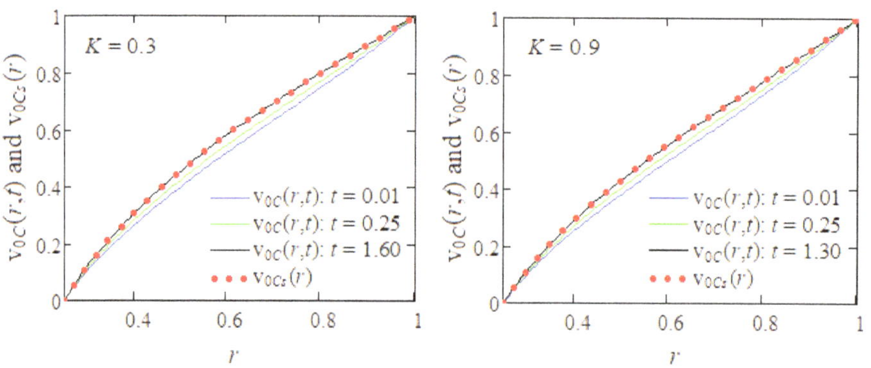

Figure 6. Convergence of starting velocity $v_{0C}(r,t)$ to its steady component $v_{0Cs}(r)$ given by Equation (44) for $\alpha = 0.4,\ M = 0.8,\ K = 0.3$ and $K = 0.9$ and increasing values of t.

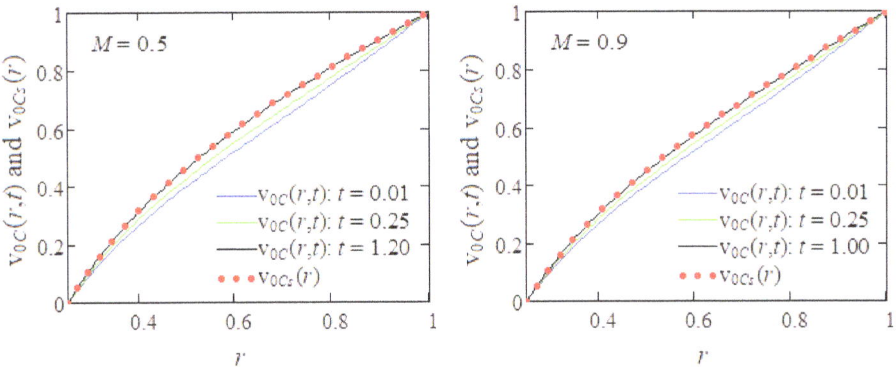

Figure 7. Convergence of starting velocity $v_{0C}(r,t)$ to its steady component $v_{0Cs}(r)$ given by Equation (48) for $\alpha = 0.4,\ K = 0.3,\ M = 0.5$ and $M = 0.9$ and increasing values of t.

Table 1. Comparative values of the steady velocity $v_{0Cs}(r)$ and starting velocity $v_{0C}(r,t)$ at dimensionless times 1.6, 1.3 and 1.2, 1 from Figures 6 and 7, respectively.

r	**Figure 6**				**Figure 7**			
	v_{0C}	v_{0Cs}	v_{0C}	v_{0Cs}	v_{0C}	v_{0Cs}	v_{0C}	v_{0Cs}
	t=1.6		**t=1.3**		**t=1.2**		**t=1**	
$r_0 = 0.25$	0	0	0	0	0	0	0	0
0.30	0.118	0.119	0.111	0.112	0.121	0.122	0.116	0.117
0.35	0.218	0.219	0.206	0.208	0.223	0.225	0.214	0.217
0.40	0.306	0.307	0.289	0.291	0.312	0.315	0.300	0.304
0.45	0.384	0.385	0.363	0.366	0.391	0.395	0.377	0.382
0.50	0.454	0.456	0.431	0.434	0.462	0.467	0.447	0.452
0.55	0.520	0.521	0.495	0.498	0.528	0.534	0.511	0.517
0.60	0.580	0.582	0.555	0.558	0.590	0.595	0.572	0.578
0.65	0.638	0.640	0.612	0.615	0.647	0.653	0.629	0.636
0.70	0.693	0.695	0.668	0.671	0.702	0.707	0.685	0.691
0.75	0.747	0.748	0.723	0.726	0.755	0.760	0.739	0.744
0.80	0.798	0.800	0.778	0.780	0.806	0.810	0.791	0.796
0.85	0.849	0.850	0.832	0.834	0.856	0.859	0.844	0.848
0.90	0.900	0.900	0.887	0.889	0.904	0.907	0.896	0.898
0.95	0.950	0.900	0.943	0.944	0.952	0.954	0.948	0.949
1.00	1.000	1.000	1.000	1.000	1.000	1.000	1.000	1.000

In conclusion, when such fluids move in a magnetic field electromagnetic forces arise, affecting the flow and models shearing flow situations which are useful in applications like lubrication, polymer processing, and geophysical flows. Even though it sounds highly theoretical, the present study can be useful to models real systems in engineering, geophysics, biomedical, metalurgical processes, polymer and chemical processes, lubrication systems and environmental engineering.

6. Conclusions

Axial Couette flows of ECISGFs through a porous medium in two cylindrical domains have been analytically and numerically investigated in the presence of a magnetic field. The fluid motion is produced by an infinite circular cylinder that slides along its symmetry axis with the time-dependent velocity $Vf(t)$. General analytic expressions were established for the dimensionless velocity field $v(r,t)$ using the finite Hankel transform only. These expressions can be used to derive the fluid velocity for any motion of this kind of the respective fluids and the problems in discussion are completely solved. For illustration, some flows with engineering applications have been considered and the corresponding velocity fields have been used to bring to light some characteristics of the fluid behavior.

The main findings that have been obtained by means of this study are:

- MHD axial Couette flows of ECISGFs through a porous medium in two cylindrical domains were analytically investigated. The fluid motion was induced by a circular cylinder that moves along its symmetry axis with the time-dependent velocity $Vf(t)$.
- General expressions have been established for the corresponding dimensionless velocity fields. They can generate exact solutions for any motion of this type of the respective fluids and the motion problems in discussion are completely solved.
- For illustration, some particular cases have been considered and the corresponding velocity fields were derived. For validation, the steady components of these velocities were presented in different forms and their equivalence was graphically proved.
- As application, since the considered motions become steady in time, the necessary time to reach the steady state was graphically determined. That time, which is important for experimental researchers, declines for increasing values of K or M.
- Consequently, the steady state for such motions of ECISGFs is earlier touched in the presence of a porous medium or magnetic field. It is also rather touched for flows between infinite circular cylinders than in an infinite circular cylinder.
- The fluid flows faster in the absence of a porous medium or magnetic field.

In completion, we provide an important equation for the non-trivial shear stress that can be useful for readers to solve similar motion problems in which the shear stress is prescribed on the boundary. In such cases they can use the fact that the dimensionless shear stress $\tau(r,t)$ satisfies the governing equation

$$\begin{aligned}\frac{\partial\tau(r,t)}{\partial t} &= \left(1+\alpha\frac{\partial}{\partial t}\right)\left[\frac{\partial^2\tau(r,t)}{\partial r^2}+\frac{1}{r}\frac{\partial\tau(r,t)}{\partial r}-\frac{1}{r^2}\tau(r,t)\right]\\ &-M\tau(r,t)-K\left(1+\alpha\frac{\partial}{\partial t}\right)\tau(r,t),\end{aligned} \quad (52)$$

that can be easily solved using the Hankel transforms. As soon as the shear stress $\tau(r,t)$ is determined for a given motion, the corresponding velocity field can be easily obtained solving the linear differential Equation (8).

Author Contributions: Conceptualization, C.F. and D.V.; Methodology, C.F. and D.V.; Software, D.V.; Validation, C.F. and D.V.; Writing—review and editing, C.F. and D.V.; Supervision, C.F. and D.V.; project administration, D.V. All authors have read and agreed to the published version of the manuscript.

Funding: This research received no external funding.

Data Availability Statement: Data are contained within the article.

Acknowledgments: The authors would like to express their gratitude to reviewers for the careful assessment and fruitful suggestions and recommendations.

Conflicts of Interest: The authors declare no conflicts of interest.

Appendix A

$$u_H(r_n) = H\{u(r)\} = \int_0^1 r u(r) J_0(r r_n) dr, \; u(r) = 2\sum_{n=1}^{\infty} \frac{J_0(r r_n)}{[J_1(r_n)]^2} u_H(r_n), \tag{A1}$$

$$\int_0^1 r\left\{\frac{d^2u(r)}{dr^2} + \frac{1}{r}\frac{du(r)}{dr}\right\} J_0(r r_n) dr = r_n u(1) J_1(r_n) - r_n^2 u_H(r_n), \tag{A2}$$

$$u_{H0}(r_n) = H\{u(r)\} = \int_{r_0}^1 r u(r) A(r, r_n) dr, \; u(r) = \frac{\pi^2}{2}\sum_{n=1}^{\infty} \frac{r_n^2 J_0^2(r_n) A(r, r_n)}{J_0^2(r_0 r_n) - J_0^2(r_n)} u_{H0}(r_n), \tag{A3}$$

$$\int_{r_0}^1 r\left[\frac{d^2u(r)}{dr^2} + \frac{1}{r}\frac{du(r)}{dr}\right] A(r, r_n) dr = \frac{2}{\pi}\left[\frac{J_0(r_0 r_n)}{J_0(r_n)} u(1) - u(r_0)\right] - r_n^2 u_{H0}(r_n), \tag{A4}$$

$$\int_{r_0}^1 r A(r, r_n) dr = \frac{2}{\pi}\frac{J_0(r_0 r_n)}{r_n^2 J_0(r_n)}, \tag{A5}$$

$$\int_{r_0}^1 r^3 A(r, r_n) dr = \frac{2}{\pi}\frac{J_0(r_0 r_n)}{r_n^2 J_0(r_n)} + \frac{2}{\pi}\frac{(4 - r_0^2 r_n^2) J_0(r_n) - 4 J_0(r_0 r_n)}{r_n^4 J_0(r_n)}, \tag{A6}$$

The last identity has been determined using the known results

$$\frac{d}{dz}\left[z^3 J_1(z) - 2z^2 J_2(z)\right] dz = z^3 J_0(z), \; \frac{d}{dz}\left[z^3 Y_1(z) - 2z^2 Y_2(z)\right] dz = z^3 Y_0(z). \tag{A7}$$

References

1. Ting, T.W. Certain unsteady flows of second grade fluids. *Arch. Ration. Mech. Anal.* **1963**, *14*, 1–26. [CrossRef]
2. Rajagopal, K.R. Longitudinal and torsional oscillations of a rod in a non-Newtonian fluid. *Acta Mech.* **1983**, *49*, 281–285. [CrossRef]
3. Bandelli, R.; Rajagopal, K.R. Start-up flows of second grade fluids in domains with one finite dimension. *Int. J. Non-Linear Mech.* **1995**, *30*, 817–839. [CrossRef]
4. Fetecau, C. Starting solutions for the motion of a second grade fluid due to longitudinal and torsional oscillations of a circular cylinder. *Int. J. Eng. Sci.* **2006**, *44*, 788–795. [CrossRef]
5. Erdogan, M.E.; Imrak, C.E. On the comparison of the methods used for the solutions of the governing equation for unsteady unidirectional flows of second grade fluids. *Int. J. Eng. Sci.* **2007**, *45*, 786–796. [CrossRef]
6. Bano, Z.; Shah, N.F.; Islam, S.; Haroon, T. Exact solution to longitudinal and torsional oscillations of an electrically conducting second grade fluid. *Int. J. Phys. Sci.* **2011**, *6*, 2939–2943. [CrossRef]
7. Fetecau, C.; Mocanu, C.; Samiulhaq; Fetecau, C. Exact solutions for motions of second grade fluids induced by an infinite cylinder that applies arbitrary shear stresses to fluid. *J. Comput. Theor. Nanosci.* **2012**, *17*, 266–270. [CrossRef]
8. Javaid, M.; Imran, M.A.; Khan, I.; Nisar, K.S. Natural convection flow of a second grade fluid in an infinite vertical cylinder. *Sci. Rep.* **2020**, *10*, 8327. [CrossRef] [PubMed]
9. Erdogan, M.E. On unsteady motions of a second-order fluid over a plane wall. *Int. J. Non-Linear Mech.* **2003**, *38*, 1045–1051. [CrossRef]
10. Erdogan, M.E.; Imrak, C.E. Effects of the side walls on unsteady flow of a second grade fluid over a plane wall. *Int. J. Non-Linear Mech.* **2008**, *43*, 779–782. [CrossRef]
11. Yao, Y.; Liu, Y. Some unsteady flows of a second grade fluid over a plane wall. *Nonlinear Anal. Real World Appl.* **2010**, *11*, 4442–4450. [CrossRef]
12. Sultan, Q.; Nazar, M.; Ahmad, I.; Ali, U. Flow of second-grade fluid between two walls induced by rectified sine pulses shear stress. *J. Mech.* **2015**, *31*, 573–582. [CrossRef]

13. Imran, M.; Tahir, M.; Nazar, M.; Kamran, M. Some Couette flows of a second-grade fluid due to tangential stresses. *Sci. Int.* **2015**, *27*, 829–834.
14. Baranovskii, E.S.; Artemov, M.A. Steady flows of second grade fluids subject to stick-slip boundary conditions. In Proceedings of the 23rd International Conference Engineering Mechanics, Svratka, Czech Republic, 15–18 May 2017; pp. 110–113.
15. VeeraKrishna, M.; Reddy, G.S. Unsteady MHD reactive flow of second grade fluid through porous medium in a rotating parallel plate channel. *J. Anal.* **2018**, *27*, 103–120. [CrossRef]
16. Fetecau, C.; Vieru, D. General solutions for some MHD motions of second-grade fluids between parallel plates embedded in a porous medium. *Symmetry* **2023**, *15*, 183. [CrossRef]
17. Baranovskii, E.S. Analytical solutions to the unsteady Poiseuille flow of a second grade fluid with slip boundary conditions. *Polymers* **2024**, *16*, 179. [CrossRef] [PubMed]
18. Kanuri, J.V.; Chandra Sekhar, K.V.; Brahmanandam, P.S.; Ramanaiah, J.V. Analytical solutions of Poiseuille flow of second-grade fluid. *J. Nav. Archit. Mar. Eng.* **2024**, *21*, 67–77. [CrossRef]
19. Jamil, M.; Zafarullah, M. MHD flows of second grade fluid through the moving porous cylindrical domain. *Eur. J. Pure Appl. Math.* **2019**, *12*, 1149–1175. [CrossRef]
20. Fetecau, C.; Akhtar, S.; Forna, N.C.; Moroşanu, C. General solutions for MHD motions of second grade fluids through a circular cylinder filled with porous medium. *Symmetry* **2025**, *17*, 319. [CrossRef]
21. Hamza, S.E.E. MHD flow of an Oldroyd-B fluid through porous medium in a circular channel under the effect of time dependent pressure gradient. *Am. J. Fluid Dyn.* **2017**, *7*, 1–11. [CrossRef]
22. Sneddon, I.N. *Fourier Transforms*; McGRAW-Hill: New-York, NY, USA, 1951.
23. Debnath, L.; Bhatta, D. *Integral Transforms and Their Applications*, 2nd ed.; Chapman & Hall/CRC: Boca Raton, FL, USA, 2007.

symmetry

MDPI

Article

On the Horizontal Divergence Asymmetry in the Gulf of Mexico

Tianshu Zhou [1,2], Jin-Han Xie [1,2,*] and Dhruv Balwada [3]

1 Department of Mechanics and Engineering Science, College of Engineering, Peking University, Beijing 100871, China; zts_coe_2020@stu.pku.edu.cn
2 State Key Laboratory for Turbulence and Complex Systems, Peking University, Beijing 100871, China
3 Lamont-Doherty Earth Observatory, Columbia University, New York, NY 10964, USA; dbalwada@ldeo.columbia.edu
* Correspondence: jinhanxie@pku.edu.cn

Abstract: Due to the geostrophic balance, horizontal divergence-free is often assumed when analyzing large-scale oceanic flows. However, the geostrophic balance is a leading-order approximation. We investigate the statistical feature of weak horizontal compressibility in the Gulf of Mexico by analyzing drifter data (the Grand LAgrangian Deployment (GLAD) experiment and the LAgrangian Submesoscale ExpeRiment (LASER)) based on the asymptotic probability density function of the angle between velocity and acceleration difference vectors in a strain-dominant model. The results reveal a notable divergence at scales between 10 km and 300 km, which is stronger in winter (LASER) than in summer (GLAD). We conjecture that the divergence is induced by wind stress with its curl parallel to the Earth's rotation.

Keywords: homogeneous turbulence; isotropic turbulence; oceanic surface flow; ocean surface drifter; skewness of horizontal divergence

1. Introduction

In large-scale oceanic flows, the Coriolis effect and the pressure gradient dominate the horizontal momentum balance [1,2]. When the Coriolis parameter is a constant, geostrophic flow is horizontally non-divergent. This horizontal non-divergent condition simplifies many theoretical analyses. For instance, the third-order structure function theory with the incompressible condition can be used to study the direction of energy cascades in the atmosphere [3] and ocean [4,5].

Geostrophic balance describes the oceanic mesoscale to the leading order of small Rossby number, while at submeso and smaller scales, horizontal divergence can be of leading order [6] and even mesoscale motion can have weak horizontal divergence. Second-order structure functions can be constructed to detect the amplitude of this compressibility [7], but the sign of horizontal velocity divergence is unknown.

We go beyond the velocity structure function and study the combined information from the velocity difference vector $\delta\boldsymbol{u}$ and acceleration difference vector $\delta\boldsymbol{a}$. In three-dimensional (3D) turbulence, the ensemble average of their dot product, denoted as $\langle\delta\boldsymbol{u}\cdot\delta\boldsymbol{a}\rangle$, is found to be a constant under a fixed energy cascading rate with its sign corresponding to the direction of cascade in the inertial range, which is justified by experiments [8]. Thus, it is natural to study the angle θ between $\delta\boldsymbol{u}$ and $\delta\boldsymbol{a}$. In a simple model with constant strain rates and without rotation, the probability density function (p.d.f.) of $\cos\theta$ is related to the ratio of the eigenvalues of the velocity strain rate tensor [9]. The

 https://doi.org/10.3390/sym17010136

p.d.f. of $\cos\theta$ under this model fits well with the p.d.f. from experiments [9]. Therefore, it is possible to obtain information about the eigenvalues of the strain rate tensor from the p.d.f. of $\cos\theta$, which has not been realized in former studies.

This paper treats the oceanic surface flow as a projection of 3D flow. In this sense, the eigenvalues of the strain rate tensor reveal the compressibility of the surface flow. By analyzing drifter data in the Gulf of Mexico, we aim to detect the amplitude and sign of the horizontal divergence in oceanic surface flow. This will be realized through an asymptotically derived relation between the p.d.f. of $\cos\theta$ and the ratio of eigenvalues of the strain rate tensor.

The structure of the rest of this paper is as follows: Section 2 introduces the theoretical background and methods for analyzing the drifter data. Then we show the main results in Section 3. Finally, we discuss the phenomenon found from the results and propose the possible physics in Section 4. Details of data processing and error estimation will be shown in Appendices A and B. More fitting results are displayed in Appendix C.

2. Theories and Methods

We built a model that captures the relation between the angle of acceleration and velocity difference vectors and the divergence at a certain scale, which is inspired by Gibert et al. (2012) [9]. Since drifters may be distributed at strain-dominant regions around vortices in a diffusion process [10], we consider that at each scale, random strains dominantly impact pair separation. Thus, we keep the model to be irrotational.

2.1. Random Strain Model

We can start from a simple case where the flow has constant strain rates and does not rotate. Consider a 3D incompressible flow, then the velocity gradient tensor $\boldsymbol{M}$ is constant and symmetric with zero trace:

$$M_{ij} = \frac{\partial u_i}{\partial x_j} = \frac{\partial u_j}{\partial x_i} = M_{ji} = \text{const.} \tag{1}$$

In general, the velocity can be integrated from its gradient tensor $\boldsymbol{M}$, which is $\boldsymbol{u} = \boldsymbol{u}_0 + \int \boldsymbol{M} \cdot \mathrm{d}\boldsymbol{r}$. Additionally, the acceleration can be derived from velocity, i.e., $\boldsymbol{a} = \partial \boldsymbol{u}/\partial t + \boldsymbol{u} \cdot \nabla \boldsymbol{u}$ [1], where $\partial \boldsymbol{u}/\partial t$-related terms are zero in a statistically steady flow. For simplicity, we consider a linear velocity profile and the corresponding acceleration can be expressed as follows [9]:

$$u_i = M_{ij} x_j, \tag{2a}$$

$$a_i = u_j \frac{\partial u_i}{\partial x_j} = M_{ij} u_j. \tag{2b}$$

Here, we chose the Cartesian coordinates $O\text{-}xyz$, where the x-axis points to the East, the y-axis points to the North and the z-axis is vertically upwards. The indices $i,\ j$ range from 1 to 3, and each pair of repeating indices represents the corresponding summation from 1 to 3. The velocity is also denoted as $\boldsymbol{u} = u\mathbf{i} + v\mathbf{j} + w\mathbf{k}$, where $\mathbf{i},\ \mathbf{j},\ \mathbf{k}$ represent the unit vectors of the axes $x,\ y,\ z$, respectively.

We can perform an orthogonal diagonalization since $\boldsymbol{M}$ is assumed symmetric. Choosing the principal coordinates $O\text{-}\xi\eta\zeta$, we obtain the following:

$$\boldsymbol{M} = \begin{pmatrix} \lambda_1 & & \\ & \lambda_3 & \\ & & \lambda_2 \end{pmatrix} = \lambda_1 \begin{pmatrix} 1 & & \\ & -(1+\alpha) & \\ & & \alpha \end{pmatrix}, \tag{3}$$

where $\lambda_1 \geq \lambda_2 \geq \lambda_3$ are the eigenvalues of $\boldsymbol{M}$. Here, we applied the incompressibility condition as follows [9]:

$$\nabla \cdot \boldsymbol{u} = \mathrm{tr}\boldsymbol{M} = \sum_{i=1}^{3} \lambda_i = 0, \tag{4}$$

and $\alpha = \lambda_2/\lambda_1$.

For the large-scale oceanic flow, due to the dominant geostrophic balance, the principal axes ξ and η lie close to the x-y plane. Therefore, it is reasonable to assume $|\alpha| \ll 1$ and the ζ-axis is parallel to the z-axis. Under these assumptions, the horizontal velocity divergence is not zero:

$$\nabla_z \cdot \boldsymbol{u} = \frac{\partial u}{\partial x} + \frac{\partial v}{\partial y} = -\frac{\partial w}{\partial z} = -\lambda_2, \tag{5}$$

where $\nabla_z := \mathbf{i}\partial_x + \mathbf{j}\partial_y$ is the horizontal gradient.

Thus, a divergent flow would correspond to a negative λ_2 and a negative α, and a convergent flow would correspond to a positive λ_2 and a positive α.

Considering two-dimensional (2D) drifter pair dispersion in the ocean, the relative position vector between two drifters can be expressed as follows:

$$\delta\boldsymbol{r} = r_0(\mathbf{e}_\xi \cos\delta\phi + \mathbf{e}_\eta \sin\delta\phi), \tag{6}$$

where $\mathbf{e}_\xi$ and $\mathbf{e}_\eta$ are the unit vectors of the axes ξ and η, r_0 is the distance between the drifters, and $\delta\phi$ represents the angle between $\delta\boldsymbol{r}$ and $\mathbf{e}_\xi$. When $\delta\boldsymbol{r}$ is fixed, the velocity and acceleration difference vectors can be expressed as follows:

$$\delta\boldsymbol{u} = \begin{pmatrix} \lambda_1 & & \\ & \lambda_3 & \\ & & \lambda_2 \end{pmatrix} \cdot \delta\boldsymbol{r} = r_0\lambda_1 \begin{pmatrix} \cos\delta\phi \\ -(1+\alpha)\sin\delta\phi \\ 0 \end{pmatrix}, \tag{7}$$

and

$$\delta\boldsymbol{a} = \begin{pmatrix} \lambda_1^2 & & \\ & \lambda_3^2 & \\ & & \lambda_2^2 \end{pmatrix} \cdot \delta\boldsymbol{r} = r_0\lambda_1^2 \begin{pmatrix} \cos\delta\phi \\ (1+\alpha)^2\sin\delta\phi \\ 0 \end{pmatrix}. \tag{8}$$

Let θ be the angle between $\delta\boldsymbol{u}$ and $\delta\boldsymbol{a}$. We can obtain the exact expression of $\cos\theta$:

$$\cos\theta = \frac{\delta\boldsymbol{u} \cdot \delta\boldsymbol{a}}{|\delta\boldsymbol{u}||\delta\boldsymbol{a}|} = \frac{\cos^2\delta\phi - (1+\alpha)^3\sin^2\delta\phi}{\sqrt{\cos^2\delta\phi + (1+\alpha)^2\sin^2\delta\phi}\sqrt{\cos^2\delta\phi + (1+\alpha)^4\sin^2\delta\phi}}, \tag{9}$$

which is independent of r_0 and λ_1.

Oceanic flows can be different when the scale changes. To apply the model to real oceanic flows, we assume λ_i ($i = 1, 2, 3$) and α to be functions of the scale, i.e., functions of r_0 in a 2D drifter pair dispersion. We further assume that the angle $\delta\phi$ is a random variable (r.v.), to capture the changing direction of the principal axes and drifter pairs. When the flow is assumed isotropic, α is independent of $\delta\phi$. Considering Equation (9), for a fixed scale, the quantity $\cos\theta$ as a function of $\delta\phi$ is also a r.v. Our aim is to obtain α from the distribution of $\cos\theta$ under different scales.

2.2. Probability Distribution of $\cos\theta$

The exact expression of $\cos\theta$ as a function of $\delta\phi$ (9) has a complicated form which is not easily applicable to data analysis. As has been discussed above, the dominant geostrophic balance implies $|\alpha| \ll 1$, so we can expand $\cos\theta$ into a series of α through Taylor expansion:

$$\cos\theta = \sum_{k=0}^{\infty} \alpha^k f_k(\delta\phi). \tag{10}$$

At the first order, we obtain the following:

$$\cos\theta = \cos 2\delta\phi + \frac{3}{2}\alpha(\cos^2 2\delta\phi - 1). \tag{11}$$

It is reasonable to reset $\cos 2\delta\phi$ instead of $\delta\phi$ to be the independent variable. We introduce the following notations for convenience:

$$x := \cos 2\delta\phi, \quad y := \cos\theta. \tag{12}$$

The reverse function of (11) exists when $|\alpha| \leq 1/3$, and the 1st-order relations become

$$y = x - \frac{3}{2}\alpha(1 - x^2), \tag{13a}$$

$$x = y + \frac{3}{2}\alpha(1 - y^2). \tag{13b}$$

To the second order, these relations become

$$\begin{aligned} y &= x - \frac{3}{2}\alpha(1 - x^2) + \alpha^2(1 - x^2)\left(\frac{3}{4} - \frac{19}{8}x\right), \\ x &= y + \frac{3}{2}\alpha(1 - y^2) - \alpha^2(1 - y^2)\left(\frac{3}{4} + \frac{17}{8}y\right), \end{aligned} \tag{14}$$

where the inverse function $x = x(y)$ here can be derived by the matching of expansions in a perturbative method. The results of our study will show that α is on the order of 10^{-2} for oceanic surface flow, making the 1st-order expansion a good approximation.

The probability distribution of $\cos\theta$ can be expressed by the distribution of $\delta\phi$ (or $\cos 2\delta\phi$) in an asymptotic form. Consider a simple case when $\delta\phi$ is uniformly distributed over $[0,\ 2\pi]$. The p.d.f. of $x = \cos 2\delta\phi$ was derived in the 2D incompressible case with $\alpha = 0$ [9]:

$$P(x) = \frac{1}{\pi\sqrt{1 - x^2}}, \quad x \in (-1,\ 1). \tag{15}$$

To the 1st order of α, we derive the expression of p.d.f. of $\cos\theta$ as follows:

$$\tilde{P}(y) = \frac{1 - \frac{3}{2}\alpha y}{\pi\sqrt{1 - y^2}}, \quad y \in (-1,\ 1). \tag{16}$$

This expression introduces an asymmetric part with α as the leading order, which inspires us to describe the influence of non-zero α by even-odd decomposition.

2.3. *Procedure for Analyzing Oceanic Surface Flow*

Hinted by (16), we decompose the p.d.f.s into the even and odd parts

$$\begin{aligned} P_{\text{even}}(x) &= \frac{P(x) + P(-x)}{2}, \qquad \tilde{P}_{\text{even}}(y) = \frac{\tilde{P}(y) + \tilde{P}(-y)}{2}, \\ P_{\text{odd}}(x) &= \frac{P(x) - P(-x)}{2}, \qquad \tilde{P}_{\text{odd}}(y) = \frac{\tilde{P}(y) - \tilde{P}(-y)}{2}. \end{aligned} \tag{17}$$

The analysis will be based on the following assumptions:

Assumption 1. *Assuming that in the even-odd decomposition* $P(x) = P_{\text{even}}(x) + P_{\text{odd}}(x)$, $P_{\text{even}}(x) \gg P_{\text{odd}}(x)$. *So we can introduce a bookkeeping parameter* ϵ *to mark the smallness of the odd part of the p.d.f., i.e.,*

$$P(x) = P_{\text{even}}(x) + \epsilon P_{\text{odd}}(x), \quad 0 < \epsilon \ll 1. \tag{18}$$

Assumption 2. $\alpha = \lambda_2/\lambda_1$ *is small in oceanic flows, i.e.,* $|\alpha| \ll 1$. *Hence the discussion in Section 2.2 holds valid.*

Assumption 3. $|\alpha| \gg \epsilon$, *i.e., the odd p.d.f. of* $\cos\theta$ *is dominated by* α *instead of* ϵ. *Thus,* α *measures the skewness of p.d.f. of the angular distribution of drifter pairs.*

The p.d.f. of $y = \cos\theta$ can be derived through a simple relation: $\tilde{P}(y) = P(x(y))|\mathrm{d}x/\mathrm{d}y|$, where $\mathrm{d}x/\mathrm{d}y = 1 - 3\alpha y$ always holds non-negative under Assumption 2. Denoting $x = x_0(y) + \alpha x_1(y)$ with $x_0(y) = y$ and $x_1(y) = 3(1-y^2)/2$, we obtain the following:

$$\begin{aligned}
\tilde{P}(y) &= P(x(y))\frac{\mathrm{d}x}{\mathrm{d}y} = P(x_0 + \alpha x_1)\left(\frac{\mathrm{d}x_0}{\mathrm{d}y} + \alpha\frac{\mathrm{d}x_1}{\mathrm{d}y}\right) \\
&= \left[P(x_0) + \alpha x_1 P'(x_0)\right]\left(\frac{\mathrm{d}x_0}{\mathrm{d}y} + \alpha\frac{\mathrm{d}x_1}{\mathrm{d}y}\right) \\
&= P(x_0)\frac{\mathrm{d}x_0}{\mathrm{d}y} + \alpha\frac{\mathrm{d}}{\mathrm{d}y}[x_1(y)P(x_0(y))] \\
&= P(y) + \frac{3}{2}\alpha\frac{\mathrm{d}}{\mathrm{d}y}[(1-y^2)P(y)],
\end{aligned} \tag{19}$$

where we omitted the second-order term. Combined with Assumption 1 and Assumption 3, we obtain the even-odd decomposition for $\tilde{P}(y)$:

$$\tilde{P}(y) = \underbrace{P_{\text{even}}(y)}_{\text{even}} + \underbrace{\frac{3}{2}\alpha\frac{\mathrm{d}}{\mathrm{d}y}\left[(1-y^2)P_{\text{even}}(y)\right]}_{\text{odd}}, \tag{20}$$

where the skewness of p.d.f. of the angular distribution is captured by α.

Considering the decomposition in Equation (20), we can use the even part to express the odd part, thus obtaining the α by fitting. Here, we use the least square method for fitting.

After obtaining the odd and even parts of the p.d.f. from data, we introduce the loss function as follows:

$$L(\alpha) = \frac{1}{n}\sum_{i=1}^{n}\left[\frac{3}{2}\alpha q'(y_i) - p(y_i)\right]^2, \tag{21}$$

where

$$p(y) = \tilde{P}_{\text{odd}}(y), \tag{22a}$$
$$q(y) = (1-y^2)\tilde{P}_{\text{even}}(y), \tag{22b}$$

then we obtain α by minimizing $L(\alpha)$:

$$\alpha = \frac{2\sum_{i=1}^{n} p(y_i)q'(y_i)}{3\sum_{i=1}^{n}[q'(y_i)]^2}. \tag{23}$$

Here, the sequence $\{y_k\}_{k=1}^{n}$ represents the data points, and it is chosen to be a finite arithmetic sequence with a common difference in Δy.

In our model, α is closely related to horizontal compressibility by Equation (5). If α itself is a r.v. with its p.d.f. denonted as $f(\alpha)$, assuming that $f(\alpha)$ is negligible when

$|\alpha| > 1/3$, then the relation between x and y is 1-to-1; therefore, the α in Equation (20) can be understood as its expectation $\int \alpha f(\alpha)\, d\alpha$. Thus, we may expect to obtain a statistical average of horizontal velocity divergence, i.e., the asymmetry towards divergence or convergence, by the sign of α.

3. Results

In this study, we analyze two data sets of drifter motion from the Grand LAgrangian Deployment (GLAD) experiment and the LAgrangian Submesoscale ExpeRiment (LASER). Drifters were deployed in the Gulf of Mexico in Summer, July to August 2012 (GLAD), [11] and in Winter, January to February 2016 (LASER) [12]. The locations of drifters are tracked using the Global Positioning System (GPS) within a nominal position error less than 10 m [5]. We think these two sets of data can reflect the velocity of oceanic surface flows precisely at the scales studied here.

The p.d.f. of $\cos\theta$ to the zeroth order is (15) when $\delta\phi$ is uniformly distributed over $[0, 2\pi]$. Therefore, we can do a rough comparison between the actual p.d.f. of $\cos\theta$ and (15). Choose the data in LASER and $|\delta \boldsymbol{r}|$ around 33 km as an example, and the result is shown in Figure 1. Details of data processing are written in Appendix A. We find that Equaiton (15) can already qualitatively describe the p.d.f. of $\cos\theta$, i.e., $\cos\theta$ distributing around ± 1 has the maximum probability, while the p.d.f. changes little and lies below 0.5 around $\cos\theta = 0$. The dimensionless quantity α, which may represent the compressibility by Equation (5), appears in the first-order expansion. Then, a rough estimation of α can be conducted according to Equation (16).

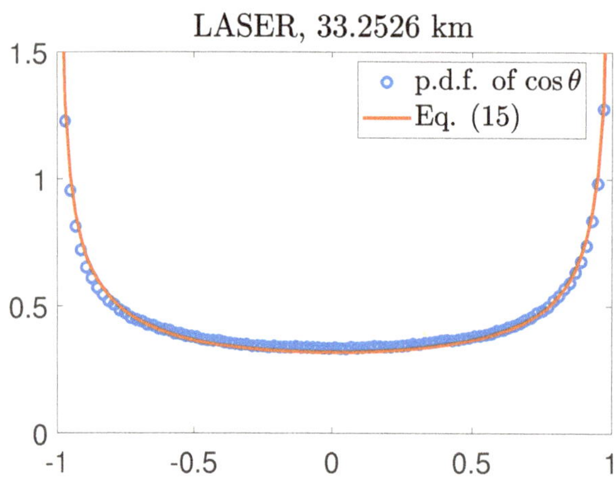

Figure 1. Comparison between the actual p.d.f. of $\cos\theta$ and Equation (15) in LASER, at the scale around 33 km.

Section 2.3 provides a method to obtain α with a weaker assumption to the distribution of $\delta\phi$, and we can take an example to show the whole process. Figure 2 shows the fitting process and its performance for $|\delta \boldsymbol{r}|$ around 33 km in LASER. We first decompose the p.d.f. of $\cos\theta$ into the even and odd parts, as shown in Figure 2a and c, respectively. Note that in Figure 2a, q relates to the even part of p.d.f. of $\cos\theta$ through (22b). Then, using the discrete format in Appendix A, we take the derivative of q, shown in Figure 2b. Thus, we fit p using q' to obtain α following the least-square procedure shown in Equations (21)–(23). The red fitting curve in Figure 2c well captures p. Figure 2d shows the comparison between the p.d.f. of $\cos\theta$ obtained from data and the theoretical expression (20) obtained from our fitting procedure.

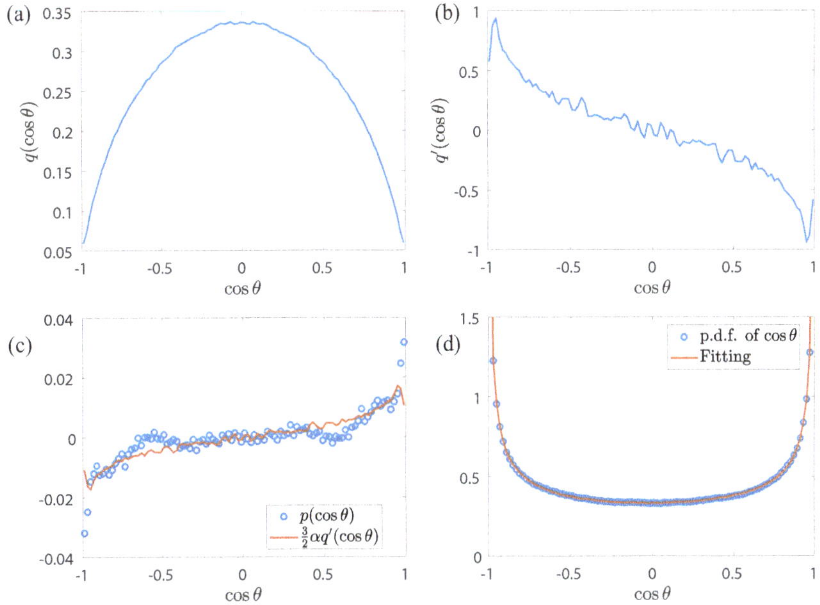

Figure 2. Fitting process in LASER at the scale around 33 km. (**a**) The even part of the p.d.f. of $\cos\theta$ multiplied by $(1-\cos^2\theta)$, denoted as $q(\cos\theta)$. (**b**) The derivative of q, using the discrete format in (A1). (**c**) The odd part of the p.d.f. of $\cos\theta$, denoted as $p(\cos\theta)$, and its fitting curve $\frac{3}{2}\alpha q'(\cos\theta)$, where α is obtained by (23). Here, $\alpha = -0.0123$. (**d**) The comparison between the original data and fitting of p.d.f. of $\cos\theta$.

The above process can be carried out in other scales of $|\delta \boldsymbol{r}|$, in GLAD and LASER. More fitting results are shown in Appendix C. Then we can obtain α as a function of scale, which is shown in Figure 3. Details of data processing and error estimation are written in Appendices A and B, which will tell us that Figure 3 is reliable when the scale is above 3 km. We can find that α is on the order of 10^{-2} in general, and stays negative when the scale ranges from about 10 km to about 300 km. This range of scale is larger in LASER than in GLAD, and so is the amplitude of negative α.

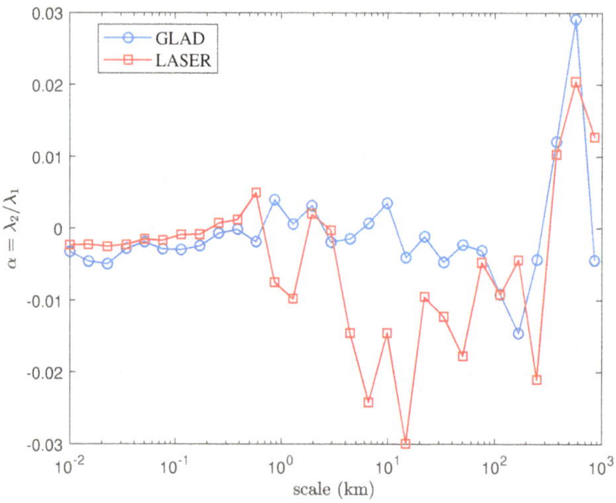

Figure 3. The obtained α-$|\delta \boldsymbol{r}|$ curve. Blue lines and red lines are used for GLAD and LASER, respectively. α is of $O(10^{-2})$ or less in general. The value of it is significantly negative at scales between 10 km and 300 km, and this phenomenon is more prominent in LASER than in GLAD. Note that the data in very small scales (less than 500 m) and very large scales (larger than 800 km) may be subject to insufficient data with a probably large error.

4. Discussion

Figure 3 shows that α is negative (implying an average divergent flow) at most of the scales and positive (convergent flows) at some scales around 1–10 km. The negative α is more prominent in LASER than in GLAD. This at least verifies that the surface flow is weakly compressible, as α is small and non-zero, and may involve some patterns in 3D flows. Additionally, compared with theories of 3D homogeneous isotropic turbulence [13] with $\langle \lambda_1 \lambda_2 \lambda_3 \rangle < 0$, which may imply a positive α). Here, λ_i refers to the eigenvalues of the symmetric part of the velocity gradient tensor, and the negative α implies the uniqueness of oceanic flow. We do not know the exact reason behind this negative α, but we propose two possible explanations in the following sections.

4.1. Kinematics: Drifter Concentration Caused by Mesoscale Vortices and Surface Compressibility

Cressman et al. (2004) [14,15] analyzed the motion of small particles on the surface of water in a square tank, and simulated the velocity distribution numerically. They claimed that the surface flow is a compressible flow with low speed (lower than the speed of sound). Structures of source and sink will exist on the surface, as the motion is still 3D in essence. Particles will be trapped in narrow convergent regions (with negative 2D divergence of velocity) and will distribute around vortices with a larger scale.

For oceanic flows, the leading-order vortices are mesoscale eddies (with scales ranging from several kilometers to several hundreds of kilometers). It is possible that these mesoscale vortices are accompanied by narrow convergent fronts, so that the drifters will accumulate around these vortices [12]. Meanwhile, the vortices correspond to divergent regions due to mass conservation. This qualitative view agrees with the suggestion of divergence (negative α) at mesoscales ($\sim$10–100 km), with some suggestion of convergence (positive α) at submesoscale fronts ($\sim$1–10 km). We further speculate that the negative α seen at even smaller scales may correspond to the divergent zones of Langmuir cells.

4.2. Dynamics: Weak Compressibility Caused by Ekman Pumping

The Coriolis force is non-negligible for large scales in oceanic flows. The balance between the Coriolis force and internal friction force is significant near the surface, resulting in the boundary layer between the ocean and the atmosphere, which is called the Ekman layer. For the Ekman layer on the surface of the ocean, friction at the bottom is zero (from definition), and wind stress $\boldsymbol{\tau}_{\mathrm{T}}$ will happen at the top. The z-component w of the velocity stays zero at the top, but non-zero at the bottom [1]:

$$w_{\mathrm{B}} = \frac{1}{\rho_0}\left(\mathrm{curl}_z \frac{\boldsymbol{\tau}_{\mathrm{T}}}{f} + \nabla \cdot \boldsymbol{M}_g\right). \tag{24}$$

where curl_z denotes the z-component of the curl, $\boldsymbol{M}_g = \int_{\mathrm{Ek}} \rho_0 \boldsymbol{u}_g \mathrm{d}z$, and $\boldsymbol{u}_g$ is the horizontal component of the velocity under geostrophic balance. The term $\nabla \cdot \boldsymbol{M}_g = -(\beta/f)\boldsymbol{M}_g \cdot \mathbf{j}$ represents the divergence of geostrophic transport, which is relatively small in Equation (24) [1].

Therefore, wind stress with a curl consistent with the rotation of the Earth (i.e., positive curl in the northern hemisphere) will cause the up-welling and surface divergence, which is likely to contribute to the negative α. The Gulf of Mexico is a region associated with negative wind stress curl on average [16], which could not explain the observed negative α (divergence) at scales between 10 km and 300 km. However, this region is also subject to hurricanes, which are cyclones with a positive wind stress curl. In particular, the LASER experiment experienced a strong hurricane, which may explain the more negative α detected in the LASER dataset.

5. Conclusions

We constructed a pure-strain model with a parameter α to analyze the drifter data, where α can reflect the existence of compressibility in a 2D flow, and the horizontal divergence asymmetry in 2D turbulence. We find that α stays negative at the mesoscales, representing a divergent flow, and we conjecture that mesoscale eddies interspersed with convergent fronts or a wind stress with a cyclonic wind stress curl arising from hurricanes may explain this observed divergence. Since particles and pollution tend to concentrate in regions with negative horizontal divergence, our results potentially lead to better models for pollution transports.

This study analyzed two data sets, and our model better matches LASER than GLAD (shown in Appendix B). This difference may be caused by the difference in data volume, or our assumptions, such as homogeneity and isotropy, are better satisfied in LASER than in GLAD. The inhomogeneity and anisotropy in GLAD were already discussed in other studies [17].

A major caveat of this work is that we assumed the flow to be pure-strain but oceanic flows are known to have comparably strong vorticity as well [18]. Thus, this strong assumption on our flow model may impact some of the results and conclusions. Also, our model is probably invalid when the velocity gradient tensor changes significantly with time, and the assumption of linearity is restrictive. Regardless, we provide an interesting and novel proof of concept for analyzing ocean observations, which is worthy of further investigation.

Author Contributions: Conceptualization, methodology, formal analysis, T.Z., J.-H.X. and D.B.; data processing: T.Z. and D.B.; writing—original draft preparation, T.Z.; writing—review and editing, T.Z., J.-H.X. and D.B. All authors have read and agreed to the published version of the manuscript.

Funding: T.Z. and J.-H.X. acknowledge financial support from the National Natural Science Foundation of China (NSFC) under grant numbers 12272006, 42361144844 and 12472219 and from the Laoshan Laboratory under grant numbers LSKJ202202000, LSKJ202300100. D.B. acknowledges financial support from the U.S. National Science Foundation under grant number NSF-OCE-2242110.

Data Availability Statement: The surface drifter data are available at https://data.gulfresearchinitiative.org/ (accessed on 12 January 2025).

Conflicts of Interest: The authors declare no conflicts of interest.

Appendix A. Details of Data Processing

The data used in this research covers the time period from day 10 to day 52 in GLAD and LASER experiments. The positions (latitudes and longitudes) and velocities of the drifters are contained in the original data. Acceleration is computed by the central differencing scheme, and the time interval between neighbouring data points is 15 min originally. The actually used data takes 30 min as the time interval.

The p.d.f. of $\cos\theta$ (defined by the first equals sign in Equation (9)) is generated at different scales of $|\delta\boldsymbol{r}|$ (i.e., the distance between drifters). We choose a sequence $\{r_k\}$ as the representative scales in advance. This is a geometric sequence with a ratio of 1.5, so it is equidistant in the logarithmic coordinate and is consistent with a former study [5]. Meanwhile, we tolerate a relative error of 0.2 when categorizing each actual $|\delta\boldsymbol{r}|$ into a certain r_k, i.e., $|\delta\boldsymbol{r}|$ is considered on the scale of r_k if and only if $||\delta\boldsymbol{r}| - r_k|/r_k < 0.2$. Then almost all the data can be used without repetition when the scale is between the first and the last term of $\{r_k\}$. The first term is set as $r_1 = 10$ m and the last term is about 852 km, so there are 29 terms of $\{r_k\}$ in total, all consistent with [5]. The examples in Figures 1, 2, and A1, where the scale is chosen as 33 km, correspond to r_{21} to be precise.

As is mentioned in Section 2.3, we choose a sequence $\{y_k\}$ with a common difference in Δy to represent different values of $\cos\theta$ when generating the p.d.f. of $\cos\theta$. To obtain the p.d.f., we compute a discrete probability mass function (p.m.f.) of the event sequence $\{\cos\theta \in [y_k - \Delta y/2,\ y_k + \Delta y/2)\}$, where $\bigcup_{k=1}^{n}[y_k - \Delta y/2,\ y_k + \Delta y/2) = [-1,\ 1)$. The extreme case $\cos\theta = 1$ will fall under the last interval. Divide the p.m.f. by Δy, and we can obtain the p.d.f. for analysis. In the main text, Δy is set as 0.02 by default.

The function $q(y)$ in Section 2.3 is also defined on discrete points $\{y_k\}$. The derivative of $q(y)$ can be defined by a discrete format with second-order accuracy, i.e.,

$$q'(y_i) := \begin{cases} \dfrac{4q(y_2) - 3q(y_1) - q(y_3)}{2\Delta y}, & i = 1, \\ \dfrac{q(y_{i+1}) - q(y_{i-1})}{2\Delta y}, & 2 \leq i \leq n-1, \\ \dfrac{3q(y_n) - 4q(y_{n-1}) + q(y_{n-2})}{2\Delta y}, & i = n. \end{cases} \quad \text{(A1)}$$

Appendix B. Error Estimation of the Fitting

Appendix B.1. Noise in the Generated p.d.f.

We generated p.d.f. of $\cos\theta$ from drifter data directly. But these functions are not smooth, and contain irregular oscillation, which is called noise. The existence of noise will reduce the effect and reliability of our fitting. Observation tells us that noise will probably decrease (i.e., Signal to Noise Ratio (SNR) will increase) as the scale and data volume increase. For instance, there are more valid drifters in LASER (956) than in GLAD (297), and accordingly, the p.d.f.s appear smoother in LASER than in GLAD.

Moreover, SNR will also increase when Δy appropriately increases. An example is taken for 33 km in LASER, shown in Figure A1. Obvious oscillations can be seen when

$\Delta y = 0.02$, and this phenomenon vanishes gradually as Δy grows up. However, changes in Δy will result in changes in the obtained α, and the amplitude of this change also depends on the scale and data volume.

We can plot the α-$|\delta \boldsymbol{r}|$ curves at different values of Δy, which are shown in Figure A2. There should not be big changes in α if our method is proper enough. According to Figure A2, small scales correspond to a bigger amplitude of change in α and thus, a lower reliability compared with the larger scales. The demarcation approximately locates at 3 km.

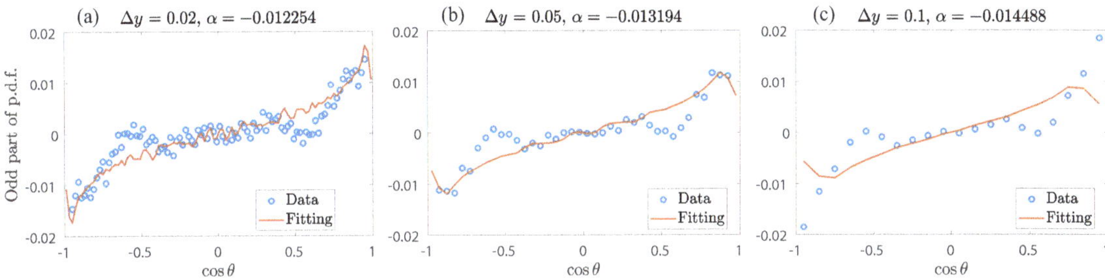

Figure A1. Odd part of the p.d.f. and the fitting curve when Δy equals (**a**) 0.02, (**b**) 0.05 and (**c**) 0.1, at the scale around 33 km in LASER.

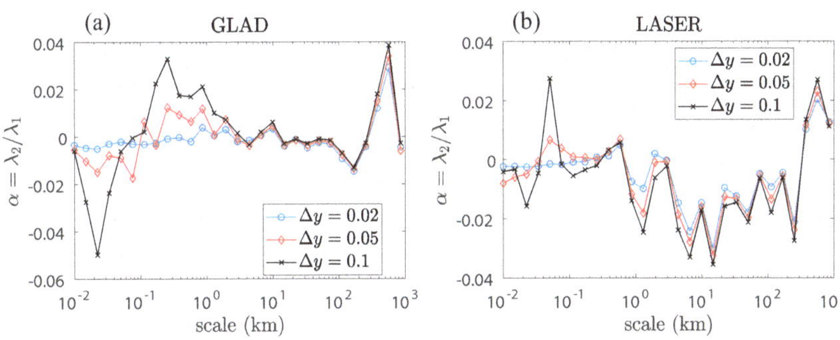

Figure A2. The $\alpha - |\delta \boldsymbol{r}|$ curves at different values of Δy in (**a**) GLAD and (**b**) LASER. The values of Δy are set as 0.02, 0.05, and 0.1, plotted in blue, red and black, respectively.

Appendix B.2. Root Mean Square Error

In the former section, we found that the fitting is effective and reliable at mesoscale and large scale. But we have not checked whether the fitting matches the p.d.f. well enough, and this is independent from the existence of noise. Therefore, we set $\Delta y = 0.1$ to increase SNR, and define the bias of fitting as follows:

$$\sigma_{\mathrm{Bias}} = \sqrt{\frac{1}{n'} \sum_{i=1}^{n'} \left[p(y_i') - p^{\mathrm{fit}}(y_i') \right]^2}, \tag{A2}$$

where n' denotes the number of data points y_i' when $\Delta y = 0.1$, $p(y_i')$ and $p^{\mathrm{fit}}(y_i') = \frac{3}{2}\alpha q'(y_i')$ denote the odd part of the p.d.f. of $\cos\theta$ and the fitting curve. Figure A3 shows the σ_{Bias}-$|\delta \boldsymbol{r}|$ curves. The value of σ_{Bias} reduces to a stable platform when the scale is over 500 m.

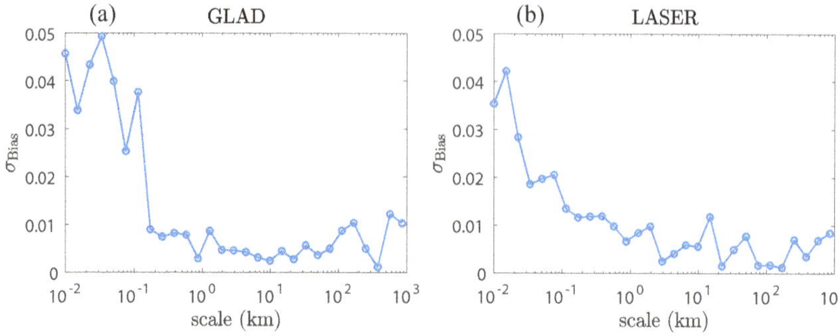

Figure A3. $\sigma_{\text{Bias}} - |\delta \boldsymbol{r}|$ curves in (**a**) GLAD and (**b**) LASER.

In addition, the amplitude of the odd part of the p.d.f. of $\cos\theta$ will vary as the scale changes. A quantity describing the relative error needs to be introduced, i.e.,

$$e_{\text{Bias}} = \frac{\sigma_{\text{Bias}}}{\sqrt{\frac{1}{n'}\sum_{i=1}^{n'} p^2(y_i')}}. \tag{A3}$$

The $e_{\text{Bias}} - |\delta \boldsymbol{r}|$ curves are shown in Figure A4. We can find that the fitting matches better in LASER than in GLAD, and the error is indeed smaller in larger scales (roughly larger than 3 km). The difference between the two data sets may be resulted by the difference in data volume, or GLAD is less likely to accord with our model.

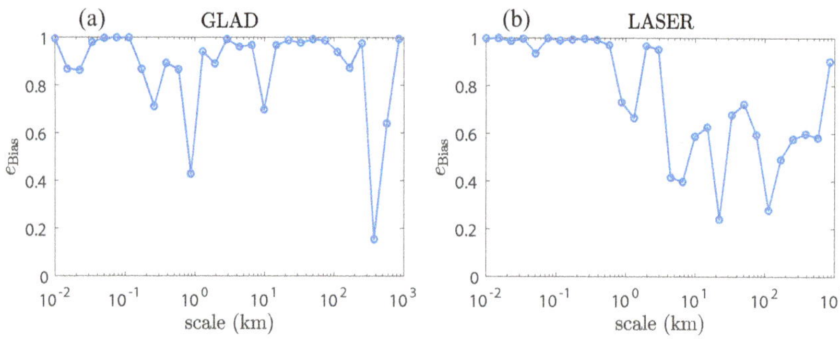

Figure A4. $e_{\text{Bias}} - |\delta \boldsymbol{r}|$ curves in (**a**) GLAD and (**b**) LASER.

Appendix C. More Fitting Results

Here, we select more fitting results in GLAD and LASER, as are shown in Figures A5 and A6. The value of Δy is set as 0.02. In GLAD, obvious oscillations of data and the fitting curve appear when the scale is small (Figure A5a). The fitting curves are approximately horizontal at mesoscale (Figure A5b–d). This is consistent with the result that the amplitude of α is smaller in GLAD than in LASER at mesoscale, as can be seen in Figure 3. Large scales have a better fitting result (Figure A5e). The oscillation of the fitting curve around $\cos\theta = 0$ in Figure A5e may be a result of insufficient data volume. In LASER, oscillation is less obvious than in GLAD when the scale is small, while the curves appear horizontal and capture little features of the data (Figure A6a,b). The fitting is fine for mesoscale and large scale (Figure A6c–e).

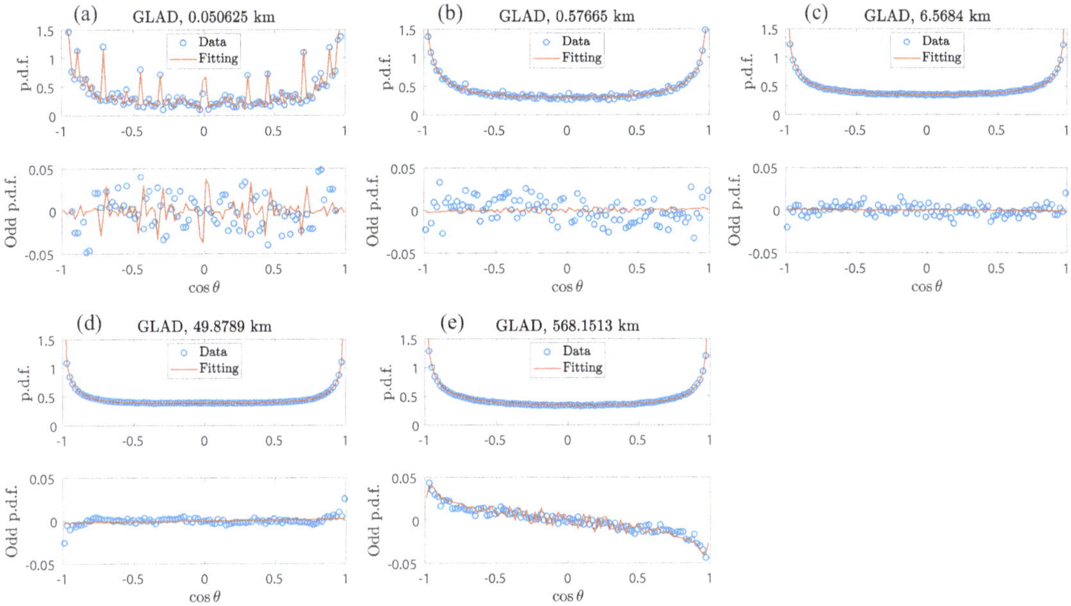

Figure A5. Fitting results of p.d.f. of $\cos\theta$ at different scales, using the data from GLAD. Five cases are shown here at scales around (**a**) 51 m, (**b**) 577 m, (**c**) 6.6 km, (**d**) 50 km and (**e**) 568 km. Each single case contains two sub-figures, plotting the whole and the odd part of the p.d.f. (blue circles) and the fitting curve (red curve).

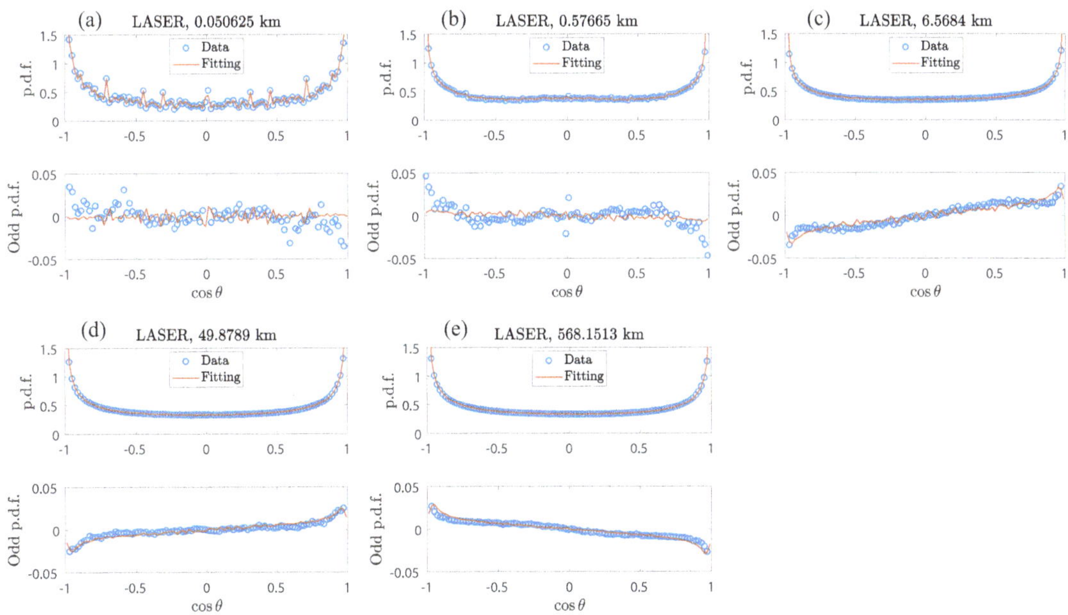

Figure A6. Fitting results of p.d.f. of $\cos\theta$ at different scales, using the data from LASER. Five cases are shown here at scales around (**a**) 51 m, (**b**) 577 m, (**c**) 6.6 km, (**d**) 50 km and (**e**) 568 km. Each single case contains two sub-figures, plotting the whole and the odd part of the p.d.f. (blue circles) and the fitting curve (red curve).

References

1. Vallis, G.K. *Atmospheric and Oceanic Fluid Dynamics*; Cambridge University Press: Cambridge, UK, 2017.
2. Vallis, G.K. *Essentials of Atmospheric and Oceanic Dynamics*; Cambridge University Press: Cambridge, UK, 2019.
3. Cho, J.Y.; Lindborg, E. Horizontal velocity structure functions in the upper troposphere and lower stratosphere: 1. Observations. *J. Geophys. Res. Atmos.* **2001**, *106*, 10223–10232. [CrossRef]
4. Qiu, B.; Nakano, T.; Chen, S.; Klein, P. Bi-directional energy cascades in the Pacific Ocean from equator to subarctic gyre. *Geophys. Res. Lett.* **2022**, *49*, e2022GL097713. [CrossRef]
5. Balwada, D.; Xie, J.H.; Marino, R.; Feraco, F. Direct observational evidence of an oceanic dual kinetic energy cascade and its seasonality. *Sci. Adv.* **2022**, *8*, eabq2566. [CrossRef] [PubMed]
6. McWilliams, J.C. Submesoscale currents in the ocean. *Proc. R. Soc. A Math. Phys. Eng. Sci.* **2016**, *472*, 20160117. [CrossRef] [PubMed]
7. Balwada, D.; LaCasce, J.H.; Speer, K.G. Scale-dependent distribution of kinetic energy from surface drifters in the Gulf of Mexico. *Geophys. Res. Lett.* **2016**, *43*, 10–856. [CrossRef]
8. Ott, S.; Mann, J. An experimental investigation of the relative diffusion of particle pairs in three-dimensional turbulent flow. *J. Fluid Mech.* **2000**, *422*, 207–223. [CrossRef]
9. Gibert, M.; Xu, H.; Bodenschatz, E. Where do small, weakly inertial particles go in a turbulent flow? *J. Fluid Mech.* **2012**, *698*, 160–167. [CrossRef]
10. Solomon, T.; Tomas, S.; Warner, J. Chaotic mixing of immiscible impurities in a two-dimensional flow. *Phys. Fluids* **1998**, *10*, 342–350. [CrossRef]
11. Poje, A.C.; Özgökmen, T.M.; Lipphardt, B.L., Jr.; Haus, B.K.; Ryan, E.H.; Haza, A.C.; Jacobs, G.A.; Reniers, A.; Olascoaga, M.J.; Novelli, G.; et al. Submesoscale dispersion in the vicinity of the Deepwater Horizon spill. *Proc. Natl. Acad. Sci. USA* **2014**, *111*, 12693–12698. [CrossRef] [PubMed]
12. D'Asaro, E.A.; Shcherbina, A.Y.; Klymak, J.M.; Molemaker, J.; Novelli, G.; Guigand, C.M.; Haza, A.C.; Haus, B.K.; Ryan, E.H.; Jacobs, G.A.; et al. Ocean convergence and the dispersion of flotsam. *Proc. Natl. Acad. Sci. USA* **2018**, *115*, 1162–1167. [CrossRef] [PubMed]
13. Betchov, R. An inequality concerning the production of vorticity in isotropic turbulence. *J. Fluid Mech.* **1956**, *1*, 497–504. [CrossRef]
14. Cressman, J.R.; Goldburg, W.I.; Schumacher, J. Dispersion of tracer particles in a compressible flow. *Europhys. Lett.* **2004**, *66*, 219. [CrossRef]
15. Cressman, J.R.; Davoudi, J.; Goldburg, W.I.; Schumacher, J. Eulerian and Lagrangian studies in surface flow turbulence. *New J. Phys.* **2004**, *6*, 53. [CrossRef]
16. Yu, L. Sea surface exchanges of momentum, heat, and freshwater determined by satellite remote sensing. *Encycl. Ocean Sci.* **2009**, *2*, 202–211.
17. Huntley, H.S.; Lipphardt, B., Jr.; Kirwan, A., Jr. Anisotropy and inhomogeneity in drifter dispersion. *J. Geophys. Res. Ocean.* **2019**, *124*, 8667–8682. [CrossRef]
18. Balwada, D.; Xiao, Q.; Smith, S.; Abernathey, R.; Gray, A.R. Vertical fluxes conditioned on vorticity and strain reveal submesoscale ventilation. *J. Phys. Oceanogr.* **2021**, *51*, 2883–2901. [CrossRef]

symmetry

MDPI

Article

Asymmetry of Two-Dimensional Thermal Convection at High Rayleigh Numbers

Jian-Chao He [1], Yun Bao [2] and Xi Chen [1,*]

[1] Institute of Fluid Mechanics, Beijing University of Aeronautics and Astronautics, Beijing 100191, China; hejch7@buaa.edu.cn
[2] School of Aeronautics and Astronautics, Sun Yat-sen University, Shenzhen 518107, China; stsby@mail.sysu.edu.cn
* Correspondence: chenxi97@outlook.com

Abstract: While thermal convection cells exhibit left–right and top–bottom symmetries at low Rayleigh numbers (Ra), the emergence of coherent flow structures, such as elliptical large-scale circulation in Rayleigh–Bénard convection (RBC), breaks these symmetries as the Rayleigh number increases. Recently, spatial double-reflection symmetry was proposed and verified for two-dimensional RBC at a Prandtl number of 6.5 and Ra values up to 10^{10}. In this study, we examined this new symmetry at a lower Prandtl number of 0.7 and across a wider range of Rayleigh numbers, from 10^7 to 10^{13}. Our findings reveal that the double-reflection symmetry is preserved for the mean profiles and flow fields of velocity and temperature for $Ra < 10^9$, but it is broken at higher Rayleigh numbers. This asymmetry at high Ra values is inferred to be induced by a flow-pattern transition at $Ra = 10^9$. Together with the previous study, our results demonstrate that the Prandtl number has an important influence on the symmetry preservation in RBC.

Keywords: thermal convection; direct numerical simulation; symmetry

1. Introduction

Turbulence, which is prevalent in both nature and industrial systems [1,2], is a complex phenomenon in fluids that is characterized by chaotic changes in velocity and pressure. One example of these complex phenomena is the Rayleigh–Bénard convection (RBC). This phenomenon occurs in a system that is heated from the bottom and cooled from the top with two constant temperatures (T_{bot} and T_{top}), while the sidewalls remain adiabatic. The fluid in this system is driven by buoyancy and gravity. The cold fluid sinks while the hot fluid rises. They encounter and mix in the center of the cell, leading to complex flow patterns. The geometry of such a system can be arbitrary, including cubic, rectangular, or cylindrical configurations. The characteristics of RBC are determined by three key parameters: the Rayleigh number $Ra = \frac{\alpha g \Delta T H^3}{\nu\kappa}$, the Prandtl number $Pr = \frac{\nu}{\kappa}$, and the aspect ratio $\Gamma = \frac{D}{H}$. Here, α is the thermal-expansion coefficient, ν is the viscosity, κ is the thermal-diffusion coefficient, g is the gravitational acceleration, H is the system's height, and D is the system's width or diameter. The temperature difference between the horizontal conducting plates at the bottom and top of the system is represented by $\Delta T = T_{\text{bot}} - T_{\text{top}}$.

Over the years, research on RBC has encompassed various aspects, including the system's heat transfer, the emergence of coherent structures, and fluid dynamics. In early studies, the instability and chaos in RBC systems attracted much attention [3–6]. After the moderately high-Ra experiments reported by Castaing et al. in 1989 [7], the properties of turbulence in RBC were widely investigated by researchers, including considerations of velocity and thermal fluctuations [8,9], kinetic and thermal dissipation [10–12], boundary layers [13,14], and small-scale turbulence [15,16]. Turbulent RBC shows different flow patterns with different Ra–Pr phases [17,18], with some coherent structures emerging, i.e., large-scale circulation (LSC) and plumes [12]. In particular, the flow patterns experience a transition with increasing Ra, leading to different statistical behaviors [18]. These properties are closely related to the heat transfer of RBC. For instance, thermal plumes are transported upward or downward by LSC, facilitating the transfer of heat flux; however, alterations to the boundary conditions or constraints on the flow motion can lead to significant changes in the system's heat transfer behavior [19–21]. Furthermore, heat transfer is influenced by the cell's aspect ratio and shape, both of which have garnered significant attention [22–26]. The heat transfer is quantified by the Nusselt number (Nu), and its scaling with Ra and Pr has been discussed in numerous reports [27–33]; however, no unified conclusions have yet been reached on this topic.

Symmetry in RBC systems has also been investigated in many studies [34–37]. Symmetry is an important concept in physics as it reflects the invariance within a system [38], and it is associated with a pattern that satisfies invariance under a specific transformation rule. Generally speaking, if there is an invariant quantity—one that remains unchanged under transformation—then a corresponding symmetry exists. With symmetric boundary conditions and symmetric geometry, symmetry is expected in the properties of the flow within an RBC system. At very low Ra values, an RBC system will maintain a motionless state with a uniform vertical temperature gradient. Above a critical Rayleigh number Ra_c, a convective motion appears and swirl structures are formed [35]. Then, as the Ra is increased, the flow state of the RBC experiences convection, oscillation, chaos, transition, and turbulence [39]. In these states, the symmetry of an RBC system also experiences different states, such as axisymmetric convection and non-axisymmetric motions [35]. With cryogenic helium gas as the working fluid, for moderately high Ra values in the turbulent state, it was found that the flow is statistically symmetric for $Ra > 10^{11}$ in 3D cylinders with aspect ratios (Γ) of 0.5 and 1.0 [34]. In experiments with water ($Pr = 5.3$), the mean velocity field is also azimuthally symmetric in a cylinder with $\Gamma = 0.5$, while no such symmetry is observed in a cylinder with $\Gamma = 1.0$ [36,37]. For a large Pr, the local boundary layers have been found to be either symmetric [40] or asymmetric [41–43]. In a 2D RBC, elliptical LSC forms within the cell at moderately high Ra values [11,44], as shown in Figure 1; therefore, the left–right and top–bottom symmetry of the system will be broken. These results indicate that the symmetry might be broken or restored in different cells or those with different Ra–Pr phases; in other words, symmetry plays a crucial role in the potential transitions exhibited by this system. Recently, the x–y double-reflection symmetry in thermal convection at Re values between 10^7 and 10^{10} has been proposed, and the symmetry under this reflection was validated [45]. This symmetry for other Pr values or higher Ra values is unknown. For this reason, we sought to examine double-reflection symmetry for higher Ra values at $Pr = 0.7$.

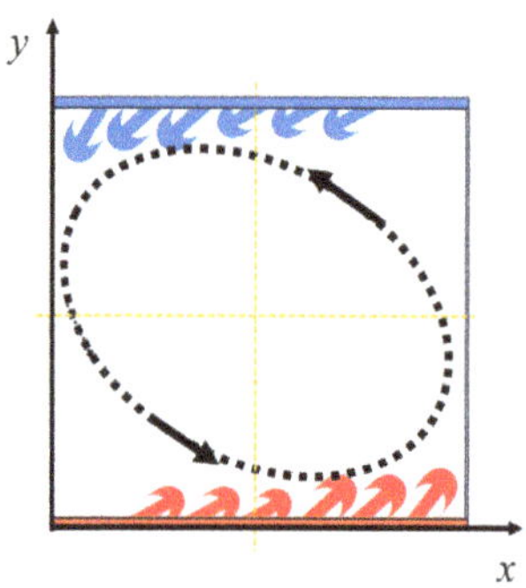

Figure 1. Schematic illustration of a turbulent RBC. This system is confined with two adiabatic side walls and two conducting plates, with heating from the bottom and cooling from the top. The major structures include thermal boundary layers, large-scale circulation (black ellipse), and thermal plumes. The yellow dashed lines represent the symmetry axes of the RBC system.

In this work, the x–y double-reflection symmetry in a 2D RBC with Ra values ranging from 10^7 to 10^{13} and $Pr = 0.7$ was investigated using direct numerical simulation (DNS). In the simulation model, the aspect ratio of the cell was unity, and it was confined by adiabatic walls. The Ra range in the previous study was only up to 10^{10}, and it was necessary to verify the symmetry at higher Ra. In addition, the Pr effect in the double reflection symmetry was unknown. Therefore, the Pr in this paper was set as 0.7, which is different from the $Pr = 6.5$ set in the previous study. The remainder of this paper is structured as follows. The numerical setup of the DNS is described in Section 2, the temperature and velocity flow fields are presented in Section 3, the verification of the double-reflection symmetry using the data is discussed in Section 4, and the conclusions are presented in Section 5.

The RBC system is considered to be incompressible, with the temperature treated as an active scalar. In general, the Oberbeck–Boussinesq approximation is used to simplify the Navier–Stokes equations [14,18,46], i.e., $\rho(T) = \rho(T_0)[1 - \alpha(T - T_0)]$, while various transport coefficients (e.g., α, ν, and κ) are treated as constants. Under this approximation, the effect of temperature on the velocity field is realized through the buoyancy term. Using the scaling factors $H, \Delta T, U = \sqrt{\alpha g H \Delta T}$ and $\tau = \sqrt{H/\alpha g \Delta T}$ [14], the governing equations can be written in a dimensionless form as follows:

$$\nabla \cdot \mathbf{u} = 0 \tag{1}$$

$$\frac{\partial \mathbf{u}}{\partial t} + (\mathbf{u} \cdot \nabla)\mathbf{u} = -\nabla p + \frac{1}{\sqrt{Ra/Pr}}\nabla^2 \mathbf{u} + \theta \hat{y}, \tag{2}$$

$$\frac{\partial \theta}{\partial t} + (\mathbf{u} \cdot \nabla)\theta = \frac{1}{\sqrt{Ra \cdot Pr}}\nabla^2 \theta, \tag{3}$$

where $\mathbf{u} = (u, v)$ is the dimensionless velocity, p is the dimensionless pressure, θ is the dimensionless temperature, and $\hat{y}$ is a vertical unit vector whose direction is opposite to that of gravity. For the thermal boundary conditions, the temperatures of the two conducting plates were kept constant, while the two sidewalls were adiabatic:

$$\theta(y = 0) = 0.5, \quad \theta(y = 1) = -0.5 \tag{4}$$

$$\frac{\partial \theta}{\partial x}(x = 0) = 0, \quad \frac{\partial \theta}{\partial x}(x = 1) = 0. \tag{5}$$

No-slip velocity boundary conditions were imposed on all the conducting plates and sidewalls:

$$u(y = 0) = 0, \quad u(y = 1) = 0, \quad u(x = 0) = 0, \quad u(x = 1) = 0 \tag{6}$$

$$v(y = 0) = 0, \quad v(y = 1) = 0, \quad v(x = 0) = 0, \quad v(x = 1) = 0. \tag{7}$$

2. Equations and Numerical Settings

The simulation scheme applied in this work uses a finite-difference method known as the parallel direct method of DNS [47]. This process for solving the Navier–Stokes equations are illustrated in Figure 2. Specifically, the process of solving the Poisson equation is also shown there. In the flowchart, $f(\mathbf{u}^n, \theta^n) = -(\mathbf{u}^n \cdot \nabla)\mathbf{u}^n + \frac{1}{\sqrt{Ra/Pr}}\nabla^2 \mathbf{u}^n + \theta^n \hat{y}$ and $g(\mathbf{u}^{n+1}, \theta^n) = -(\mathbf{u}^{n+1} \cdot \nabla)\theta^n + \frac{1}{\sqrt{Ra \cdot Pr}}\nabla^2 \theta^n$. A fully explicit projection method, with second-order accuracy in both space and time, was used to solve the governing equations. In the first step, the pressure p^{n+1} was unknown, and the Poisson equation for the pressure was solved in the third step. An explicit second order Runge–Kutta scheme was employed in the time-marching. The Poisson equation for pressure was solved using the parallel diagonal dominant (PDD) algorithm in combination with fast Fourier transforms (FFTs). Finally, the velocity $\mathbf{u}^{n+1}$ was corrected with u^* and pressure p^{n+1}, and the temperature θ^{n+1} was calculated with $\mathbf{u}^{n+1}$ and θ^n.

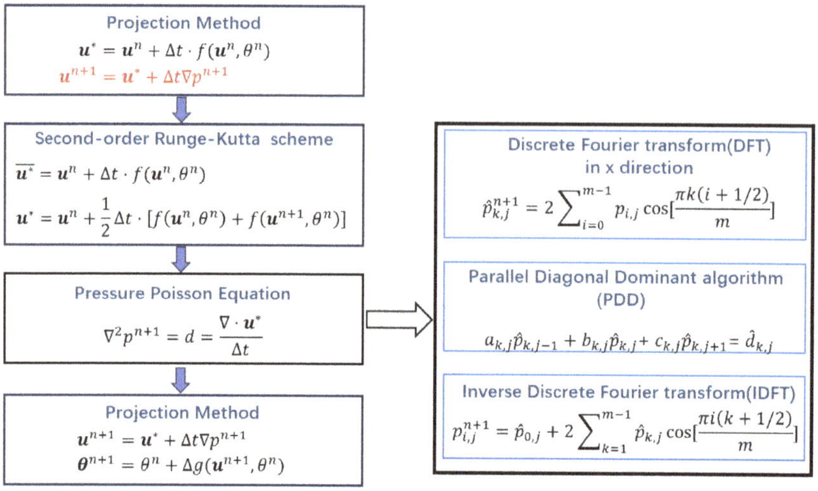

Figure 2. Numerical process for solving the Navier–Stokes equations (**left**) and the pressure Poisson equation (**right**) .

The code was written in Fortran using parallel technologies, including MPI and OpenMP. We used this code for the work we reported in previous publications that related to investigating flow patterns, the scaling transition of thermal dissipation, and boundary layers [14]. In our simulations, staggered grids were employed. The grids were uniform in the horizontal direction (x), while they were un-uniform in the vertical direction (y). To accurately capture the intense fluctuations near the top and bottom plates, the grid points were clustered closer to these regions.

The key parameters and grid for the cases in this work are outlined in Table 1. To resolve the smallest turbulence scales—i.e., the Kolmogorov scale $\eta_K = HPr^{1/2}/[Ra(Nu-1)]^{1/4}$ and the Batchelor scale $\eta_B = \eta_K Pr^{-1/2}$—the grid in the vertical direction (y) was refined in the boundary layer, and the number of grid points (N_{BL} in Table 1) in the thermal boundary layer satisfied the condition derived by Shishkina et al. [48]:

$$N_{\theta,\mathrm{BL}} \geq \sqrt{2}aNuPr^{-0.5355+0.033\log Pr}, \tag{8}$$

$$N_{v,\mathrm{BL}} \geq \sqrt{2}aNu^{1/2}Pr^{-0.1785+0.011\log Pr}, \tag{9}$$

where $a \approx 0.482$ and $N_{\theta,\mathrm{BL}}$ and $N_{v,\mathrm{BL}}$ are the minimum numbers of grid points in the thermal boundary layer and the viscous boundary layer, respectively. Note that Equations (8) and (9) were only valid for $3 \times 10^{-4} \leq Pr \leq 1$; the conditions for other Pr values can be found in Ref. [48] and are not discussed here. The dimensionless time step Δt was set to be less than 1/1000 of the Kolmogorov time scale $\tau_K = \sqrt{Pr/(Nu-1)}$ to accurately capture the intense fluctuations. The simulations were executed on the Tianhe-2 supercomputer, requiring millions of core hours to complete.

For cases with low Ra values ($\leq 2 \times 10^8$), the number of grid points was relatively small, allowing the statistical time t_{avg} to reach up to 1000 free-fall times (t_{f}), with t_{f} calculated as $t_{\mathrm{f}} = \sqrt{H/\alpha g \Delta T}$. For higher Ra values, the number of grid points rapidly increased, and the cost of the simulation also increased, resulting in a shorter statistical time. The value of t_{avg} was set to at least $200t_{\mathrm{f}}$ before $Ra = 10^{13}$ to ensure convergence of the data. It was noted that Nu is a key response variable for assessing the state of the system and is commonly used to verify convergence. For the statistical analysis, Nu was calculated as the average of data from two successive long-term statistical segments. It was found that the relative differences between the Nu values from these segments were within 1%. For the highest Ra case, t_{avg} was only $50t_{\mathrm{f}}$, but the relative difference was about 3.6%.

Table 1. Details of the numerical simulations employed for the present simulations. Here, Nu is calculated as $Nu = \sqrt{RaPr}\langle v\theta \rangle_{\Omega,t} - \langle \partial_y \theta \rangle_{\Omega,t}$, where $\langle\rangle_{\Omega,t}$ represents the time–space average over the entire system; N_x and N_y represent the grid resolution in the horizontal and vertical directions, respectively; and N_{BL} is the number of grid points in the thermal boundary layer of a conducting plate.

Ra	Pr	Γ	$N_x \times N_y$	t_{avg}	Nu	N_{BL}
1×10^7	0.70	1.0	512 × 576	1000	11.38	33
2×10^7	0.70	1.0	512 × 576	1000	14.33	27
5×10^7	0.70	1.0	512 × 576	1000	19.71	20
1×10^8	0.70	1.0	512 × 576	1000	25.22	16
2×10^8	0.70	1.0	512 × 576	1000	31.24	13
5×10^8	0.70	1.0	1024 × 1152	1000	42.13	26
1×10^9	0.70	1.0	1024 × 1152	400	50.53	21
2×10^9	0.70	1.0	1024 × 1152	400	59.83	18
5×10^9	0.70	1.0	1024 × 1152	400	78.25	14
1×10^{10}	0.70	1.0	1536 × 1728	600	95.50	30
2×10^{10}	0.70	1.0	2048 × 2304	400	118.23	36
5×10^{10}	0.70	1.0	2048 × 2304	400	158.12	28
1×10^{11}	0.70	1.0	2048 × 2304	400	198.43	22
2×10^{11}	0.70	1.0	2560 × 2880	200	243.14	40
5×10^{11}	0.70	1.0	2560 × 2880	200	326.14	30
1×10^{12}	0.70	1.0	4096 × 4608	200	404.86	51
2×10^{12}	0.70	1.0	4096 × 4608	300	502.11	41
5×10^{12}	0.70	1.0	5120 × 5760	200	666.74	57
1×10^{13}	0.70	1.0	5120 × 5760	50	839.71	48

3. Flow Fields

3.1. Instantaneous Flow Fields

The instantaneous velocity and temperature fields are presented in Figure 3. In the top row of Figure 3, the high-speed fluid is represented in red and the low-speed fluid is represented in blue. In the bottom row of Figure 3, red is used to represent the high-temperature fluid, while blue represents the low-temperature fluid. Note that, in the 2D RBC, the direction of the LSC was arbitrary as it could be either clockwise or counter-clockwise. Similar to the approach taken in Ref. [13], the direction of the LSC was unified to be counter-clockwise in this work, and this did not affect the analysis.

In cases with low Ra values, a large elliptical swirl (red on the outside and blue on the inside) fills almost the entire system, and there are usually two small swirls beside it (see Figure 3a,b top-right and bottom-left corners). Generally, this large elliptical swirl is referred to as LSC, and the small counter-rotating swirls are referred to as corner rolls. In contrast, the top-left and bottom-right corners were almost blue, meaning that the motion of the fluid in these regions was very slow. The four corners were filled by cold and hot fluid, as shown in Figure 3e,f, as a result of the emission of thermal plumes. The motion of the thermal plumes was primarily constrained by the LSC, resulting in an elliptical distribution of the mean-temperature fluid in the bulk region. This flow pattern has also been observed at lower Ra values in some previous works [11,44]. Due to the presence of the LSC and corner rolls, the flow pattern remained almost steady for the low Ra values in the current range up to 5×10^8.

For relatively high Ra values ($\geq 10^9$ in this paper), the red areas were less stable, and a large-scale swirl moved around in the system, as shown in Figure 3c,d. Additionally, the corner rolls became less prominent and were able to move away from their original positions. More small swirls (shown by vectors) appeared in the system, making the flow pattern more complex. These swirls varied in scale, which is related to the energy inverse cascade in 2D turbulence [49]. The small swirls were generated by the shearing action between the large-scale swirl and the boundary layers. As these swirls moved throughout the system with the large-scale swirl, they mixed with the larger ones. Meanwhile, the motion of the plumes was disorganized, as shown in Figure 3g,h, and the plumes were mainly carried by the small swirls. Moreover, the size of the plumes decreased as the *Ra* increased, as has been previously reported [14]. The flow pattern was markedly different from that observed at lower *Ra* values, as can be seen in Figure 3a,b,e,f. This transition in a flow pattern was also reported by Gao et al., who conducted a more detailed investigation using stability analysis [18]. This transition also lead to changes in the thermal dissipation [14] and fluctuations [9].

In these instantaneous fields, the top–bottom and left–right symmetry of the system was, evidently, broken. Next, we examined the time-averaged temperature and velocity fields.

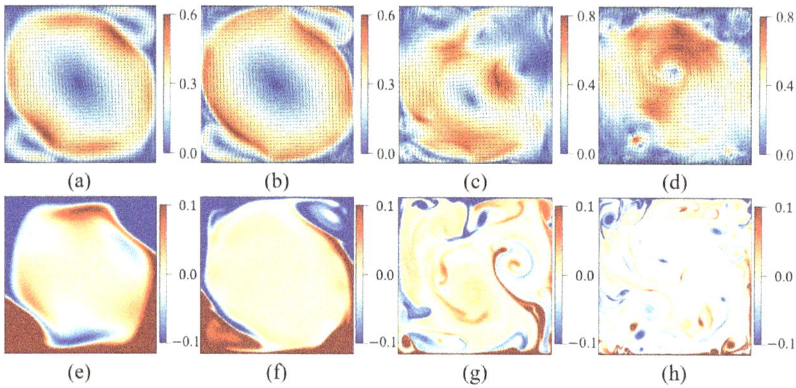

Figure 3. Instantaneous snapshots of the velocity fields $\mathcal{V}(x,y) = \sqrt{u^2+v^2}$ (top row) and temperature fields $\theta(x,y)$ (bottom row) for (**a**,**e**) $Ra = 10^7$, (**b**,**f**) $Ra = 10^8$, (**c**,**g**) $Ra = 10^9$, and (**d**,**h**) $Ra = 10^{10}$. Note that the color map for the velocity field in each case was set from the minimum value to maximum value of that case, while the color maps for the temperature fields were set from -0.1 to 0.1 for better visualization of the plumes. The vectors indicate velocity, with their lengths representing its magnitude.

3.2. Time-Averaged Flow Fields

The time-averaged velocity and temperature fields are shown in Figure 4. As shown in Figure 4a, there were only two corner rolls, in the top-right and bottom-left corners, while there were no swirls in the other two corners. The high-speed flow formed a large swirl, which is the LSC mentioned above. The maximum velocity was located in the region between the LSC and corner roll. As shown in Figure 4e, the time-averaged temperature field revealed a yellow ellipse, indicating a mean temperature distribution of zero in most parts of the cell. The thermal plumes were primarily carried by the LSC or were confined in the corner rolls, leading to the high- and low-temperature regions that correspond to the LSC shown in Figure 4a.

For the $Ra = 10^8$ in Figure 4b,f, the flow pattern and temperature fields were very similar to those for $Ra = 10^7$; however, the flow motion in top-left and bottom-right corners became more complex, suggesting the appearance of additional swirls in these regions. Moreover, the velocity magnitude remained small, indicating that these swirls had little influence on the heat transfer. For these two cases, the time-averaged velocity fields were similar to the instantaneous velocity fields, suggesting that the flow motion was relatively steady for *Ra* values ranging from 10^7 to 5×10^8.

As shown in Figure 4c,d, a large red annulus almost filled the system together with four low-speed corner rolls. At $Ra = 10^{10}$, the shapes of these corner rolls appeared different, and their sizes were smaller. This distribution was quite different from that of the instantaneous velocity fields that are shown in Figure 3. The time-averaged velocity field indicated that a large-scale swirl persisted in the cell while many smaller swirls were present, making it challenging to identify the largest swirl. Reviewing the fields in

Figure 3c,d, the red region resembles an annulus and it was surrounded by several swirls. Therefore, the LSC still exists in the system at high Ra values, but it was difficult to identify. For the temperature fields in Figure 4g,h, the fluctuations of the plumes were eliminated in the bulk region. The high- (or low-) temperature region in the bottom-left (or top-right) corner disappeared as the flow in these two corner rolls was not steady, unlike those in Figure 4a,b. The hot and cold layers near the conducting plates were thermal boundary layers. In previous studies [14], the boundary layer was divided into three regions (from left to right): the impacting region, the shearing region, and the ejecting region. In these three regions, the motions of the fluid were different, and their properties changed with different Ra values.

The mean fields in Figure 4 indicate that the top–bottom and left–right symmetries were absent. The flow fields for $Ra \leq 5 \times 10^8$ were similar to the results for $Pr = 6.5$, which were also found in a recent study [45], although the effects of the flow transitions were not included. Therefore, we applied double reflection to the flow fields, as shown in the next section, to investigate the symmetry at higher Ra values.

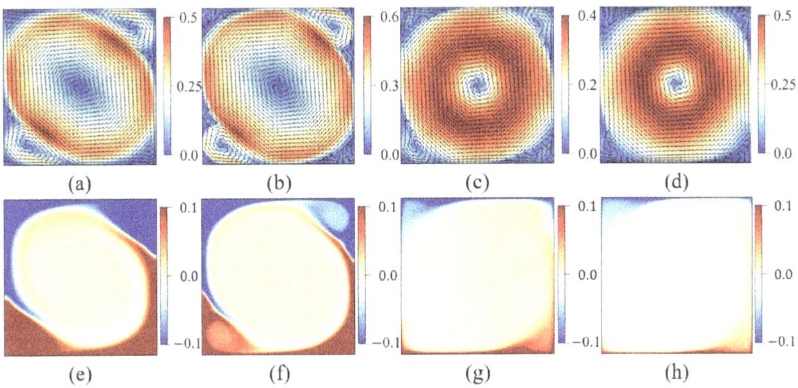

Figure 4. The time-averaged velocity fields $\mathbb{V}(x, y) = \sqrt{\langle u \rangle_t^2 + \langle v \rangle_t^2}$ (top row) and temperature fields (bottom row) $\theta(x, y)$ at (**a**,**e**) $Ra = 10^7$, (**b**,**f**) $Ra = 10^8$, (**c**,**g**) $Ra = 10^9$, and (**d**,**h**) $Ra = 10^{10}$. The velocity color map was set from the minimum to maximum value in each case, while the temperature color maps each ranged from -0.1 to 0.1. Note that the vectors only represent the direction of the velocity, and their length did not correspond to the magnitude. To show the velocity directions more clearly, the vectors were each normalized to unity.

4. Double-Reflection Symmetry and Data Verification

In this section, we will examine the symmetry by applying an x–y reflection at the center of the system. If a reflection is performed over $x = 1/2$ and is followed by another reflection over $y = 1/2$, then

$$t^* = t, \quad x^* = 1 - x, \quad y^* = 1 - y, \quad \mathbf{u}^* = -\mathbf{u}, \quad p^* = p, \quad \theta^* = -\theta, \tag{10}$$

where the superscript * indicates transformed variables. Equations (1)–(3) can then be written as follows:

$$\nabla^* \cdot \mathbf{u}^* = 0 \tag{11}$$

$$\frac{\partial \mathbf{u}^*}{\partial t^*} + (\mathbf{u}^* \cdot \nabla)\mathbf{u}^* = -\nabla p^* + \frac{1}{\sqrt{Ra/Pr}} \nabla^{*2} \mathbf{u}^* + \theta^* \hat{y}^* \tag{12}$$

$$\frac{\partial \theta^*}{\partial t^*} + (\mathbf{u}^* \cdot \nabla)\theta^* = \frac{1}{\sqrt{Ra \cdot Pr}} \nabla^2 \theta^*. \tag{13}$$

Comparing Equations (11)–(13) with Equations (1)–(3), it can be seen that their forms are the same; it is, thus, clear that the governing equations are symmetric with the reflections applied in both the x and y directions.

4.1. Flow Fields

As shown in this section, the time-averaged temperature $\langle\theta(x,y)\rangle_t$ and velocity magnitude fields $\mathbb{V} = \sqrt{\langle u\rangle_t^2 + \langle v\rangle_t^2}$ were transformed using double reflection, where $\langle\rangle_t$ indicates time averaging. For brevity, we use $\Theta(x,y)$ to denote $\langle\theta(x,y)\rangle_t$; $U(x,y)$ to denote $\langle u(x,y)\rangle_t$; and $V(x,y)$ to denote $\langle v(x,y)\rangle_t^2$. Taking the temperature field $\Theta(x,y)$ as an example, the process of the double-reflection transformation is as follows:

$$\Theta(x,y) \to \Theta(1-x,y) \to -\Theta(1-x,1-y) = \Theta^*(x^*,y^*). \tag{14}$$

This process at $Ra = 10^8$ is directly shown in Figure 5. The process was similar for other physical fields, as shown in Equation (10). From Figure 5a,c, it can be seen that the differences in the temperature fields before and after the transformation were not visually apparent; the two plots are almost identical, showing a dominant LSC accompanied by two corner rolls. Therefore, the relative difference between the two fields was defined to quantify the quality of the double reflection as follows:

$$D(\Theta) = |\Theta(x,y) - \Theta^*(x^*,y^*)|/\Delta\Theta \times 100\%, \tag{15}$$

where $\Delta\Theta = \Theta_{\text{bot}} - \Theta_{\text{top}} = 1$. The relative difference between the two temperature fields is shown clearly in Figure 5d. If the difference $D(\Theta)$ across most of the region is very small, the double-reflection symmetry is validated. The symmetry of the velocity magnitude field $\mathbb{V}$ will also be examined in a similar way, with the relative difference $D(\mathbb{V})$ defined as follows:

$$D(\mathbb{V}) = |\mathbb{V} - \mathbb{V}^*|/\mathbb{V}_{\max} \times 100\%, \tag{16}$$

where $\mathbb{V}^* = \sqrt{U^{*2} + V^{*2}}$ and $\mathbb{V}_{\max}$ is the maximum velocity magnitude.

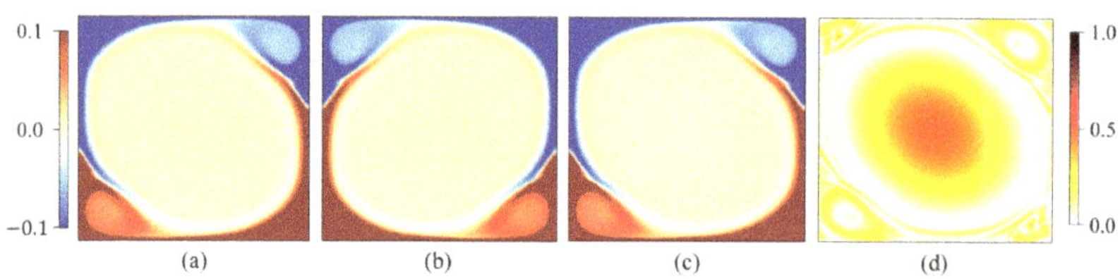

Figure 5. Time-averaged temperature fields at $Ra = 10^8$ during the transformation process: (**a**) original temperature field $\Theta(x,y)$; (**b**) temperature field after horizontal reflection $\Theta(1-x,y)$; (**c**) the temperature field after both horizontal and vertical reflections $\Theta^*(x^*,y^*) = \Theta(1-x,1-y)$; and (**d**) the relative difference $D(\Theta)$ between $\Theta(x,y)$ and $\Theta^*(x^*,y^*)$.

For $Ra < 10^9$, the mean velocity fields were symmetrical after the double reflection, with the relative difference $D(\mathbb{V})$ remaining below 1%, as shown in Figure 6a,b. The maximum value of $D(\mathbb{V})$ was found in the region between the LSC and the corner rolls, where the velocity fluctuations were the most intense [50].

In contrast, for $Ra \geq 10^9$, the distribution of the difference $D(\mathbb{V})$ changed, as shown in Figure 6c,d. The value of $D(\mathbb{V})$ in the center of the cell was much larger, exceeding 10%. The value of $D(\mathbb{V})$ also increased in the middle of the conducting plates (the shearing regions), as well as in the bottom-left and top-right regions, which were dominated by impacting plumes. These distributions indicate that the double-reflection symmetry was broken in the mean velocity fields for $Ra \leq 10^9$ at $Pr = 0.7$. This asymmetry corresponds to the flow-pattern transition mentioned above and reflects the change in the flow fields, as shown in Figures 3 and 4. In the flow state for $Ra \leq 10^9$, more of the plumes moved around the center of the cell, inducing more intense fluctuations. Therefore, $D(\mathbb{V})$ became larger in the center of the cell.

The mean temperature fields exhibited symmetry after a double reflection for $Ra < 10^9$, with $D(\Theta)$ values remaining below 1% in Figure 6e,f; however, for $Ra \geq 10^9$, the $D(\Theta)$ values increased throughout the system, which is similar to the behavior seen in $D(\mathbb{V})$. The maximum $D(\Theta)$ value was around 5%, as shown in Figure 6g,h, and this was located in the ejecting region. In this region, thermal plumes were primarily emitted, leading to stronger thermal fluctuations [50]. For the mean temperature fields, the symmetry of double reflection held for $Ra < 10^9$ within the current range, while the symmetry was broken for $Ra \geq 10^9$. According to [18], the flow pattern transition occurred at about $Ra = 1.1 \times 10^9 Pr^{1.41}$. In our previous study for $Pr = 6.5$, the highest Ra was 10^{10}, while the transition Ra for this Pr was, approximately, 1.54×10^{10}. Under the transition Ra, the double-reflection symmetry held. In this paper, the highest Ra was 10^{13}, which was significantly higher than the transition Ra for $Pr = 0.7$ (6.65×10^8 approximately). As a result, asymmetry was observed in this regime.

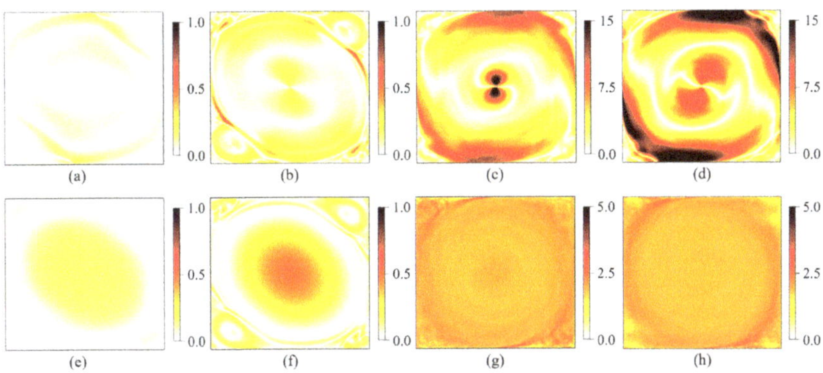

Figure 6. Top row: the relative difference $D(\mathbb{V})$ between $\mathbb{V}$ and $\mathbb{V}^*$. Bottom row: the relative difference $D(\Theta)$ between $\Theta(x,y)$ and $\Theta^*(x^*,y^*)$. (**a**,**e**) $Ra = 10^7$, (**b**,**f**) $Ra = 10^8$, (**c**,**g**) $Ra = 10^{11}$, and (**d**,**h**) $Ra = 10^{12}$. Note the varying color maps: 0 to 1% in (**a**,**b**,**e**,**f**), 0 to 15% in (**c**,**d**); and 0 to 5% in (**g**,**h**).

4.2. Mean Velocity and Temperature Profiles

In this section, the symmetry at the low Ra values was examined in more detail using the horizontal profiles of the mean temperature (Θ) and velocity components (U, V). Two horizontal sections at different heights, $y = 0.01$ and $y = 0.1$, were selected to study the double-reflection symmetry of the velocity and temperature profiles. The $y = 0.01$ section was closer to the boundary layer, while the $y = 0.1$ section was situated in the bulk region.

The mean profile at $y = 0.99$ was transformed to correspond with the mean profile at $y = 0.01$ in accordance with Equation (10). If the two profiles collapsed onto one another, the symmetry of the double reflection held; otherwise, the symmetry was considered broken. Similarly, the profile at $y = 0.9$ was transformed to match the profile at $y = 0.1$.

The profiles for $Ra = 10^7$ and 10^8 were compared, as shown in Figure 7. It can be seen that the profile $U(x, y = 0.01)$ closely matches $-U(x, y = 0.99)$, confirming the symmetry after the double reflection. Similarly, the profiles of $V(x, y = 0.01)$ and $\Theta(x, y = 0.01)$ also aligned with their counterparts following double reflection (see Figure 7, left-hand panel). Furthermore, the profiles at $y = 0.1$ and $y = 0.9$ were nearly identical, validating the symmetry of the double reflection. Although the profiles for other Ra values are not shown here and they varied with Ra, the double-reflection symmetry consistently held for all Ra values below 10^9 in this study.

It was also evident that these profiles exhibited asymmetry in the x direction. Using the profiles at $y = 0.1$ as an example, we can explain the source of the left–right asymmetry. On the left-hand side, the corner roll generated a negative horizontal velocity U and a positive vertical velocity V (see Figure 4a,b). In contrast, on the right-hand side, the absence of a corner roll resulted in velocity components that differed from those on the left side of the cell. The plateaus observed in Figure 7b,f were induced by the shear from the elliptical LSC, and they were not symmetrical about the center axis at $x = 0.5$. Therefore, the left–right symmetry was broken in these cases.

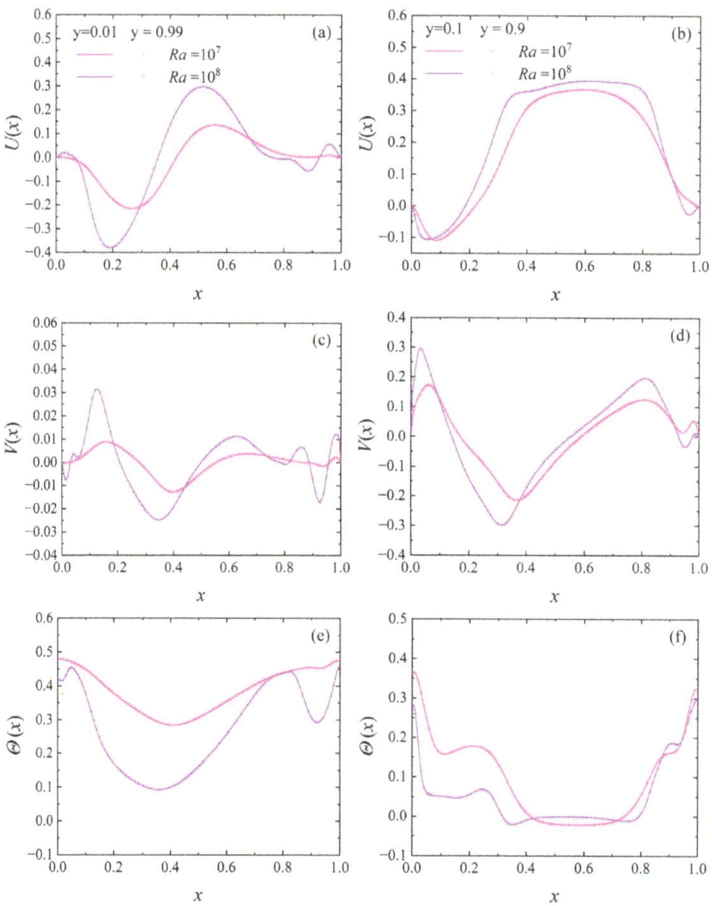

Figure 7. Streamwise variations of the (**a**,**b**) mean horizontal velocity $U(x)$, (**c**,**d**) mean vertical velocity $V(x)$, and (**e**,**f**) mean temperature profiles $\Theta(x)$. In the left-hand panels, the solid lines represent the original profiles before transformation at $y = 0.01$, while the plus symbols represent the double-reflection-transformed profiles from data at $y = 0.99$ (dashed lines). Similarly, in the right-hand panels, the solid lines represent the original profiles at $y = 0.1$, while the plus symbols represent the double-reflection-transformed profiles from the data at $y = 0.9$. The pink lines and corresponding plus symbols represent the data for $Ra = 10^7$, while the purple lines and corresponding plus symbols represent the data for $Ra = 10^8$.

At high Ra values ($\geq 10^9$), the double-reflection symmetry was broken, as indicated in Figure 8. For the velocity components U and V, as well as the temperature Θ, the trends of the profiles near the bottom plate ($y = 0.01, 0.1$) were similar to those near the top plate

($y = 0.99, 0.9$). However, the profiles near the bottom plate (represented by plus symbols) did not coincide with those near the top plate (shown by solid lines). We can see clear discrepancies in Figure 8, and these discrepancies increased with increasing Ra.

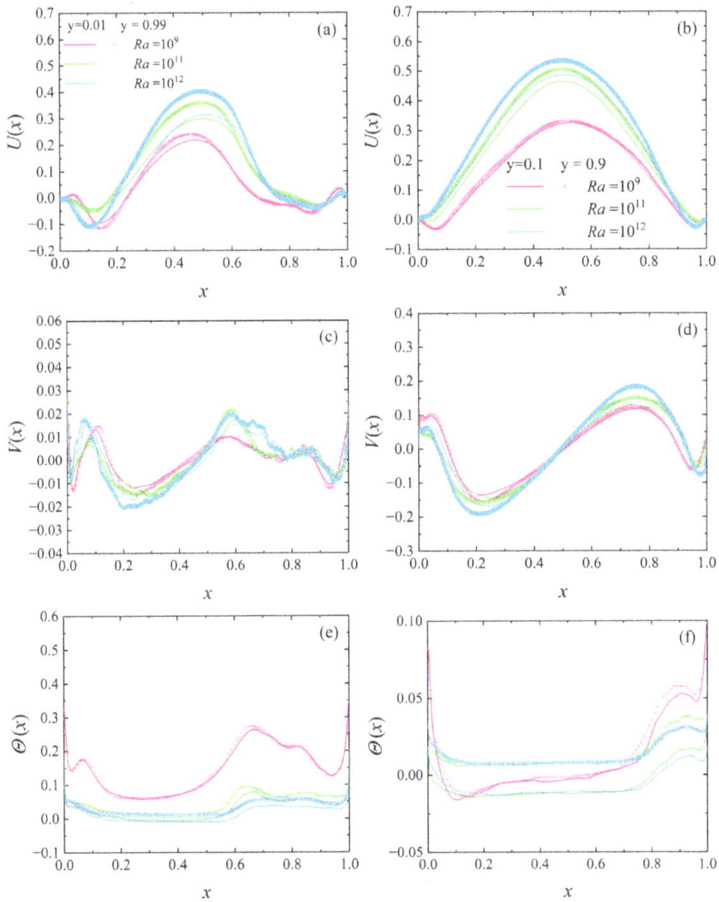

Figure 8. Streamwise variations of the (**a**,**b**) mean horizontal velocity $U(x)$, (**c**,**d**) mean vertical velocity $V(x)$, and (**e**,**f**) mean temperature profiles $\Theta(x)$. In the left-hand panels, the solid lines represent the original profiles before transformation at $y = 0.01$, while the plus symbols represent the double-reflection-transformed profiles from the data at $y = 0.99$ (dashed lines). Similarly, in the right-hand panels, the solid lines represent the original profiles at $y = 0.1$, while the plus symbols represent the double-reflection-transformed profiles from data at $y = 0.9$. The colors pink, green, and blue represent the data for $Ra = 10^9$, 10^{11}, and 10^{12}, respectively.

4.3. Boundary Layer Characteristics

As shown in this section, the wall units of the boundary layers were compared to examine the double-reflection symmetry. The wall units were calculated for two regions (as shown in Figure 1): the top-left region ($y = 1, 0 \leq x \leq 0.5$) and the bottom-right region ($y = 0, 0.5 \leq x \leq 1$).

Following Ref. [46], the friction velocity in the top-left (TL) region u_τ^T or in the bottom-right (BR) region u_τ^B were calculated as

$$u_\tau^T = \left(\frac{Pr}{Ra}\right)^{1/4} \left\langle \left|\frac{\partial u}{\partial y}\right|_{y=1}^{1/2} \right\rangle_{t,0\le x\le 0.5}, \quad u_\tau^B = \left(\frac{Pr}{Ra}\right)^{1/4} \left\langle \left|\frac{\partial u}{\partial y}\right|_{y=0}^{1/2} \right\rangle_{t,0.5\le x\le 1}, \tag{17}$$

where $\langle\rangle_{t,a\le x\le b}$ denotes the time–space average over the region $a \le x \le b$. The wall units of temperature θ_τ^T and θ_τ^B were also investigated, and these were calculated as follows [46]:

$$\theta_\tau^T = -\frac{1}{u_\tau\sqrt{Ra\cdot Pr}} \left\langle \frac{\partial\theta}{\partial y}\bigg|_{y=1} \right\rangle_{t,0\le x\le 0.5}, \quad \theta_\tau^B = -\frac{1}{u_\tau\sqrt{Ra\cdot Pr}} \left\langle \frac{\partial\theta}{\partial y}\bigg|_{y=0} \right\rangle_{t,0.5\le x\le 1}. \tag{18}$$

As shown in Figure 9a,b, the wall units u_τ and θ_τ in the TL and BR regions were identical before $Ra = 10^9$; however, after $Ra = 10^9$, the u_τ in the TL region decreased with increasing Ra, while the u_τ in the BR region was slightly smaller than that in the TL. Additionally, the temperature wall unit θ_τ showed a clear discrepancy between the TL and BR regions. Since θ_τ was calculated using u_τ in Equation (18), a decrease in u_τ led to an increase in θ_τ.

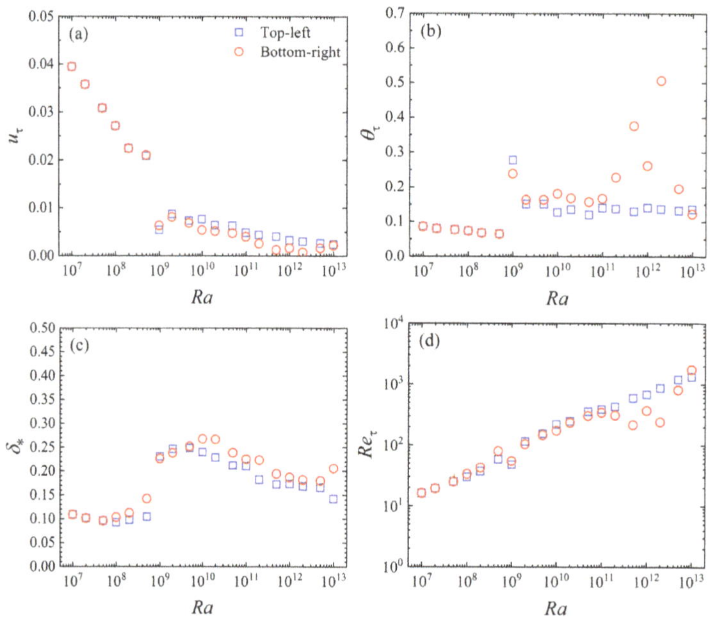

Figure 9. Variation within the Ra of (**a**) the friction velocity u_τ, (**b**) the wall unit temperature θ_τ, (**c**) the boundary layer thickness δ_*, and (**d**) the friction Reynolds number Re_τ. The blue squares represent the data for the top-left region, and the red circles represent the data for the bottom-right region.

We further investigated the boundary layer thickness in the TL and BR regions. The boundary layer thickness δ_*^T was defined as the distance between the top wall and the maximum value of $\langle|U|\rangle$ in the domain $(0 \le x \le 0.5, 0.5 \le y \le 1.0)$. Similarly, δ_*^B is the distance from the bottom wall to the maximum value of $\langle|U|\rangle$ in the domain $(0.5 \le x \le 1.0, 0 \le y \le 0.5)$. Consequently, two friction Reynolds numbers Re_τ were defined as follows:

$$Re_\tau^T = u_\tau^T\delta_*^T\sqrt{Ra/Pr}, \quad Re_\tau^B = u_\tau^B\delta_*\sqrt{Ra/Pr}. \tag{19}$$

The boundary layer thickness δ_* remained almost constant for $Ra \leq 5 \times 10^7$, but it began to diverge from $Ra = 1 \times 10^8$, as shown in Figure 9c. A sharp increase appeared at $Ra = 10^9$, after which δ_* was consistently larger until $Ra = 10^{10}$. Although the variations in δ_*^T and δ_*^B with respect to Ra were similar, they showed some differences for $Ra \geq 10^{10}$. The values of Re_τ were almost the same for $Ra = 10^{11}$, as shown in Figure 9d, but, for the highest five cases, the discrepancy was clear, meaning that the double-reflection symmetry was broken.

Based on these results for the TL and BR regions, we can conclude that the flow in RBC remained symmetrical after an x–y reflection for low Ra values in the current cases, i.e., $Pr = 0.7$ and $10^7 \leq Ra \leq 5 \times 10^8$. However, the symmetry was broken above $Ra = 10^9$. There was a slight deviation between the wall units of the TL and BR regions, indicating that the flow was different after double reflection. This transition at $Ra = 10^9$ was correlated with the flow-pattern transition discussed above [18].

5. Conclusions

In this study, we investigated the symmetry after an x–y double reflection using the DNS data of a 2D RBC in a confined square cell for Ra values ranging from 10^7 to 10^{13} at $Pr = 0.7$. As illustrated by the instantaneous and mean flow fields, both the top–bottom and left–right symmetries were broken within the current range, as was expected. We examined the mean temperature and velocity fields, the horizontal profiles at two specific heights, and the boundary layer properties in the TL and BR regions, including the friction velocity, friction temperature, boundary layer thickness, and friction Reynolds number. To assess the validity of the symmetry, the relative differences between the original and transformed quantities were calculated. It was revealed that the double-reflection symmetry held for $Ra < 10^9$ but was broken at higher Ra values within the range considered. In contrast to a previous study [45], which found double-reflection symmetry for Ra values up to 10^{10} at $Pr = 6.5$, the present results indicate that the symmetry was broken at $Ra = 10^9$ when the Pr value was decreased to 0.7. Therefore, the effect of Pr was crucial for determining the symmetry properties in a RBC. The observed asymmetry may be related to the flow-pattern transitions previously reported [18], leading to stronger fluctuations in the flow fields. The current findings demonstrate the parameter range for the double-reflection symmetry. The application in the latter in the boundary settings could potentially accelerate the simulations of RBC cases and reduce computational costs, but this will have to be explored in the future.

Author Contributions: Conceptualization, Y.B. and J.-C.H.; methodology, Y.B. and J.-C.H.; software, Y.B. and J.-C.H.; validation, J.-C.H.; formal analysis, X.C. and J.-C.H.; investigation, X.C. and J.-C.H.; resources, Y.B. and X.C.; data curation, Y.B. and J.-C.H.; writing—original draft preparation, X.C. and J.-C.H.; writing—review and editing, X.C., Y.B., and J.-C.H.; visualization, J.-C.H. and Y.B.; supervision, X.C. and Y.B.; project administration, X.C.; funding acquisition, X.C. All authors have read and agreed to the published version of the manuscript.

Funding: X. Chen appreciates the funding support provided by the National Key Research and Development Program of China (2022YFF0610805), the National Natural Science Foundation of China (Grant Nos. 12072012 and 92252201), and the Fundamental Research Funds for the Central Universities.

Data Availability Statement: The data that support the findings of this study are available from the corresponding author upon reasonable request.

Acknowledgments: J.-C. He and Y. Bao are grateful to S.-D. Huang and Z.-Y. Gao from the Southern University of Science and Technology for their helpful discussions regarding the transitions of flow patterns.

Conflicts of Interest: The authors declare no conflicts of interest.

Abbreviations

The following abbreviations are used in this manuscript:

RBC	Rayleigh–Bénard convection
DNS	Direct numerical simulation
Ra	Rayleigh number
Pr	Prandtl number
Re	Reynold number
Nu	Nusselt number
LSC	Large-scale circulation
TL	Top left
BR	Bottom right

References

1. Schumacher, J.; Sreenivasan, K.R. Colloquium: Unusual dynamics of convection in the Sun. *Rev. Mod. Phys.* **2020**, *92*, 041001. [CrossRef]
2. Jiménez, J. Cascades in wall-bounded turbulence. *Annu. Rev. Fluid Mech.* **2012**, *44*, 27–45. [CrossRef]
3. Rosenblat, S. Thermal convection in a vertical circular cylinder. *J. Fluid Mech.* **1982**, *122*, 395–410. [CrossRef]
4. Behringer, R. Rayleigh-Bénard convection and turbulence in liquid helium. *Rev. Mod. Phys.* **1985**, *57*, 657. [CrossRef]
5. Morris, S.W.; Bodenschatz, E.; Cannell, D.S.; Ahlers, G. Spiral defect chaos in large aspect ratio Rayleigh-Bénard convection. *Phys. Rev. Lett.* **1993**, *71*, 2026. [CrossRef]
6. Bodenschatz, E.; Pesch, W.; Ahlers, G. Recent developments in Rayleigh-Bénard convection. *Annu. Rev. Fluid Mech.* **2000**, *32*, 709–778. [CrossRef]
7. Castaing, B.; Gunaratne, G.; Heslot, F.; Kadanoff, L.; Libchaber, A.; Thomae, S.; Wu, X.Z.; Zaleski, S.; Zanetti, G. Scaling of hard thermal turbulence in Rayleigh-Bénard convection. *J. Fluid Mech.* **1989**, *204*, 1–30. [CrossRef]
8. Xie, Y.C.; Hu, Y.B.; Xia, K.Q. Universal fluctuations in the bulk of Rayleigh–Bénard turbulence. *J. Fluid Mech.* **2019**, *878*, R1. [CrossRef]
9. Labarre, V.; Fauve, S.; Chibbaro, S. Heat-flux fluctuations revealing regime transitions in Rayleigh-Bénard convection. *Phys. Rev. Fluids* **2023**, *8*, 053501. [CrossRef]
10. Bhattacharya, S.; Samtaney, R.; Verma, M.K. Scaling and spatial intermittency of thermal dissipation in turbulent convection. *Phys. Fluids* **2019**, *31*, 075104. [CrossRef]
11. Xu, A.; Shi, L.; Xi, H.D. Statistics of temperature and thermal energy dissipation rate in low-Prandtl number turbulent thermal convection. *Phys. Fluids* **2019**, *31*, 125101. [CrossRef]
12. Vishnu, V.T.; De, A.K.; Mishra, P.K. Statistics of thermal plumes and dissipation rates in turbulent Rayleigh-Benard convection in a cubic cell. *Int. J. Heat Mass Transf.* **2022**, *182*, 121995. [CrossRef]
13. Zhou, Q.; Sugiyama, K.; Stevens, R.J.; Grossmann, S.; Lohse, D.; Xia, K.Q. Horizontal structures of velocity and temperature boundary layers in two-dimensional numerical turbulent Rayleigh-Bénard convection. *Phys. Fluids* **2011**, *23*, 125104. [CrossRef]
14. He, J.C.; Bao, Y.; Chen, X. Turbulent boundary layers in thermal convection at moderately high Rayleigh numbers. *Phys. Fluids* **2024**, *36*, 025140. [CrossRef]
15. Lohse, D.; Xia, K.Q. Small-scale properties of turbulent Rayleigh-Bénard convection. *Annu. Rev. Fluid Mech.* **2010**, *42*, 335–364. [CrossRef]
16. Schumacher, J.; Scheel, J.D.; Krasnov, D.; Donzis, D.A.; Yakhot, V.; Sreenivasan, K.R. Small-scale universality in fluid turbulence. *Proc. Natl. Acad. Sci. USA* **2014**, *111*, 10961–10965. [CrossRef]
17. Li, X.M.; He, J.D.; Tian, Y.; Hao, P.; Huang, S.D. Effects of Prandtl number in quasi-two-dimensional Rayleigh-Benard convection. *J. Fluid Mech.* **2021**, *915*, A60. [CrossRef]
18. Gao, Z.Y.; Tao, X.; Huang, S.D.; Bao, Y.; Xie, Y.C. Flow state transition induced by emergence of orbiting satellite eddies in two-dimensional turbulent Rayleigh-Bénard convection. *J. Fluid Mech.* **2024**, *997*, A54. [CrossRef]
19. Sun, C.; Wu, J.z.; Meng, X.h.; Liu, C.x.; Xu, W.; Dong, Y.h.; Zhou, Q. Heat transfer and flow structure in centrally-confined 2-D Rayleigh-Bénard convection. *J. Hydrodyn.* **2024**, *36*, 772–780. [CrossRef]
20. Zhu, X.; Stevens, R.J.; Shishkina, O.; Verzicco, R.; Lohse, D. Scaling enabled by multiscale wall roughness in Rayleigh–Bénard turbulence. *J. Fluid Mech.* **2019**, *869*, R4. [CrossRef]
21. Huang, M.; He, X. Effect of slip length on flow dynamics and heat transport in two-dimensional Rayleigh–Bénard convection. *J. Turbul.* **2022**, *23*, 492–514. [CrossRef]
22. Ren, L.; Tao, X.; Xia, K.Q.; Xie, Y.C. Transition to fully developed turbulence in liquid-metal convection facilitated by spatial confinement. *J. Fluid Mech.* **2024**, *981*, R2. [CrossRef]
23. Shishkina, O. Rayleigh-Bénard convection: The container shape matters. *Phys. Rev. Fluids* **2021**, *6*, 090502. [CrossRef]
24. Pandey, A.; Krasnov, D.; Schumacher, J.; Samtaney, R.; Sreenivasan, K.R. Similarities between characteristics of convective turbulence in confined and extended domains. *Phys. D Nonlinear Phenom.* **2022**, *442*, 133537. [CrossRef]

25. Zhang, L.; Xia, K.Q. Heat transfer in a quasi-one-dimensional Rayleigh–Bénard convection cell. *J. Fluid Mech.* **2023**, *973*, R5. [CrossRef]
26. Wang, Q.; Wan, Z.H.; Yan, R.; Sun, D.J. Flow organization and heat transfer in two-dimensional tilted convection with aspect ratio 0.5. *Phys. Fluids* **2019**, *31*, 025102. [CrossRef]
27. Malkus, W.V. Discrete transitions in turbulent convection. *Proc. R. Soc. Lond. A Math. Phys. Sci.* **1954**, *225*, 185–195. [CrossRef]
28. Kraichnan, R.H. Turbulent thermal convection at arbitrary Prandtl number. *Phys. Fluids* **1962**, *5*, 1374–1389. [CrossRef]
29. Doering, C.R. Turning up the heat in turbulent thermal convection. *Proc. Natl. Acad. Sci. USA* **2020**, *117*, 9671–9673. [CrossRef]
30. Iyer, K.P.; Scheel, J.D.; Schumacher, J.; Sreenivasan, K.R. Classical 1/3 scaling of convection holds up to Ra = 10(15). *Proc. Natl. Acad. Sci. USA* **2020**, *117*, 201922794. [CrossRef]
31. Lohse, D.; Shishkina, O. Ultimate turbulent thermal convection. *Phys. Today* **2023**, *76*, 26–32. [CrossRef]
32. Jiang, H.; Wang, D.; Liu, S.; Sun, C. Experimental evidence for the existence of the ultimate regime in rapidly rotating turbulent thermal convection. *Phys. Rev. Lett.* **2022**, *129*, 204502. [CrossRef] [PubMed]
33. He, X.; Funfschilling, D.; Nobach, H.; Bodenschatz, E.; Ahlers, G. Transition to the ultimate state of turbulent Rayleigh-Bénard convection. *Phys. Rev. Lett.* **2012**, *108*, 024502. [CrossRef] [PubMed]
34. Niemela, J.J.; Skrbek, L.; Sreenivasan, K.R.; Donnelly, R.J. The wind in confined thermal convection. *J. Fluid Mech.* **2001**, *449*, 169–178. [CrossRef]
35. Borońska, K.; Tuckerman, L.S. Standing and travelling waves in cylindrical Rayleigh–Bénard convection. *J. Fluid Mech.* **2006**, *559*, 279–298. [CrossRef]
36. Xia, K.Q.; Sun, C.; Zhou, S.Q. Particle image velocimetry measurement of the velocity field in turbulent thermal convection. *Phys. Rev. E* **2003**, *68*, 066303. [CrossRef]
37. Sun, C.; Xi, H.D.; Xia, K.Q. Azimuthal symmetry, flow dynamics, and heat transport in turbulent thermal convection in a cylinder with an aspect ratio of 0.5. *Phys. Rev. Lett.* **2005**, *95*, 074502. [CrossRef]
38. Chen, X.; Hussain, F.; She, Z.S. Quantifying wall turbulence via a symmetry approach. Part 2. Reynolds stresses. *J. Fluid Mech.* **2018**, *850*, 401–438. [CrossRef]
39. Heslot, F.; Castaing, B.; Libchaber, A. Transitions to turbulence in helium gas. *Phys. Rev. A* **1987**, *36*, 5870. [CrossRef]
40. Lui, S.L.; Xia, K.Q. Spatial structure of the thermal boundary layer in turbulent convection. *Phys. Rev. E* **1998**, *57*, 5494. [CrossRef]
41. Wang, J.; Xia, K.Q. Spatial variations of the mean and statistical quantities in the thermal boundary layers of turbulent convection. *Eur. Phys. J. B* **2003**, *32*, 127–136. [CrossRef]
42. Pandey, A.; Verma, M.K.; Barma, M. Reversals in infinite-Prandtl-number Rayleigh-Bénard convection. *Phys. Rev. E* **2018**, *98*, 023109. [CrossRef] [PubMed]
43. Silano, G.; Sreenivasan, K.R.; Verzicco, R. Numerical simulations of Rayleigh–Bénard convection for Prandtl numbers between 10^{-1} and 10^4 and Rayleigh numbers between 10^5 and 10^9. *J. Fluid Mech.* **2010**, *662*, 409–446. [CrossRef]
44. Zhang, Y.; Zhou, Q.; Sun, C. Statistics of kinetic and thermal energy dissipation rates in two-dimensional turbulent Rayleigh-Benard convection. *J. Fluid Mech.* **2017**, *814*, 165–184. [CrossRef]
45. He, J.C.; Duan, P.Y.; Chen, X. Double-reflection symmetry of thermal convection for Rayleigh number up to 10^{10}. *Phys. Fluids* **2024**, *36*, 105113. [CrossRef]
46. Pandey, A. Thermal boundary layer structure in low–Prandtl–number turbulent convection. *J. Fluid Mech.* **2021**, *910*, A13. [CrossRef]
47. Bao, Y.; Luo, J.H.; Ye, M.X. Parallel direct method of DNS for two-dimensional turbulent Rayleigh-Benard convection. *J. Mech.* **2017**, *34*, 159. [CrossRef]
48. Shishkina, O.; Stevens, R.J.A.M.; Grossmann, S.; Lohse, D. Boundary layer structure in turbulent thermal convection and its consequences for the required numerical resolution. *New J. Phys.* **2010**, *12*, 075022. [CrossRef]
49. Zhang, Y.; Zhou, Q. Low-Prandtl-number effects on global and local statistics in two-dimensional Rayleigh–Bénard convection. *Phys. Fluids* **2024**, *36*, 015107. [CrossRef]
50. Zhang, Y.; Huang, Y.X.; Jiang, N.; Liu, Y.L.; Lu, Z.M.; Qiu, X.; Zhou, Q. Statistics of velocity and temperature fluctuations in two-dimensional Rayleigh-Bénard convection. *Phys. Rev. E* **2017**, *96*, 023105. [CrossRef]

symmetry

MDPI

Article

Blood Damage Analysis within the FDA Benchmark Nozzle Geometry at Laminar Conditions: Prediction Sensitivities to Software and Non-Newtonian Viscosity Models

Gautham Krishnamoorthy * and Nasim Gholizadeh

Department of Chemical Engineering, University of North Dakota, Grand Forks, ND 58202-7101, USA; nasim.gholizadeh@und.edu
* Correspondence: gautham.krishnamoorthy@und.edu

Abstract: There is a prevailing consensus that most Computational Fluid Dynamics (CFD) frameworks can accurately predict global variables under laminar flow conditions within the Food and Drug Administration (FDA) benchmark nozzle geometry. However, variations in derived variables, such as strain rate and vorticity, may arise due to differences in numerical solvers and gradient evaluation methods, which can subsequently impact predictions related to blood damage and non-Newtonian flow behavior. To examine this, flow symmetry indices, vortex characteristics, and blood damage—were assessed using Newtonian and four non-Newtonian viscosity models with CFD codes Ansys Fluent and OpenFOAM on identical meshes. At Reynolds number (Re) 500, symmetry breakdown and complex vortex shapes were predicted with some non-Newtonian models in both OpenFOAM and Ansys Fluent, whereas these phenomena were not observed with the Newtonian model. This contradicted the expectation that employing a non-Newtonian model would delay the onset of turbulence. Similarly, at Re 2000, symmetry breakdown occurred sooner (following the sudden expansion section) with the non-Newtonian models in both Ansys Fluent and OpenFOAM. Vortex identification based on the Q-criterion resulted in distinctly different vortex shapes in Ansys Fluent and OpenFOAM. Blood damage assessments showed greater prediction variations among the non-Newtonian models at lower Reynolds numbers.

Keywords: flow symmetry; FDA Nozzle Geometry; non-Newtonian; Q-criterion; laminar; viscosity model; blood damage; CFD

1. Introduction

Numerical simulations aimed at predicting the transition from laminar to turbulent flow (with throat Reynolds numbers (Re) ranging from 2000 to 3500) within the Food and Drug Administration (FDA) benchmark nozzle geometry (Figure 1a) face considerable challenges. However, there is a general consensus that laminar flow behavior (Re < 2000) within the FDA benchmark nozzle geometry can be accurately predicted. This has been demonstrated through various computational fluid dynamics (CFD) methodologies and has been summarized in Table 1. However, just as small variations in geometric configuration (such as eccentricity) and disturbances in inflow conditions can trigger a transition to turbulence, subtle differences in derived variable predictions, such as strain rates and vorticities resulting from differences in numerical solvers and gradient evaluation methods among different CFD solvers can result in vortex shape differences and transition onset identification. These differences in the derived variables may be further exacerbated when non-Newtonian models are employed to model the blood viscosity.

Therefore, this paper presents a numerical simulation study that highlights the challenges and complexities associated with predicting the transition from laminar to turbulent flow within the FDA benchmark nozzle geometry, focusing on Reynolds numbers (Re) ranging from 500 to 2000 employing only laminar models to simulate the flow. Additionally, prediction sensitivities to modeling blood viscosity using different non-Newtonian model formulations were explored. The importance of this study lies in its focus on the complexities and challenges of accurately predicting the transition from laminar to turbulent flow within the FDA benchmark nozzle geometry.

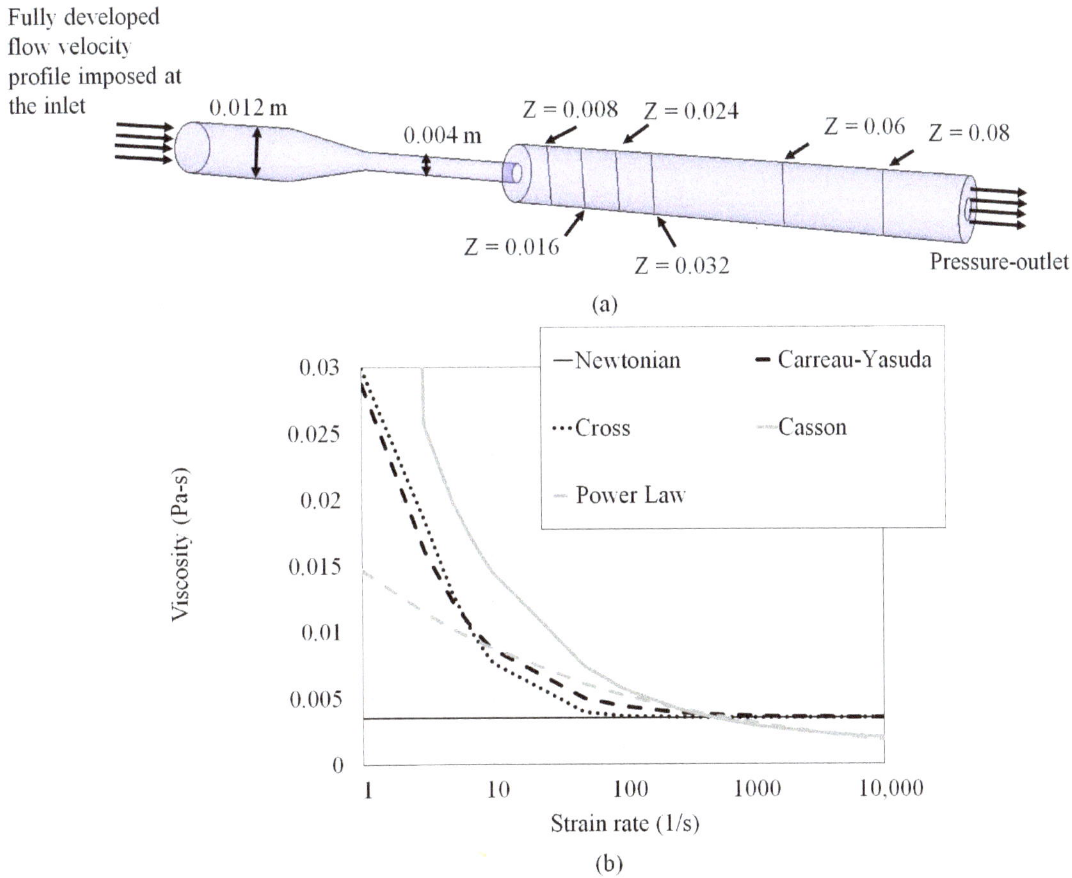

Figure 1. (**a**) Schematic Representation of FDA Benchmark Nozzle with different axial distances following the sudden expansion (Z in meters) at which the results in this study are reported; (**b**) Viscosity variations employing the different non-Newtonian models employed in this study.

Table 1. A summary of CFD studies corresponding to Re 500, 2000 within the FDA Nozzle with an emphasis on symmetry and jet breakdown after the sudden expansion region.

Authors	Publication Year	Summary
Bergersen et al. [1]	2013	In this study, at a Reynolds number (Re) of 500, simulation predictions remained stable despite numerical noise introduced at the inlet to simulate experimental uncertainties. However, at a Reynolds number of 3500, such inlet perturbations influenced the location of vortex breakdown
Bhushan et al. [2]	2013	In this study, the use of turbulence models at Reynolds number (Re) 500 led to excessive diffusion, which resulted in lower centerline velocities compared to experimental data. However, transition-sensitive unsteady RANS turbulence models can deliver accurate predictions across the entire Reynolds number range if the turbulence intensity levels at the inlet are correctly specified.

Table 1. *Cont.*

Authors	Publication Year	Summary
Chabannes et al. [3]	2017	In this study, at Re 500, solution accuracies were assessed using mesh refinement studies. At Re 2000, the jet breakdown occurred downstream of the experimentally measured location.
Delorme et al. [4]	2013	LES and Immersed Boundary Method were employed. Predictions agreed with the experimental results for the mean velocities at Re = 500, 2000 and 3500. While Re = 500 stayed symmetric throughout, coherent vortices, as identified by the λ_2 criterion, were observed close to the 0.032 m mark.
Fehn et al. [5]	2019	Flow remains laminar throughout at Re 500. At Re 2000 transition can be induced if the simulation parameters can be tweaked. Simulations were carried out using a high-order discontinuous Galerkin discretization technique.
Huang et al. [6]	2022	No jet breakdown after sudden expansion at Re 500 for unsteady simulations carried out with the Lattice Boltzmann Method. Symmetric solutions and zero centerline shear stresses persist downstream of the sudden expansion.
Jain [7]	2020	High resolution (HR) and Extreme spatial resolution (XR) of 40 μm; 20 μm is necessary for accurate results in Lattice Boltzmann Method-based simulations. Jet breakdown for the Re 2000 case occurs between 0.06 m and 0.08 m downstream of the sudden expansion. At low resolutions, the jet breaks down between 0.024 m and 0.032 m.
Manchester et al. [8]	2020	Using LES at Re 2000 using the OpenFOAM solver, laminar quantities were predicted throughout the model. In the jet breakdown region (0.06 m downstream of sudden expansion), where maximum Reynolds stresses occur, Reynolds shear stresses showed excellent agreement with measurements.
Nicoud et al. [9]	2018	LES simulations were carried out with a fourth-order accurate solver. Inlet perturbations did not induce turbulence in the laminar case (Re 500), whereas improved axial velocity predictions were seen at Re 2000.
Passerini et al. [10]	2013	Transient finite element simulations were carried out. At Re 500, the flow is axially symmetric throughout. At Re 2000, simulations predicted a jet breakdown location further downstream of the experimental and the differences were attributed to the fidelity of the numerical integration in the solver.
Pewowaruk et al. [11]	2021	Adaptive Mesh Refinement (AMR) study was carried out in a cut-cell representation of the device. Velocity contours show that symmetric profiles persist downstream of the sudden expansion at Re 500.
Sanchez, A. et al. [12]	2020	Used a high-order Spectral Element Method to carry out the simulations. At Re 500, no flow transition was observed. At Re 2000, flow did not transition to turbulence. However, by increasing the length of the domain downstream of the expansion and by incorporating some fluctuations at the inlet, a transition was eventually observed.
Stewart et al. [13]	2012	Re 500–6500 simulated. Turbulence models (k-epsilon, k-omega, and shear stress transport models) were employed in all Re > 500 simulations. Turbulence models were unable to accurately predict velocities and shear stresses in the recirculation zones downstream of the sudden expansion.
Stewart et al. [14]	2012	Summarized the results from several CFD groups. Results showed modest agreement in global and local flow behaviors. All CFD data sets contained wide degrees of velocity variation in comparison to the experiment and with each other at Re 2000 and between 0.032 m and 0.06 m downstream of the sudden expansion section.
Stiehm et al. [15]	2017	Pulsatile (time-dependent) boundary condition simulations were performed on a scaled-down (1/3rd scale) of the original FDA geometry at Re 500. Replacing the time-dependent boundary condition with an appropriate time-average boundary condition to employ in steady-state simulations was deemed to be sufficient if the interest is on obtaining time-averaged values.

Table 1. *Cont.*

Authors	Publication Year	Summary
Taylor et al. [16]	2016	Large variations in the LDV and PIV measurements of axial velocity at 0.06 m downstream of the sudden expansion due to instabilities for the Re 2000 case. The highest turbulence intensity and shear stresses were seen during jet breakdown between 0.032 m and 0.06 m downstream of the sudden expansion.
Zmijanovic et al. [17]	2017	LES simulations showed that Re 500 remains laminar throughout with no breakdown in symmetry. Results indicate a considerable impact of numerical aspects on the prediction of the location of the transition to turbulence for Re 3500.
Qiao et al. [18]	2022	Thrombus formation investigated at different Re. An inverse relationship between thrombosis and hemolysis was observed with thrombosis formation accelerated at lower Re.
Tobin et al. [19]	2020	The objective of this study is to utilize existing models to account for turbulence's chaotic nature and the interplay of fluctuating velocities and eddies within the flow. To validate these adaptations, the study compared model predictions with experimental data gathered from previous research.

1.1. Flow Symmetry within the FDA Benchmark Nozzle in the Laminar Regime (Re < 2000)

As detailed in Table 1, flow conditions at Reynolds number (Re) 500 within the FDA benchmark nozzle are expected to remain symmetric throughout. However, jet breakdown or asymmetry has been observed at Re 2000, occurring between 0.032 m and 0.06 m of the sudden expansion region. In the biomedical field, flow symmetry is crucial for achieving harmonious biological flow patterns ensuring the safety and efficacy of medical procedures and outcomes for patients. For instance, symmetric flow is essential to minimize turbulence and reduce the risk of damage to cells and vessel walls. Disruptions in flow symmetry can alter the forces and velocities within the fluid, potentially leading to cell deformation, blood clot formation (thrombosis), or cell damage (hemolysis).

Studies by Sobey and Drazin [20] demonstrate that a symmetric flow through a channel with an expanding throat becomes asymmetrical as the Reynolds number increases (Re = 384). Although the studies listed in Table 1 report a general agreement on global flow variables (such as velocity profiles) in this regime, variations in symmetry breakdown and vortex shape predictions can arise due to differences in numerical solvers and gradient evaluation methods. These variations, which are dependent on velocity derivatives like strain rate and vorticity, can occur even after achieving grid-independent results for global variables such as velocity and pressure. Recognizing these prediction variations among different CFD frameworks is important, particularly when CFD predictions are used to optimize device geometry based on the identification of stagnation zones or vortices that could promote hemolysis.

Contribution #1 of this study: To address the identified gap, this study first examines symmetry breakdown and vortex predictions for a Newtonian fluid using the commercial code Ansys Fluent [21] and OpenFOAM [22] on identical meshes. The analysis is limited to steady-state, laminar flow solvers in both frameworks to eliminate the effects of turbulent wall functions and time discretization variations.

1.2. Newtonian vs. Non-Newtonian Viscosity Model Effects in the Laminar Regime (Re < 2000)

Even within the laminar regime (Re < 2000), predicting clot formation (thrombosis) in blood-contacting devices remains challenging. From a fluid mechanics perspective, shear stresses and viscosity are critical in the thrombosis process, which involves platelet aggregation, platelet activation, and the formation of a gel-like mesh by fibrin proteins in blood plasma. The viscosity of red blood cells, which depends on hematocrit content, influences the flow and transport characteristics of platelets in recirculation zones resulting from abrupt changes in flow direction or cross-section. While various Newtonian and non-Newtonian viscosity models have been proposed to characterize blood's transport properties, the choice of model can affect flow characteristics and thrombosis prediction, as summarized in Table 2. Key conclusions from Table 2 include: non-Newtonian viscosity models have been shown to delay thrombus formation due to their shear-thinning behavior, and flow geometry and inlet boundary conditions have a more significant impact on flow characteristics than the choice of non-Newtonian model. There is no consensus on which non-Newtonian model formulation offers the highest fidelity.

Table 2. Thrombosis and transition to turbulence (TT) prediction variations in different geometries at low Reynolds Numbers (Re < 2000) resulting from the choice of non-Newtonian viscosity models.

Authors	Publication Year	Summary
Biswas et al. [23]	2016	Two mathematically defined methods, based on the velocity profile shape change and turbulent kinetic energy (TKE), were used to detect the transition to turbulence experimentally. Overall, the critical Re for turbulence onset was delayed by 20% (2316 vs. 2871) for blood compared to the Newtonian fluid.
Cebral et al. [24]	2005	Non-Newtonian viscosity models, like the Casson model, produce smaller velocity gradients due to higher local viscosity in low-flow regions. Despite this, the overall flow patterns and mean wall shear stress remain largely unchanged. The most significant factor influencing intra-aneurysmal flow patterns is the geometry of the aneurysm and its connections.
Costa et al. [25]	2022	Normalized turbulent kinetic energy (TKE) was employed to establish the critical Reynolds number for each fluid. The study found a 19% delay in TT for whole blood compared to the Newtonian fluid.
Haley et al. [26]	2021	Delays in TT between "noisy" non-Newtonian and Newtonian results were studied in an eccentric geometry. Introducing noise did not eliminate this distinction. However, as Re increased, the flow profiles under both rheologies started to converge towards similar outcomes.
Johnston et al. [27]	2004	While the Newtonian model of blood viscosity provides a good approximation in regions of mid-range to high shear, it is advisable to use the Generalized Power Law (non-Newtonian) model. This model converges to the Newtonian model in high shear ranges but offers a better approximation of wall shear stress at low shear rates.
Khan et al. [28]	2019	Blood exhibits a delayed TT compared to Newtonian fluids. However, defining a critical Reynolds number for non-Newtonian fluids accurately remains challenging in complex geometries. This difficulty is compounded by factors beyond shear thinning, including the potential influence of viscoelastic properties.
Lee, S.W. et al. [29]	2007	Prediction sensitivities to non-Newtonian models were in different geometries and flow rates. Using the Carreau–Yasuda model, it was observed that shear-thinning has a minimal effect on carotid bifurcation hemodynamics. In contrast, high shear rates, inlet boundary conditions, and geometric factors exerted more significant influences.
Razavi, A. et al. [30]	2011	In this study, flow conditions in a stenosed artery for various viscosity models, including Newtonian and six non-Newtonian models (power law, generalized power law, Carreau, Carreau–Yasuda, modified Casson, and Walburn–Schneck) were compared. The findings indicate that the power law model exhibits greater variation in wall shear stress and velocity values compared to other models. In contrast, the modified Casson and generalized power law models show a closer approximation to Newtonian behavior.

Contribution #2 of this study: This study investigates the sensitivity of symmetry breakdown and vortex shapes to four non-Newtonian (shear-thinning) viscosity models in both Ansys Fluent and OpenFOAM. The study aims to explore prediction variations across both frameworks regarding symmetry breakdown and vortex shapes as indicators of thrombus formation. Additionally, since non-Newtonian viscosities depend on strain rates, the study examines whether symmetry and vortex shape prediction variations between the two frameworks are exacerbated in non-Newtonian simulations.

1.3. Non-Newtonian Simulations of the FDA Nozzle

Non-Newtonian model studies conducted within the FDA nozzle geometry under well-characterized inflow conditions are summarized in Table 3. From this table, it is evident that while differences between Newtonian and non-Newtonian models become more pronounced at lower Reynolds numbers (Re) within this device, there is no consensus on the impact of viscosity models on shear stresses, pressure drop, and hemolysis.

Table 3. Non-Newtonian simulations of the FDA nozzle.

Authors	Publication Year	Summary
Good [31]	2023	Model differences were most evident at Re = 500. The non-Newtonian model predicted blunter upstream velocity profiles, and a greater pressure drop, with minimal differences observed at higher Reynolds numbers.
Hussein et al. [32]	2021	At Re 500, TKE and turbulent viscosity ratio were lower with Carreau in comparison to the Newtonian model. Pressure profiles were also different, attributed to larger recirculation zones at lower Re. Both differences diminished between the viscosity models at higher Re.
Trias, et al. [33]	2014	Axial velocity in the downstream region is inversely proportional to viscosity, with the highest velocity in the Newtonian model, followed by Carreau–Yasuda and then Casson. Velocity differences are within 10%. Viscosity models do not significantly affect the pressure drop. Non-Newtonian effects, while not greatly impacting velocity or pressure drop, are important for assessing blood damage, as they predict higher wall shear stresses and greater blood damage.
Zakaria et al. [34]	2019	The Newtonian model exhibits higher centerline velocities due to its constant viscosity assumption, which should be higher at lower shear rates. It also shows a higher pressure drop, likely due to more vortices. Using the lambda-2 criterion, chaotic and unstable vortices were observed with the Newtonian model. Overall, the Newtonian model produced more severe hemodynamic properties.

Contribution #3 of this study: To address this gap, this study examines the sensitivity of hemolysis predictions to non-Newtonian models within the Ansys Fluent framework.

1.4. Prediction Sensitivity to Software

While Table 1 demonstrates that various CFD methodologies can adequately represent global variables and overall flow characteristics (at Re < 2000) within the FDA nozzle, subtle differences in vortex identification may arise due to variations in gradient prediction methods among CFD frameworks. These differences are illustrated in Table 4, which compares two established CFD codes, Ansys Fluent [21] and OpenFOAM [22]. Recognizing these variations is crucial, particularly when optimizing device geometry based on the identification of stagnation zones or vortices that could promote hemolysis. Examining these prediction variations is especially important in the context of vortex identification, as highlighted in Table 5, given its increasingly significant role when CFD simulations are being employed to guide the design of medical devices.

Table 4. Literature summary comparing the predictions of Ansys Fluent and OpenFOAM.

Authors	Publication Year	Summary
Berg et al. [35]	2012	Centerline pressure predictions between Ansys Fluent and OpenFOAM show a high level of agreement, with differences being less than 1.03%. This indicates that both software packages provide reliable and consistent results for predicting centerline pressures in CFD simulations.

Table 4. *Cont.*

Authors	Publication Year	Summary
Lysenko et al. [36]	2013	In simulations of turbulent flow (Reynolds number = 3900) over a circular cylinder, AF and OF showed differences in predicting small, secondary vortices at the rear of the cylinder. These discrepancies underscore the importance of turbulence modeling and numerical methods in accurately capturing flow features and phenomena in CFD simulations.
Robertson et al. [37]	2015	In flow simulations around a sphere at Reynolds number 104, OF exhibits secondary circulations near the sphere's separation point. Variations in predictions between OF and other solvers primarily stem from differences in turbulence source term modeling rather than numerical methods. For simulations involving a backward-facing step, OF shows greater sensitivity to grid resolution. However, both OF and other solvers agree well on integral and mean flow variables, demonstrating reliable performance in predicting overall flow characteristics despite localized differences.
Jones et al. [38]	2015	Accurate CFD simulations are crucial for submarine support. The authors compared Fluent to commonly utilized OpenFOAM. Simulations on a submarine model showed up to 15% differences in drag coefficients between the two codes. Some discrepancies were resolved, but significant differences remain. This report details efforts to address these differences and suggests further work to ensure confidence in both codes.
Greifzu et al. [39]	2015	This study investigates two benchmark problems for turbulent dispersed particle-laden flow using CFD programs OpenFOAM and Ansys Fluent. The Lagrangian–Eulerian point-particle models for RANS simulations are compared in steady state and transient modes against experimental data. The results from both programs align well with experimental data. Dispersed phase results show Ansys Fluent slightly under-predicting and OpenFOAM slightly overestimating particle dispersion.

Table 5. Sensitivity of vortex identification characteristics (based on the Q-criterion adopted in this study) to CFD models in biomedical devices.

Authors	Date of Publication	Summary
Dresar et al. [40]	2019	RANS, SAS, DES and LES turbulence models were compared. While RANS and SAS did not show vortices, their predictions were insensitive to boundary mesh size.
Jarrell et al. [41]	2021	Intracellular delivery of functional macromolecules, such as DNA and RNA, across the cell microfluidic chips CFD simulations highlighted key hydrodynamic features that enhance vortex shedding and promote intracellular delivery of functional macromolecules such as DNA and RNA across cell microfluidic chips.
Menon et al. [42]	2013	The study visualized vortical structures in simulated aortic cannulation configurations using the Q-criterion to identify coherent vortex patterns in the jet wake. It compared two configurations: one arbitrarily oriented, leading to complex vortex structures and high hemolysis near the transverse aortic arch, and another based on surgeon sketches, showing simpler vortex patterns and lower hemolysis near the descending aorta. The findings highlight the importance of optimizing cannulation parameters for better hemodynamic outcomes during surgery.
Mancini et al. [43]	2020	LDA, along with Q-criterion, was employed to assess stenosis in vivo. Receiver operating characteristic analyses of power spectra revealed that the most relevant frequency bands for the detection of moderate (56–76%) and severe (86–96%) stenoses were 80–200 Hz and 0–40 Hz, respectively.

Table 5. *Cont.*

Authors	Date of Publication	Summary
Ozturk et al. [44]	2017	The study found that smaller eddies, specifically those with diameters up to about 10 μm (Kolmogorov scale eddies), are significantly associated with hemolysis. There is a clear correlation between hemolysis and the total surface area occupied by these smaller eddies. However, no such relationship was observed for larger eddies. This suggests that predicting the distribution of Kolmogorov scale eddies could help evaluate the susceptibility of medical device designs to hemolysis. Adjusting the design to increase the size of Kolmogorov scales may be beneficial in reducing the risk of hemolysis associated with these smaller turbulent structures.
Sonntag et al. [45]	2019	Intraventricular vortices were identified and visualized using the Q criterion in various studies. These vortices are linked to ventricular health.

2. Materials and Methods

Detailed dimensions and specifications of the FDA benchmark nozzle geometry are shown in Figure 1a, along with different axial distances following the sudden expansion where the results of this study are reported. Fully developed flow velocity profiles were imposed for both Newtonian and non-Newtonian scenarios by simulating a straight pipe (diameter: 0.012 m) with a constant velocity at the inlet and allowing the velocity profiles to evolve to fully developed conditions at the outlet by making the pipe sufficiently long. The velocity profiles at the outlet were then imposed at the inlet of the FDA nozzle simulations. Three hexahedral-dominant mesh configurations with cell counts—coarse [231 K], medium [439 K], and fine [966 K]—were used to discretize the domain and were employed in both Ansys Fluent and OpenFOAM. The simulations were performed at Reynolds numbers (Re) 500 and 2000, incorporating both Newtonian and non-Newtonian fluid properties. For Newtonian fluid simulations, constant density (1056 kg/m^3) and viscosity (0.0035 cp) were assumed, while non-Newtonian fluid properties were evaluated using four commonly used viscosity models: Cross, Power Law, Carreau–Yasuda (CY), and Casson, to account for variations in fluid behavior. The non-Newtonian viscosity models were implemented in Ansys Fluent as user-defined functions. The different non-Newtonian viscosity model formulations and corresponding parameters are summarized in Table 6. The shear-thinning viscosity predicted by the different models is shown in Figure 1b.

Table 6. Viscosity Models and coefficients were utilized to evaluate the non-Newtonian behavior of the fluid (blood) based on Kopernik [46].

Type	Viscosity Model	Constants
Carreau–Yasuda (CY)	$\mu = \mu_\infty + (\mu_0 - \mu_\infty)\left[1 + (\lambda\dot{\gamma})^{\alpha}\right]^{\frac{n-1}{\alpha}}$	$\mu_\infty = 0.0035$ Pa.s $\mu_0 = 0.056$ Pa.s $\lambda = 1.902$ $\alpha = 1.25$ $n = 0.22$
Casson	$\mu = \frac{\mu_\infty^2}{\dot{\gamma}} + \frac{2\mu_\infty N_\infty}{\sqrt{\dot{\gamma}}} + N_\infty^2$ $N_\infty = \frac{\sqrt{\mu_p}}{\sqrt[8]{1-Hct}}$ $\mu_\infty = \sqrt{0.625\,Hct}$	$\mu_p = 0.00145$ $Hct = 0.4$
Cross	$\mu = \mu_\infty + \frac{\mu_o - \mu_\infty}{1+\left(\frac{\dot{\gamma}}{\gamma_c}\right)^n}$	$\mu_\infty = 0.0035$ Pa.s $\mu_o = 0.0364$ Pa.s $\gamma_c = 2.63$ s^{-1} $n = 1.45$
Power Law	$\mu_{min} < \mu = \lambda\lvert\dot{\gamma}\rvert^{n-1} < \mu_{max}$	n = 0.7755 λ = 0.01467 $\mu_{max} = 0.025$ Pa.s $\mu_{min} = 0.00345$ Pa.s

3. Results

3.1. Mesh Convergence

Convergence in the axial velocity across both Ansys Fluent and OpenFOAM frameworks at the three mesh resolutions is showcased in Figures 2–5.

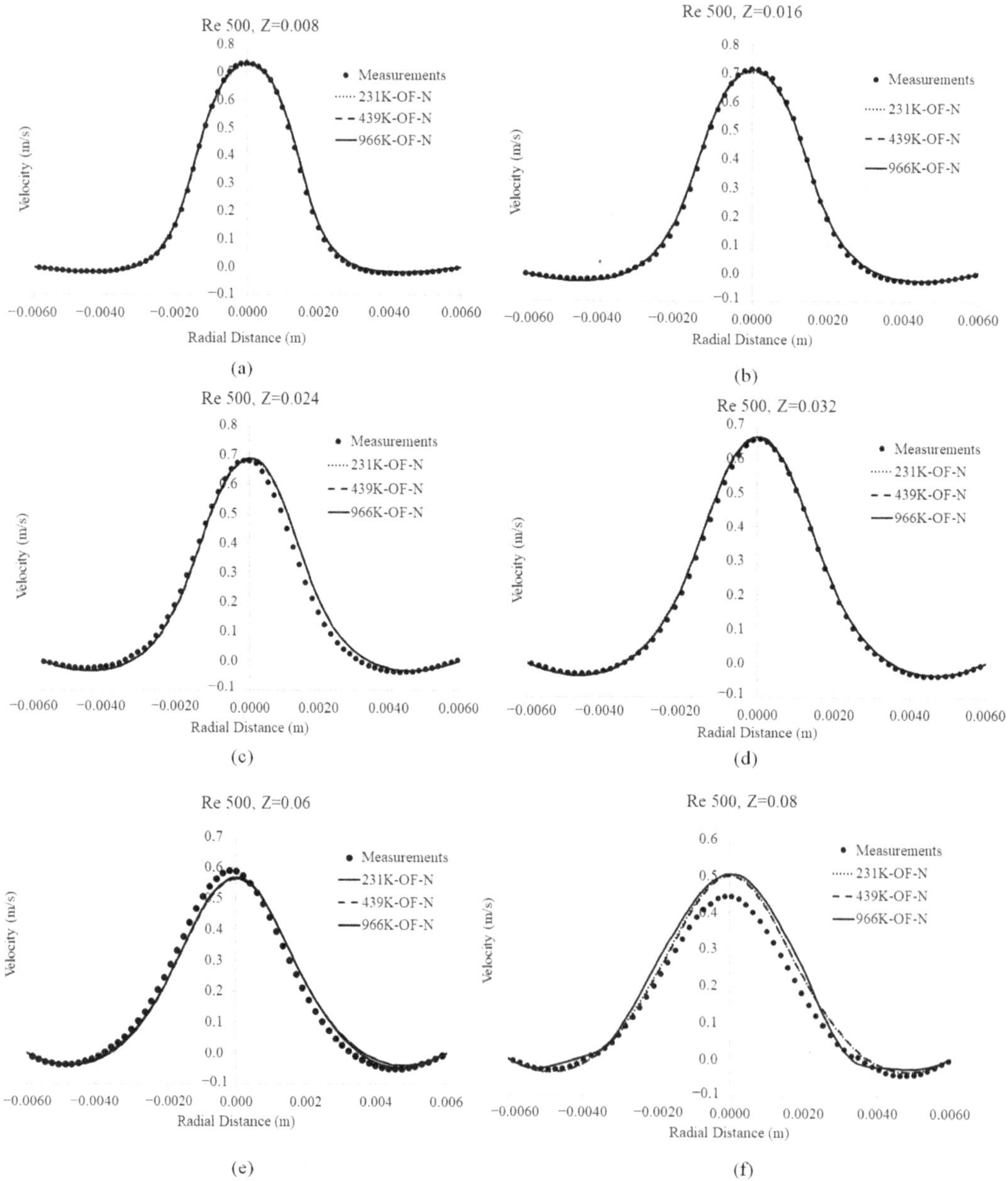

Figure 2. OpenFOAM (OF) predictions of axial velocity at different axial distances following the sudden expansion region at Re 500 employing the Newtonian (N) viscosity model at different grid resolutions (231 K, 439 K and 966 K): (**a**) Z = 0.008 m; (**b**) Z = 0.016 m; (**c**) Z = 0.024 m; (**d**) Z = 0.032 m; (**e**) Z = 0.06 m; (**f**) Z = 0.08 m.

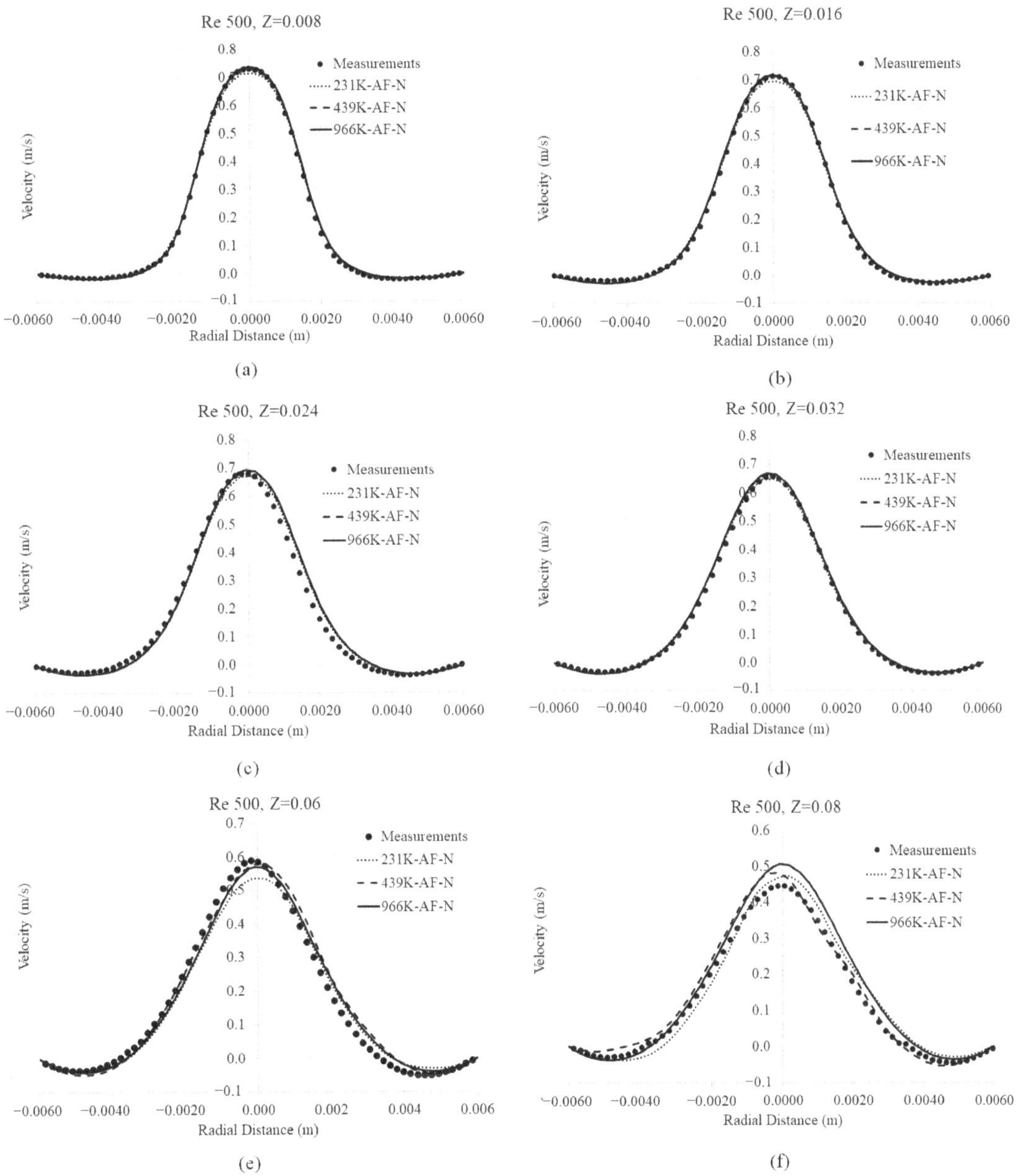

Figure 3. Ansys Fluent (AF) predictions of axial velocity at different axial distances following the sudden expansion region at Re 500 employing the Newtonian (N) viscosity model at different grid resolutions (231 K, 439 K and 966 K): (**a**) Z = 0.008 m; (**b**) Z = 0.016 m; (**c**) Z = 0.024 m; (**d**) Z = 0.032 m; (**e**) Z = 0.06 m; (**f**) Z = 0.08 m.

At Re 500, velocity predictions from both OpenFOAM (Figure 2) and Ansys Fluent (Figure 3) are in good agreement with experimental measurements [13,14] at all three mesh resolutions, with a radially symmetric profile being maintained through z = 0.06 m. At Re 500, slight differences in the velocity profiles at z = 0.06 m between the coarse and medium/fine mesh simulations are observed in Ansys Fluent; the symmetry indices, vortex identification and blood damage calculations are all reported in the finest mesh (966K) employed in this study.

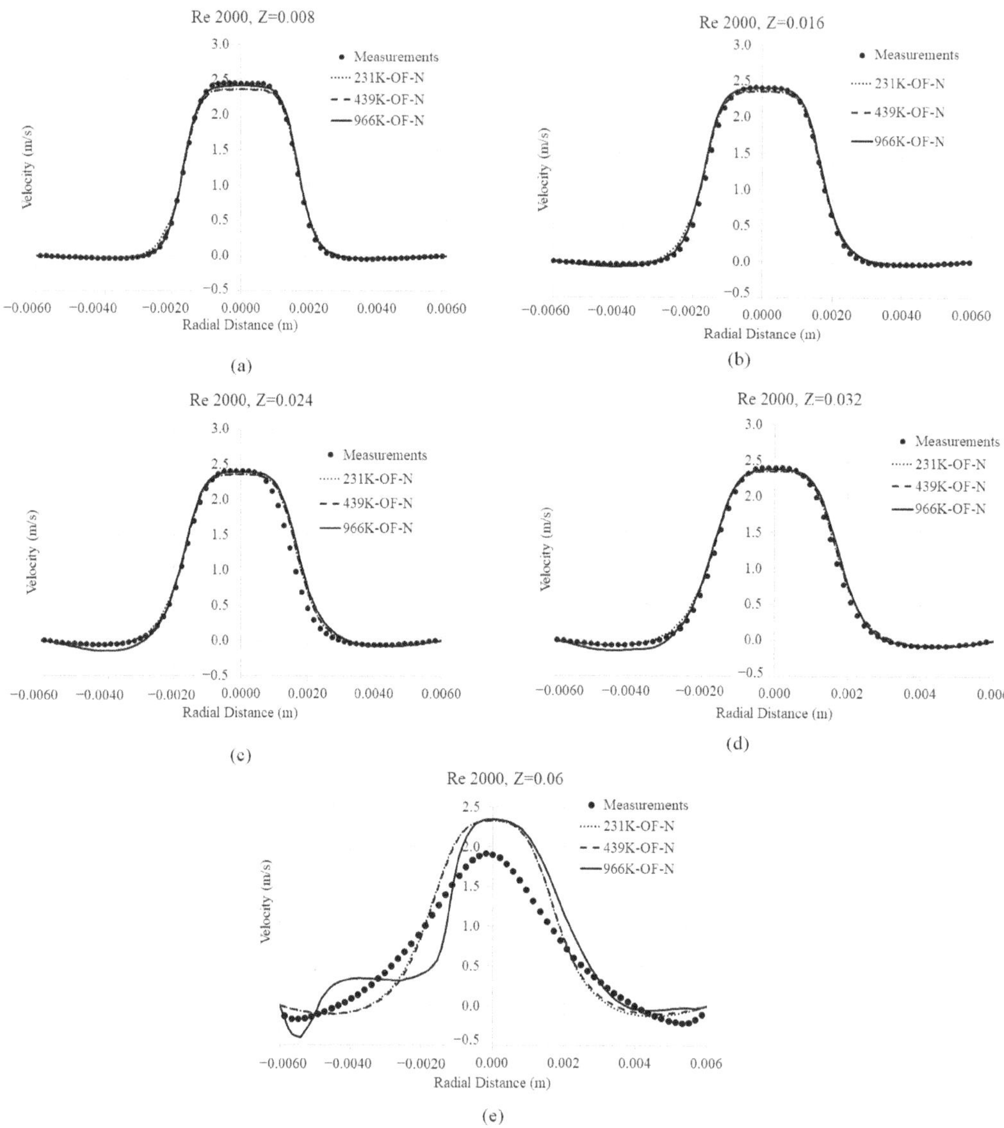

Figure 4. OpenFOAM (OF) predictions of axial velocity at different axial distances following the sudden expansion region at Re 2000 employing the Newtonian (N) viscosity model at different grid resolutions (231 K, 439 K and 966 K): (**a**) Z = 0.008 m; (**b**) Z = 0.016 m; (**c**) Z = 0.024 m; (**d**) Z = 0.032 m; (**e**) Z = 0.06 m.

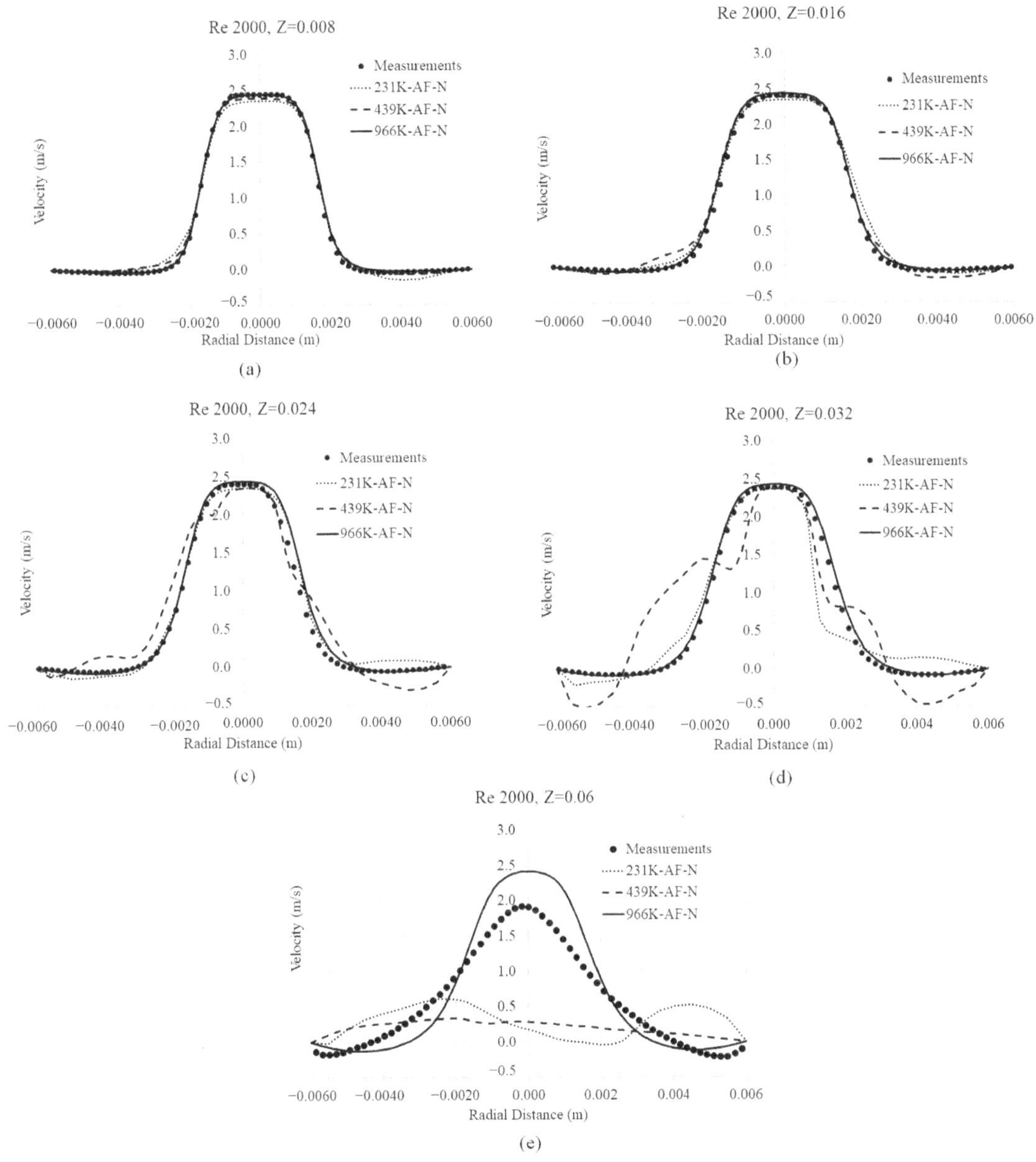

Figure 5. Ansys Fluent (AF) predictions of axial velocity at different axial distances following the sudden expansion region at Re 2000 employing the Newtonian (N) viscosity model at different grid resolutions (231 K, 439 K and 966 K): (**a**) Z = 0.008 m; (**b**) Z = 0.016 m; (**c**) Z = 0.024 m; (**d**) Z = 0.032 m; (**e**) Z = 0.06 m.

Similarly, at Re 2000, data shows good agreement with experimental results for both CFD software platforms OpenFOAM version 9 and Ansys Fluent version 2023 R2 (Figures 4 and 5). In the case of Ansys Fluent prediction, results start to deviate at Z = 0.032, whereas, in OpenFOAM, similar symmetry degeneration occurs at Z = 0.06 m.

Axial velocities (shown in Figures 2–5) indicate that grid-independent results have been obtained employing the 966 K mesh up to an axial distance of Z = 0.06 in most cases. Therefore, all subsequent analyses were performed at this mesh resolution.

3.2. Axial Velocity Prediction Sensitivity to Viscosity Models

Axial velocity prediction variations with the non-Newtonian models are shown in Figures 6 and 7. The profiles align with our expectations, indicating that non-Newtonian models predict lower peak velocities at Re 500 due to enhanced momentum diffusion resulting from higher viscosity. However, a consistent trend in peak velocity variations with non-Newtonian models is not observed across both frameworks. This inconsistency may result from differences in strain rate predictions between the frameworks, which subsequently impact the non-Newtonian viscosity. This will be examined in the context of vortex formation in Section 3.4. At Re 2000, however, the variations in axial velocity predictions between the non-Newtonian models are minimized as the importance of convection relative to diffusion becomes more significant.

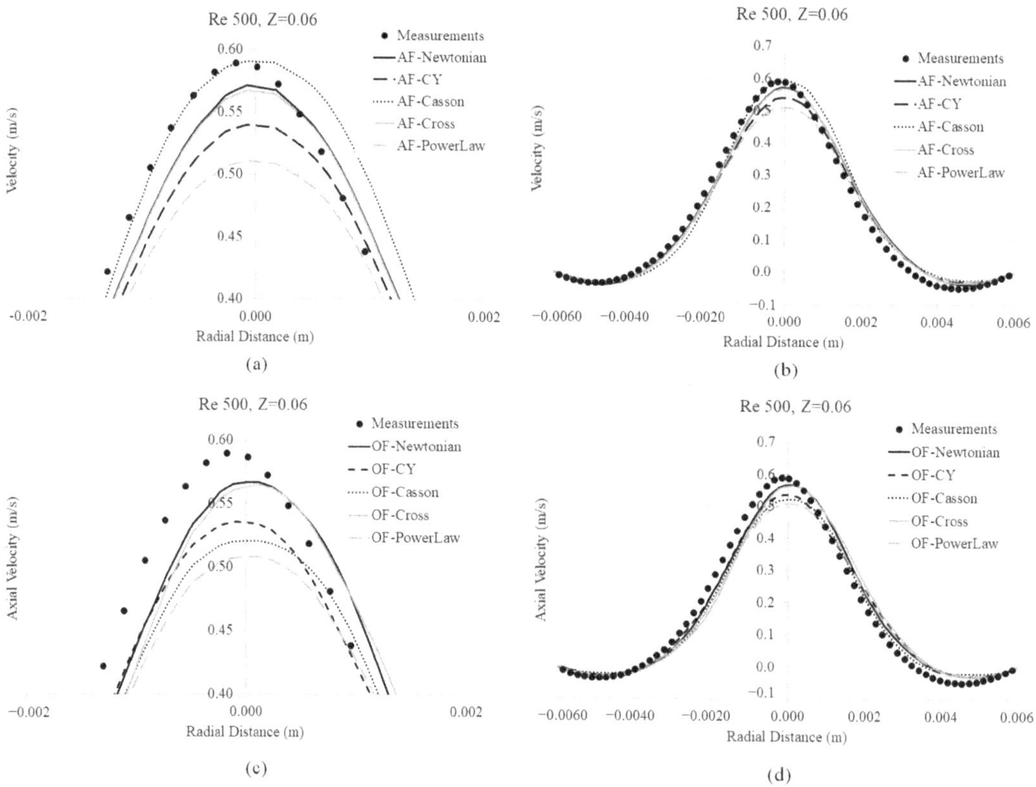

Figure 6. Axial velocity profiles (Re 500) at Z = 0.06 m following the sudden expansion using different non-Newtonian models in (**a**) Ansys Fluent (AF) (radial distances − 0.002 m to 0.002 m only to highlight differences); (**b**) Ansys Fluent (across the entire radius); (**c**) OpenFOAM (OF) (radial distances −0.002 m to 0.002 m only to highlight differences); (**d**) OpenFOAM (across the entire radius).

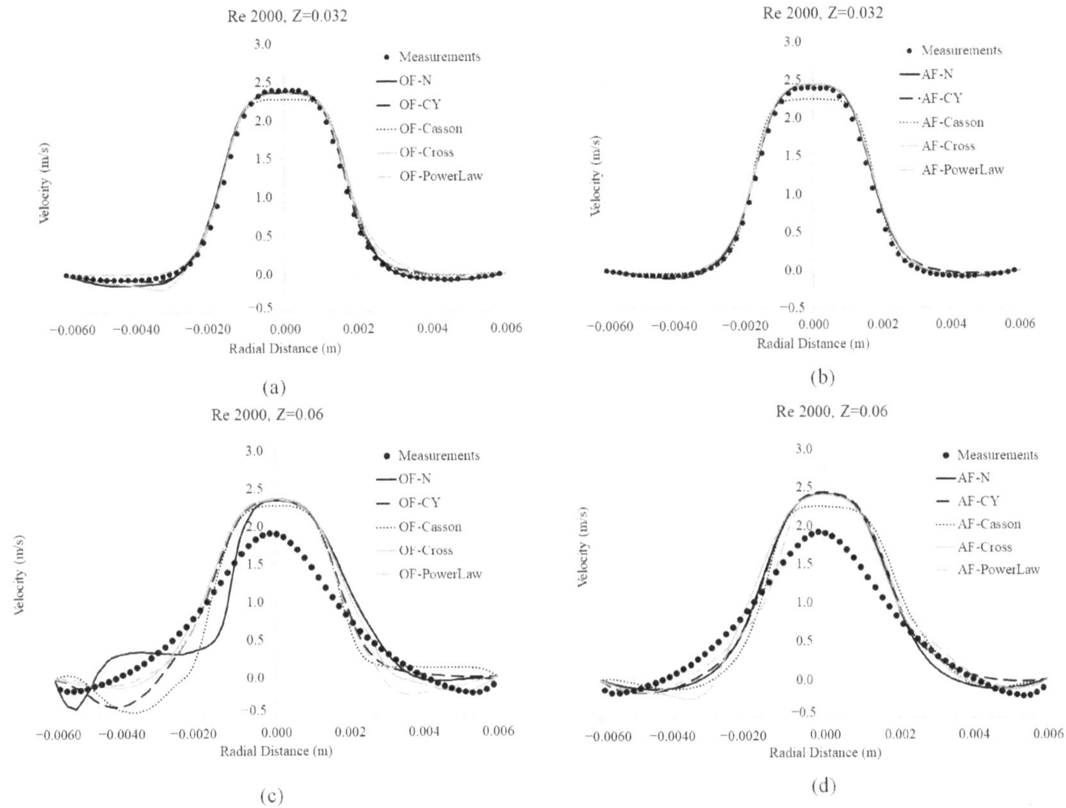

Figure 7. Axial velocity profiles (Re 2000) at different distances following the sudden expansion using different non-Newtonian models in (**a**) Ansys Fluent (Z = 0.032 m); (**b**) OpenFOAM (Z = 0.032 m); (**c**) Ansys Fluent (Z = 0.06 m); (**d**) OpenFOAM (Z = 0.06 m).

3.3. Symmetry Index (SI)

The symmetry index (SI) is a metric used to evaluate the degree of symmetry in various flow transport parameters. In this study, SI values are calculated using Equations (5) and (6) at different axial distances from the nozzle's sudden expansion section based on the velocity values along the lines shown in Figure 1a. As illustrated in Figure 8, a significant breakdown of symmetry is observed beyond a radial distance of 0.06 m (at Re 500). This breakdown indicates a point where the flow begins to deviate significantly from a symmetrical pattern, suggesting regions of potential instability or turbulence transition. At 0.06 m, SI predictions from Ansys Fluent are greater than those from OpenFOAM. Unexpectedly, while Newtonian fluid simulations maintained high symmetry across both frameworks, consistent with previous studies listed in Table 1, three non-Newtonian models in OpenFOAM (CY, Cross, Power Law) and one in Ansys Fluent (Casson) show a symmetry breakdown at 0.06 m. This is contrary to the expectation that non-Newtonian models would delay the onset of turbulence. This phenomenon is also observed at Re 2000 (Figure 9), where a symmetry breakdown occurred sooner (following the sudden expansion section) with all non-Newtonian models in both Ansys Fluent and OpenFOAM, while the Newtonian fluid in AF maintained its symmetry. However, a consistent pattern in SI variations with different non-Newtonian models is not discernible from Figures 8 and 9.

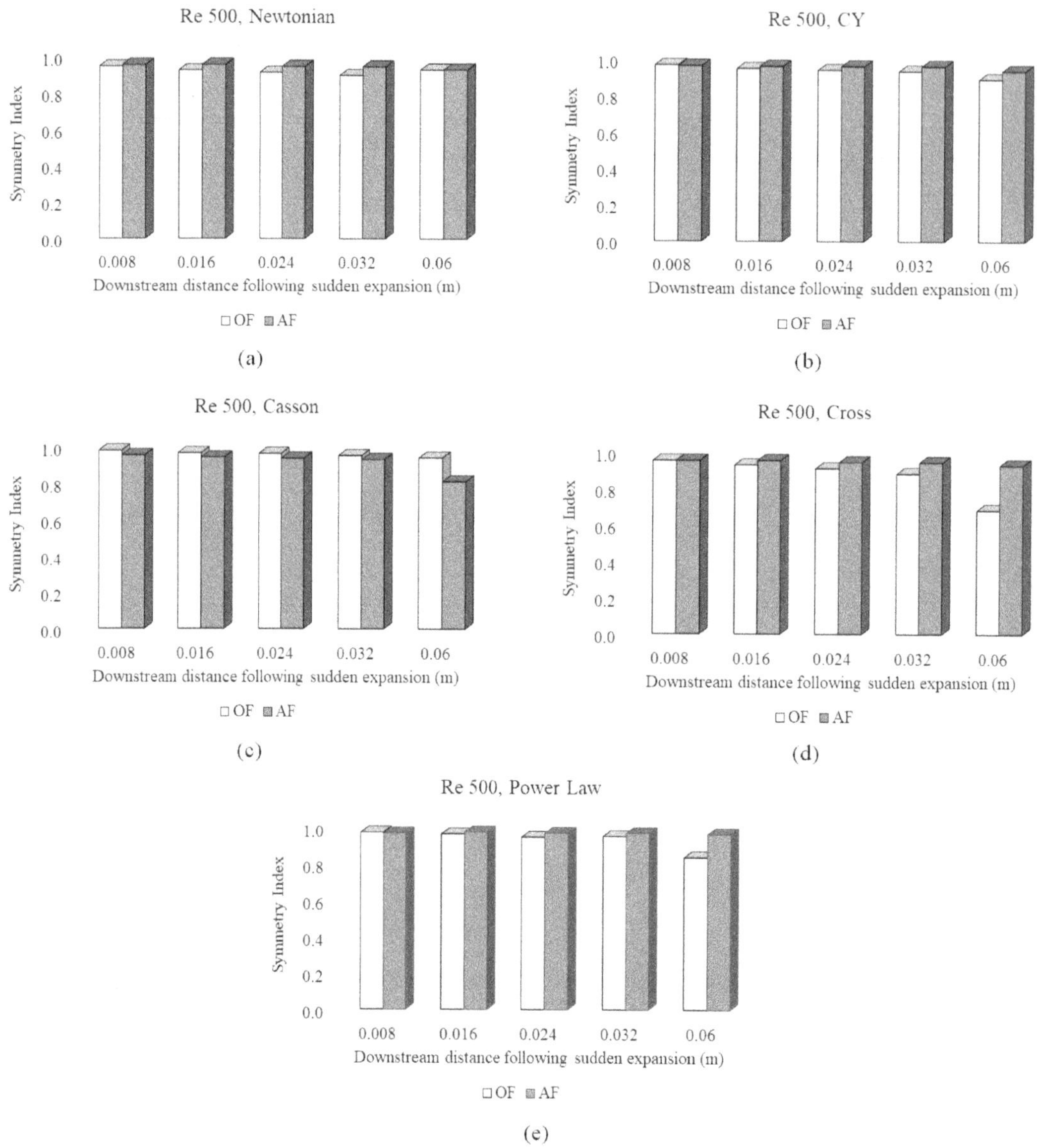

Figure 8. Symmetric index predicted from employing different viscosity models at Re 500 using OpenFOAM (OF) and Ansys Fluent (AF) (**a**) Newtonian Fluid, (**b**) Carreau–Yasuda (CY), (**c**) Casson, (**d**) Cross (**e**) Power Law.

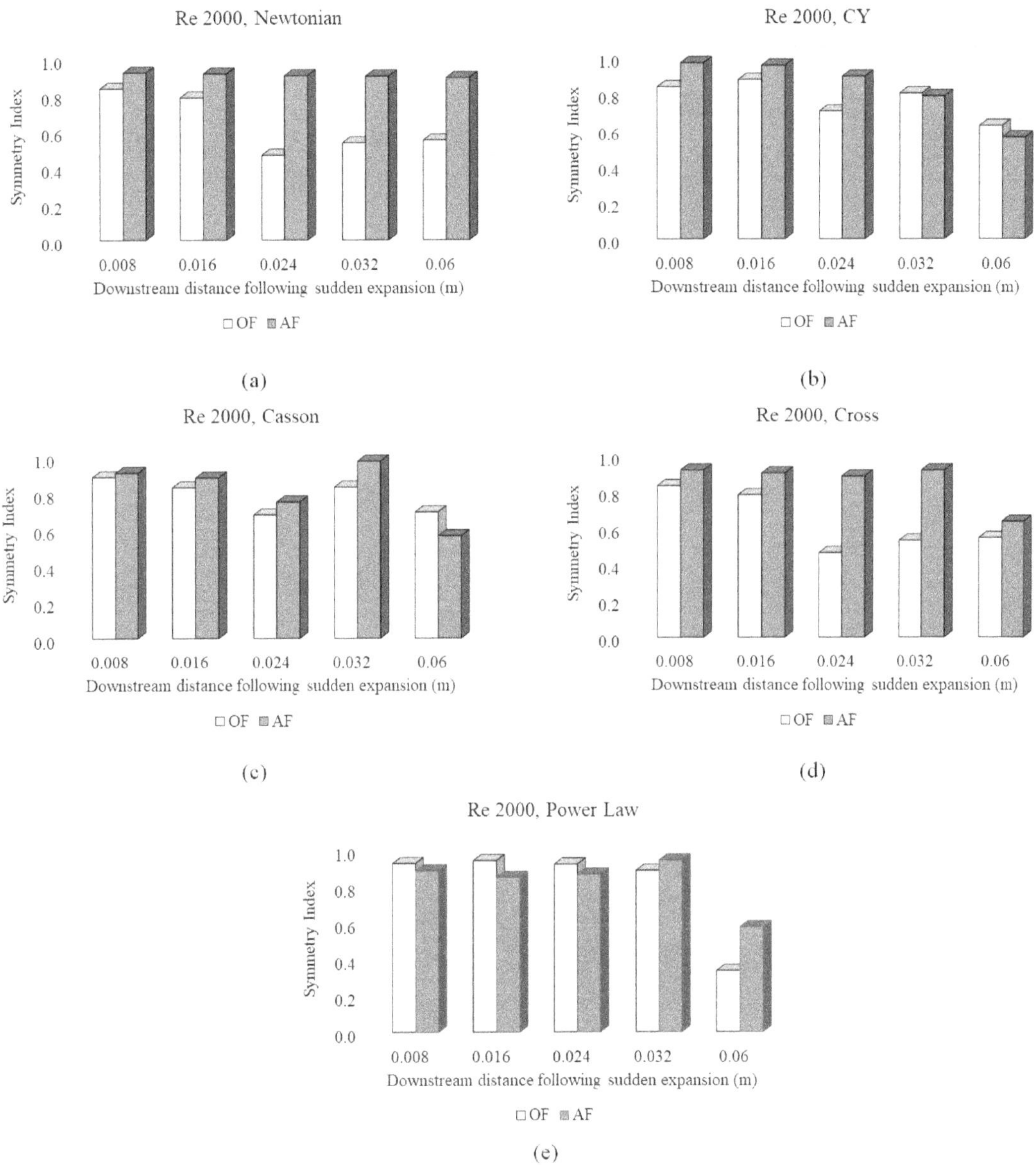

Figure 9. Symmetric index predicted from employing different viscosity models at Re 2000 using OpenFOAM (OF) and Ansys Fluent (AF) (**a**) Newtonian Fluid, (**b**) Carreau–Yasuda (CY), (**c**) Casson, (**d**) Cross (**e**) Power Law using OpenFOAM (OF) and Ansys Fluent (AF).

3.4. Q-Criterion and Vortex Shape Identification

The Q-Criterion is a metric used in fluid dynamics to evaluate turbulent flows and detect vortices. In this study, the Q-Criterion was employed to visualize and analyze vortices, providing insights into the complex dynamics of turbulent flows resulting from the interaction between medical devices and blood. The Q criterion is defined as follows:

$$Q = 0.5\left(||\Omega||^2 - ||S||^2\right) \quad (1)$$

where S is the rate of strain tensor, and Ω is the velocity tensor.

Various vortex structures were identified by creating iso-surfaces of Q at zero, as illustrated in Figures 10–13. At Re 500 (Figure 10), the Ansys Fluent-based analysis showed that all viscosity models, except for the Casson model, exhibited no vortex formation until the axial distance of Z = 0.06 m was reached. In contrast, the Casson model showed vortex formation starting at a radial distance of approximately 0.032 m. Consistent and symmetric vortex patterns were observed in OpenFOAM up to Z = 0.06 m. At Re 2000, vortex formation began at approximately 0.032 m, as shown consistently by both OpenFOAM and Ansys Fluent software outputs. At this Reynolds number, the flow is in a transitional regime and on the verge of becoming fully turbulent. This means the fluid experiences higher inertial forces compared to viscous forces, leading to the formation of vortices and other complex flow structures. The onset of vortex formation at 0.032 m marks a critical point where these instabilities become noticeable.

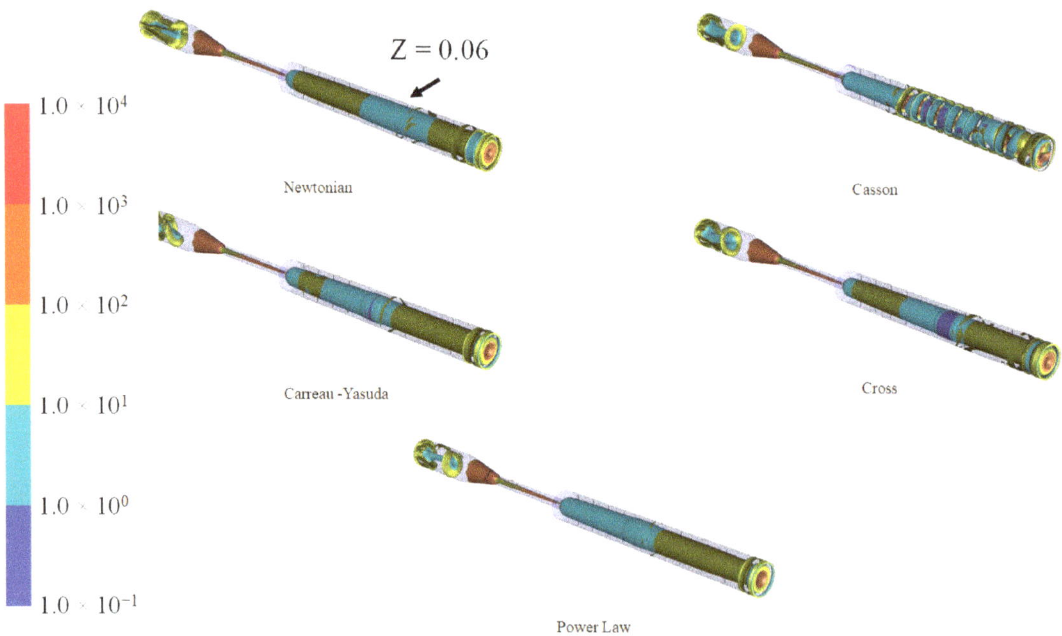

Figure 10. Three-dimensional depiction of Q-criterion-based vortex formation at Re 500 using Ansys Fluent (colored by strain rate magnitudes, 1/s).

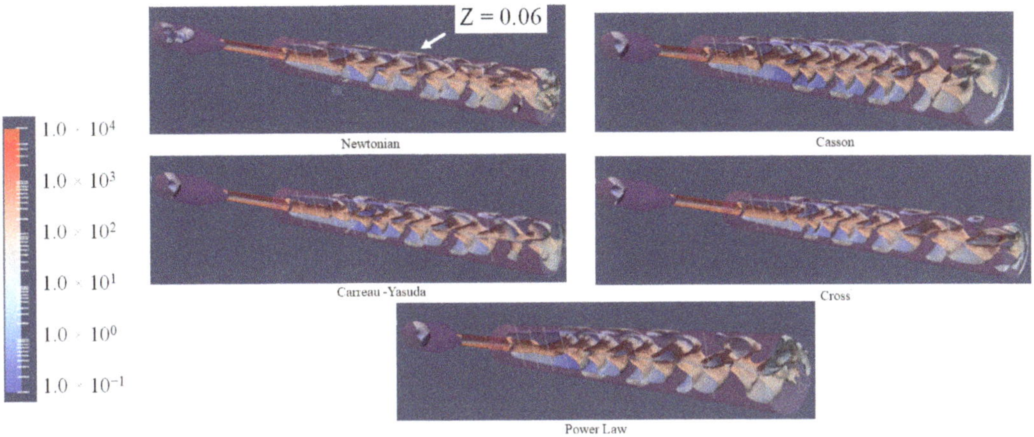

Figure 11. 3D depiction of Q-criterion-based vortex formation at Re 500 using OpenFoam (colored by strain rate magnitudes, 1/s).

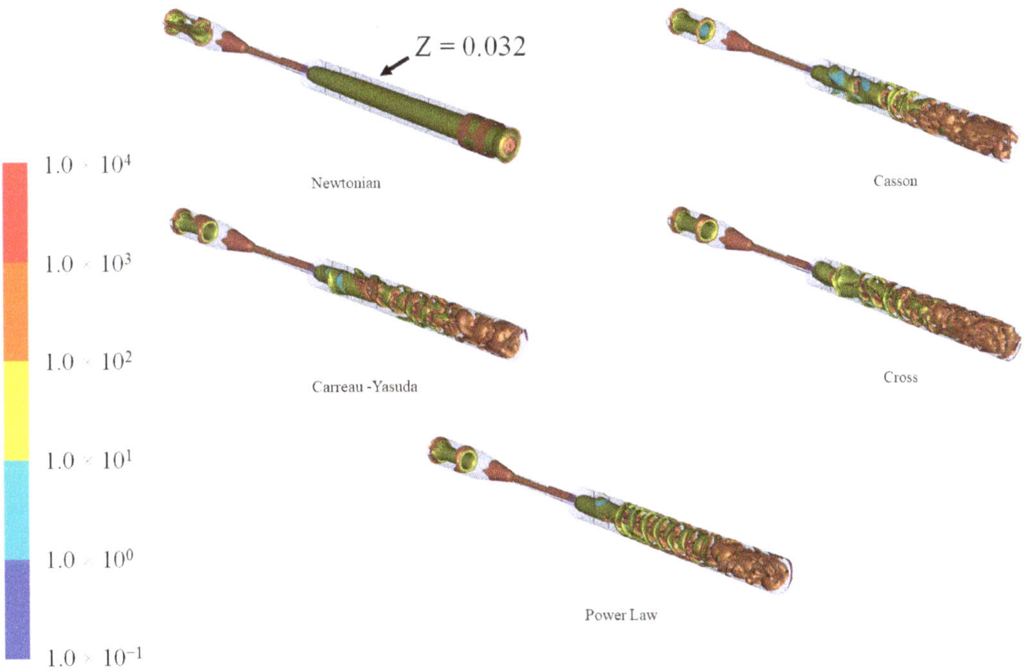

Figure 12. Three-dimensional depiction of Q-criterion-based vortex formation at Re 2000 using Ansys Fluent (colored by strain rate magnitudes, 1/s).

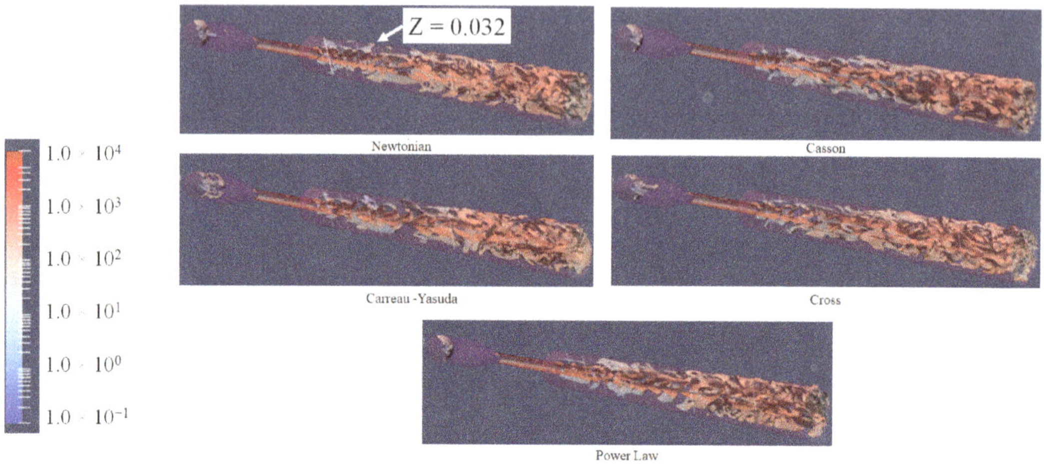

Figure 13. Three-dimensional depiction of Q-criterion-based vortex formation at Re 2000 using OpenFOAM (colored by strain rate magnitudes, 1/s).

3.5. Normalized Index of Hemolysis (NIH) Evaluation

The Normalized Index of Hemolysis (NIH) is a standardized metric used to evaluate and compare the hemolytic effects (red blood cell destruction) of blood-handling devices. This metric is crucial for developing safer and more efficient medical devices, ultimately improving patient outcomes. However, predicting NIH is particularly challenging due to the highly device-specific nature of the coefficients in viscosity models. These coefficients vary significantly based on the characteristics and configurations of the devices, making it difficult to generalize predictions across different setups.

In this study, as shown in Figure 14, the overall findings demonstrated a significant correlation between NIH values and the flow regime, whether laminar or turbulent. At Re 500, NIH values indicated that all models except Casson predict a higher NIH than the Newtonian (N) model. This observation contrasts the results of Trias et al. [33], who showed that both the Casson and Carreau–Yasuda models predicted higher NIH than the Newtonian model. Nevertheless, the NIH values predicted by the Newtonian model in our study were in reasonable agreement with those reported by Trias et al. [33], even though the values were computed using different software packages.

At Re 2000, NIH values across different Reynolds numbers appear to be nearly independent of the non-Newtonian model used. This suggests that in spite of the viscosity prediction variations among the non-Newtonian models at low strain rates, the enhancement ratio of NIH across different Reynolds numbers can be predicted fairly consistently and is independent of the non-Newtonian model. The Newtonian models, on the other hand, tend to provide conservative estimates of NIH values. This conservatism arises because Newtonian models can underestimate viscosity, leading to lower predicted NIH values. The higher viscosity predicted by other models could potentially result in more accurate and possibly higher NIH predictions, but the Newtonian model's tendency to underpredict provides a cautious upper bound for the enhancement ratio. This is a crucial consideration when assessing the potential hemolytic effects of blood-handling devices.

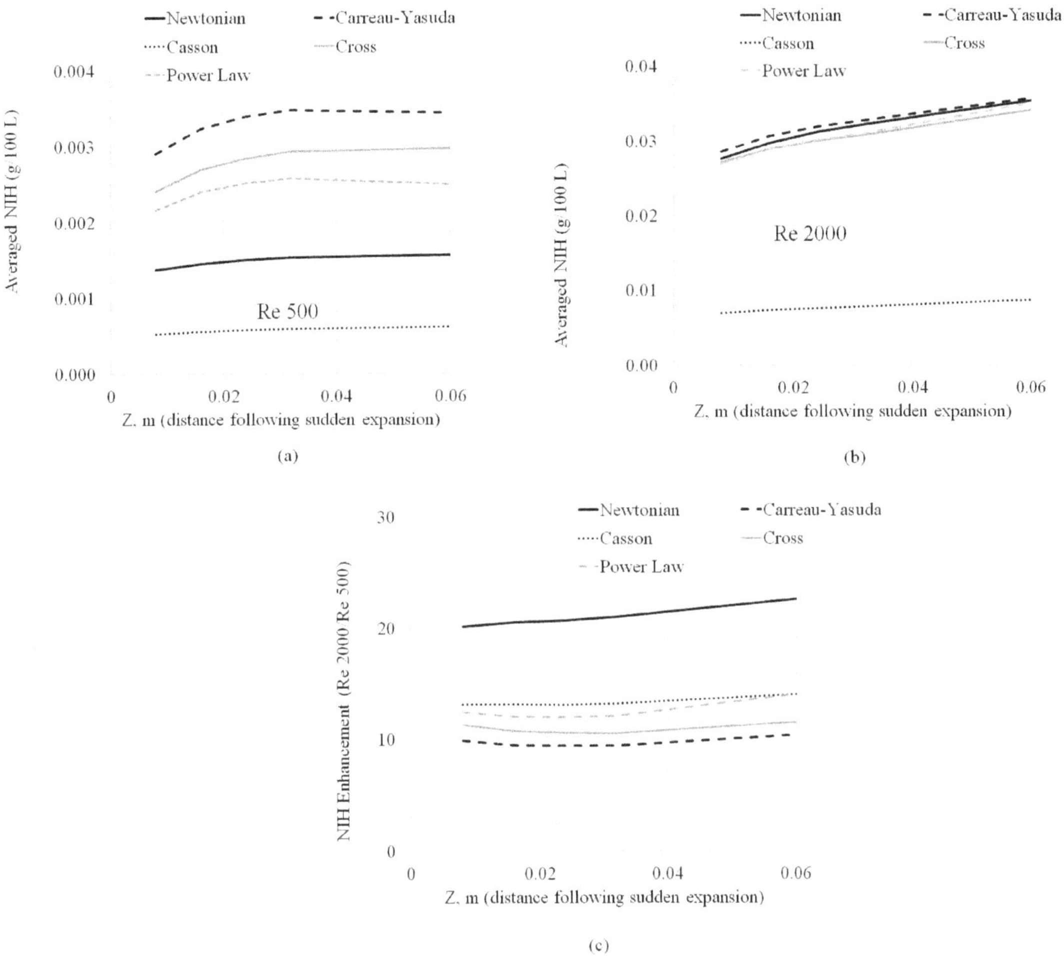

Figure 14. NIH values for evaluating hemolytic effects at (**a**) Re 500, (**b**) Re 2000 and (**c**) the ratio of NIH values at Re 2000/Re 500.

3.6. Mathematical Component

The incompressible flow Navier–Stokes equation was used to calculate velocity (v) and Shear Stress (τ). Although velocity has been the generated parameter, shear stress evaluation is more relevant and critical to medical device applications. Therefore, the shear stress values were indirectly estimated by velocity profiles. The steady state mass conservation equation is given as [21]:

$$\nabla \cdot \left(\rho \vec{v}\right) = 0 \quad (2)$$

where ρ and $\vec{v}$ represent the density and velocity vector, respectively. The momentum conservation for the fluid can be written as [21]:

$$\nabla \cdot \left(\rho \vec{v} \vec{v}\right) = -\nabla p + \nabla \cdot \left(\vec{\tau}\right) + \rho \vec{g} \quad (3)$$

where $\vec{\tau}$ is the stress tensor, p the static pressure, and $\vec{g}$ the direction of the gravitational component. The stress tensor is evaluated as follows:

$$\vec{\tau} = \mu\left[\left(\nabla\vec{v} + \nabla\vec{v}^{T}\right) - \frac{2}{3}\nabla\cdot\vec{v}I\right] \tag{4}$$

where μ is the molecular viscosity, I is the unit tensor and the second term on the right-hand side is the effect of volume dilation. The pressure velocity coupling associated with Equations (2) and (3) was handled using the SIMPLE scheme in both Ansys Fluent and OpenFOAM [22,23]. The gradients in Ansys Fluent were evaluated using the Lease-Square Cell Based option, whereas a Gauss Linear scheme was employed in OpenFOAM [22,23].

The symmetry index (*SI*) at different axial locations (lines shown in Figure 1a) was evaluated by computing the volumetric flow rates above and below the axial center employing the axial velocities shown in Figures 2–5. Equations (5) and (6) were employed to limit the *SI* values to a maximum value of one.

$$If\ Q_{Upper\ half} < Q_{Lower\ half}\ SI = \frac{Flow\ Rate\ (Upper\ half)}{Flow\ Rate\ (Lower\ half)} \tag{5}$$

$$If\ Q_{Lower\ Half} < Q_{Upper\ half}\ SI = \frac{Flow\ Rate\ (Lower\ half)}{Flow\ Rate\ (Upper\ half)} \tag{6}$$

NIH was computed by solving a scalar transport equation for linear damage D_e:

$$\frac{\partial(\rho D_e)}{\partial t} + \nabla\left(\rho D_e\vec{u}\right) = C^{\frac{1}{a}}\rho|\tau|^{\frac{b}{a}} \tag{7}$$

Equation (7) shows that D_e is a dynamic field that can be evolved together with the density and linear momentum, governed by Equations (2)–(4). The constant a, b and C on the right-hand side of Equation (7) were assigned values of: 0.785, 2.416 and 3.62×10^{-7}, respectively, following Trias et al. [33].

The normalized index of hemolysis (*NIH*) quantified the extent of blood damage resulting from shear stresses within a specific flow domain and was evaluated using Equation (8).

$$NIH = 100 * (1 - Hct) \times D_e \times k \tag{8}$$

In Equation (8), D_e was computed at each spatial location using Equation (7). *Hct* is the hematocrit count (obtained from literature at 40%), and k is the hemoglobin content of blood and was set at 160 g/L per literature [33].

4. Discussion

The research highlights significant variability in predicting flow symmetry indices and vortex characteristics when using laminar flow models across two leading CFD codes: Ansys Fluent and OpenFOAM. Although both codes employed identical mesh configurations to control mesh-related discrepancies, differences in their numerical methods and gradient evaluation approaches resulted in notable variations in predictions. These variations can be attributed to differences in mesh topologies, solver algorithms, and gradient evaluation methods, which affect how each code calculates and interprets flow dynamics, leading to discrepancies in vortex formation and flow symmetry.

Beyond the standard Newtonian fluid modeling, which aligns with FDA guidelines, the study also incorporated simulations using four different non-Newtonian viscosity models in both CFD frameworks. The goal of these initial simulations was to achieve high fidelity by ensuring that the predicted velocity profiles closely matched experimental measurements.

At Reynolds number 500 (Re 500), both Ansys Fluent and OpenFOAM predicted symmetric vortices, indicating that turbulence had not yet onset for the Newtonian fluid. This finding is consistent with existing literature. However, discrepancies arise with non-

Newtonian models. OpenFOAM showed a breakdown in symmetry for three of the four non-Newtonian models, whereas Ansys Fluent exhibited this breakdown for only one non-Newtonian model. This result contradicts the expected behavior of non-Newtonian fluids, which are anticipated to delay the onset of turbulence compared to Newtonian fluids.

Similarly, at Reynolds number 2000, a higher flow regime where turbulence is more prominent, all non-Newtonian models in both frameworks displayed a symmetry breakdown at approximately 0.032 m downstream from the sudden expansion. In contrast, the Newtonian fluid in Ansys Fluent maintained symmetry beyond 0.06 m. This suggests that the non-Newtonian models react differently to flow conditions, leading to an earlier onset of turbulence.

Visualization using Q-criterion-based iso-surfaces further highlighted these differences. Distinctly different vortex shapes were observed between Ansys Fluent and OpenFOAM, underscoring the sensitivity of vortex structure identification to numerical solver variations. This sensitivity reflects how different numerical methods can impact turbulence onset predictions using vortex characterization.

Regarding blood damage assessments, non-Newtonian models demonstrated a greater sensitivity at lower Reynolds numbers (Re 500). However, this sensitivity decreased at higher Reynolds numbers (Re 2000). However, the enhancement ratio of blood damage between Re 2000 and Re 500 was relatively consistent across non-Newtonian models, indicating that a fairly consistent assessment of how blood damage scales with Reynolds number may be obtained irrespective of the non-Newtonian model employed. In contrast, the Newtonian model estimated a higher blood damage ratio, which could be employed to establish safety boundaries when designing medical devices.

In summary, this study highlights how variations in CFD software and viscosity models can result in different predictions of fluid dynamics, which in turn can impact assessments of blood damage. This underscores the importance of understanding these differences when evaluating risks using medical device simulations.

5. Conclusions

This study delves into the complexities associated with predicting the transition from laminar to turbulent flow (Reynolds numbers (Re) 500–2000) within the FDA benchmark nozzle geometry using CFD codes. Despite a rigorous examination of numerical noise and geometric imperfections at the inflow, there is a consensus that most CFD codes can accurately predict the global variables under laminar flow conditions in this geometry. However, variations in predictions for derived variables, such as strain rate and vorticity, may arise due to differences in numerical solvers and gradient evaluation methods. These variations can subsequently impact blood flow-related phenomena such as blood damage, thrombosis, and non-Newtonian flow predictions.

In conclusion, the study highlights how prediction differences in numerical solvers can be exacerbated by the use of non-Newtonian viscosity models, leading to distinctly different vortex shapes and patterns. The results underscore the importance of understanding prediction variabilities across software frameworks when using simulations to assess risks associated with hemolysis and thrombosis, even under assumed steady-state, laminar flow conditions.

Author Contributions: N.G. contributed to this article by writing/creating the original draft. G.K. contributed to this article by writing and editing, supervising the project. Both authors contributed equally to the formulation of the problem, execution of work and data analysis. All authors have read and agreed to the published version of the manuscript.

Funding: This research received no external funding.

Data Availability Statement: All data that is relevant to this study is contained within the article. Any additional supplemental information that is needed is available upon request from the corresponding author.

Conflicts of Interest: The authors declare no conflicts of interest.

References

1. Bergersen, A.W.; Mortensen, M.; Valen-Sendstad, K. The FDA nozzle benchmark: "In theory there is no difference between theory and practice but in practice there is". *Int. J. Numer. Methods Biomed. Eng.* **2019**, *35*, 3150. [CrossRef] [PubMed]
2. Bhushan, S.; Walters, D.K.; Burgreen, G.W. Laminar, turbulent, and transitional simulations in benchmark cases with cardiovascular device features. *Cardiovasc. Eng. Technol.* **2013**, *4*, 408–426. [CrossRef]
3. Chabannes, V.; Prud'Homme, C.; Szopos, M.; Tarabay, R. High-order finite element simulations for fluid dynamics validated by experimental data from the FDA benchmark nozzle model. *arXiv* **2017**, arXiv:1701.02179.
4. Delorme, Y.T.; Anupindi, K.; Frankel, S.H. Large eddy simulation of FDA's idealized medical device. *Cardiovasc. Eng. Technol.* **2013**, *4*, 392–407. [CrossRef]
5. Fehn, N.; Wall, W.A.; Kronbichler, M. Modern discontinuous Galerkin methods for the simulation of transitional and turbulent flows in biomedical engineering: A comprehensive LES study of the FDA benchmark nozzle model. *Int. J. Numer. Methods Biomed. Eng.* **2019**, *35*, 3228. [CrossRef]
6. Huang, F.; Noël, R.; Berg, P.; Hosseini, S.A. Simulation of the FDA nozzle benchmark: A lattice Boltzmann study. *Comput. Methods Programs Biomed.* **2022**, *221*, 106863. [CrossRef]
7. Jain, K. Efficacy of the FDA nozzle benchmark and the lattice Boltzmann method for the analysis of biomedical flows in the transitional regime. *Med. Biol. Eng. Comput.* **2020**, *58*, 1817–1830. [CrossRef]
8. Manchester, E.L.; Xu, X.Y. The effect of turbulence on transitional flow in the FDA's benchmark nozzle model using large-eddy simulation. *Int. J. Numer. Methods Biomed. Eng.* **2020**, *36*, 3389. [CrossRef]
9. Nicoud, F.; Chnafa, C.; Siguenza, J.; Zmijanovic, V.; Mendez, S. Large-eddy simulation of turbulence in cardiovascular flows. In *Biomedical Technology: Modeling, Experiments, and Simulation*; Springer: Berlin/Heidelberg, Germany, 2018; pp. 147–167.
10. Passerini, T.; Quaini, A.; Veneziani, A.; Wood, N.B. Validation of an open-source framework for the simulation of blood flow in rigid and deformable vessels. *Int. J. Numer. Methods Biomed. Eng.* **2013**, *29*, 1192–1213. [CrossRef]
11. Pewowaruk, R.; Li, Y.; Rowinski, D.; Roldán-Alzate, A. Solution adaptive refinement of cut-cell Cartesian meshes can improve FDA nozzle computational fluid dynamics efficiency. *Int. J. Numer. Methods Biomed. Eng.* **2021**, *37*, 3432. [CrossRef]
12. Sánchez Abad, N.; Vinuesa, R.; Schlatter, P.; Andersson, M.; Karlsson, M. Simulation strategies for the Food and Drug Administration nozzle using Nek5000. *AIP Adv.* **2020**, *10*, 2. [CrossRef]
13. Stewart, S.F.; Paterson, E.G.; Burgreen, G.W.; Hariharan, P.; Reddy, V.; Day, S.W. Assessment of CFD performance in simulations of an idealized medical device: Results of FDA's first computational interlaboratory study. *Cardiovasc. Eng. Technol.* **2013**, *4*, 374–391. [CrossRef]
14. Stewart, S.F.; Paterson, E.G.; Burgreen, G.W.; Hariharan, P.; Reddy, V.; Day, S.W. Results of FDA's first interlaboratory computational study of a nozzle with a sudden contraction and conical diffuser. *Cardiovasc. Eng. Technol.* **2012**, *3*, 139–160. [CrossRef]
15. Stiehm, M.; Wüstenhagen, C.; Siewert, S.; Grabow, N.; Schmitz, K.P. Numerical simulation of pulsatile flow through a coronary nozzle model based on FDA's benchmark geometry. *Curr. Dir. Biomed. Eng.* **2017**, *3*, 775–778. [CrossRef]
16. Taylor, C.A.; Steinman, D.A.; Simon, H.A.; Valen-Sendstad, K.; Bergersen, A.W.; Loth, F. Predicting the quality of the FDA's nozzle benchmark. *J. Biomech.* **2016**, *49*, 2245–2251.
17. Zmijanovic, V.; Löhner, R.; Indrakanti, S.S. Large eddy simulation of transitional and turbulent flow in the FDA nozzle model. *J. Biomech. Eng.* **2017**, *139*, 041003.
18. Qiao, Y.; Luo, K. Computational prediction of thrombosis in FDA's benchmark nozzle. *Front. Physiol.* **2022**, *13*, 867613. [CrossRef]
19. Tobin, N.; Manning, K.B. Large-eddy simulations of flow in the FDA benchmark nozzle geometry to predict hemolysis. *Cardiovasc. Eng. Technol.* **2020**, *11*, 254–267. [CrossRef]
20. Sobey, I.J.; Drazin, P.G. Bifurcations of two-dimensional channel flows. *J. Fluid Mech.* **1986**, *171*, 263–287. [CrossRef]
21. ANSYS Inc. *ANSYS Fluent, Version 2023 R2*; ANSYS Inc.: Canonsburg, PA, USA, 2023. Available online: https://www.ansys.com/products/fluids/ansys-fluent (accessed on 1 January 2024).
22. The OpenFOAM Foundation. *OpenFOAM, Version 9*; The OpenFOAM Foundation: London, UK, 2021; Available online: www.openfoam.org (accessed on 1 January 2024).
23. Biswas, D.; Casey, D.M.; Crowder, D.C.; Steinman, D.A.; Yun, Y.H.; Loth, F. Characterization of transition to turbulence for blood in a straight pipe under steady flow conditions. *J. Biomech Eng.* **2016**, *138*, 071001. [CrossRef]
24. Cebral, J.R.; Castro, M.A.; Appanaboyina, S.; Putman, C.M.; Millan, D.; Frangi, A.F. Efficient pipeline for image-based patient-specific analysis of cerebral aneurysm hemodynamics: Technique and sensitivity. *IEEE Trans. Med. Imaging* **2005**, *24*, 457–467. [CrossRef]
25. Costa, R.P.; Simplice Talla Nwotchouang, B.; Yao, J.; Biswas, D.; Casey, D.; McKenzie, R.; Steinman, D.A.; Loth, F. Transition to turbulence downstream of a stenosis for whole blood and a Newtonian analog under steady flow conditions. *J. Biomech. Eng.* **2022**, *144*, 031008. [CrossRef] [PubMed]
26. Haley, A.L.; Valen-Sendstad, K.; Steinman, D.A. On delayed transition to turbulence in an eccentric stenosis model for clean vs. noisy high-fidelity CFD. *J. Biomech.* **2021**, *125*, 110588. [CrossRef] [PubMed]
27. Johnston, B.M.; Johnston, P.R.; Corney, S.; Kilpatrick, D. Non-Newtonian blood flow in human right coronary arteries: Steady-state simulations. *J. Biomech.* **2004**, *37*, 709–720. [CrossRef] [PubMed]
28. Khan, M.O.; Valen-Sendstad, K.; Steinman, D.A. Direct numerical simulation of laminar-turbulent transition in a non-axisymmetric stenosis model for Newtonian vs. shear-thinning non-Newtonian rheologies. *Flow Turbul. Combust.* **2019**, *102*, 43–72. [CrossRef]

29. Lee, S.W.; Steinman, D.A. On the relative importance of rheology for image-based CFD models of the carotid bifurcation. *J. Biomech. Eng.* **2007**, *129*, 273–278. [CrossRef]
30. Razavi, A.; Shirani, E.; Sadeghi, M.R. Numerical simulation of blood pulsatile flow in a stenosed carotid artery using different rheological models. *J. Biomech.* **2011**, *44*, 2021–2030. [CrossRef]
31. Good, B.C. The effects of non-Newtonian blood modeling and pulsatility on hemodynamics in the Food and Drug Administration's benchmark nozzle model. *Biorheology* **2023**, *59*, 1–18. [CrossRef]
32. Hussein, B.K.; Al-Azawy, M.G.; Al-Waaly, A.A.; Hamza, Z.A. Evaluation of turbulence and non-Newtonian blood rheology models through FDA nozzle. In Proceedings of the 2021 International Conference on Advance of Sustainable Engineering and its Application (ICASEA), Wasit, Iraq, 27–28 October 2021; IEEE: Toulouse, France, 2021; pp. 1–5.
33. Trias, M.; Arbona, A.; Massó, J.; Miñano, B.; Bona, C. FDA's nozzle numerical simulation challenge: Non-Newtonian fluid effects and blood damage. *PLoS ONE* **2014**, *9*, e92638. [CrossRef]
34. Zakaria, M.S.; Zainudin, S.H.; Abdullah, H.; Yuan, C.S.; Abd Latif, M.J.; Osman, K. CFD simulation of non-Newtonian effect on hemodynamics characteristics of blood flow through benchmark nozzle. *J. Adv. Res. Fluid Mech. Therm. Sci.* **2019**, *64*, 117–125.
35. Berg, P.; Janiga, G.; Thévenin, D. CFD Challenge: Solutions using the commercial solver Fluent and the open-source solver OpenFOAM. In Proceedings of the Summer Bioengineering Conference, Fajardo, PR, USA, 20–23 June 2012; American Society of Mechanical Engineers: New York, NY, USA, 2012; Volume 44809, pp. 115–116.
36. Lysenko, D.A.; Ertesvåg, I.S.; Rian, K.E. Modeling of turbulent separated flows using OpenFOAM. *Comput. Fluids* **2013**, *80*, 408–422. [CrossRef]
37. Robertson, E.; Choudhury, V.; Bhushan, S.; Walters, D.K. Validation of OpenFOAM numerical methods and turbulence models for incompressible bluff body flows. *Comput. Fluids* **2015**, *123*, 122–145. [CrossRef]
38. Jones, D.A. *CFD RANS Simulations on a Generic Conventional Scale Model Submarine: Comparison between Fluent and OpenFOAM*; Technical Note; Defence Science and Technology Organisation, Maritime Division: Fishermans Bend, VIC, Australia, 2015. Available online: https://apps.dtic.mil/sti/pdfs/ADA622263.pdf (accessed on 26 March 2024).
39. Greifzu, F.; Kratzsch, C.; Forgber, T.; Lindner, F.; Schwarze, R. Assessment of particle-tracking models for dispersed particle-laden flows implemented in OpenFOAM and ANSYS Fluent. *Eng. Appl. Comput. Fluid Mech.* **2015**, *10*, 30–43. [CrossRef]
40. Dresar, P.; Duhovnik, J. A hybrid RANS-LES computational fluid dynamics simulation of an FDA medical device benchmark. *Mechanics* **2019**, *25*, 291–298. [CrossRef]
41. Jarrell, J.A.; Sytsma, B.J.; Wilson, L.H.; Pan, F.L.; Lau, K.H.; Kirby, G.T.; Lievano, A.A.; Pawell, R.S. Numerical optimization of microfluidic vortex shedding for genome editing T cells with Cas9. *Sci. Rep.* **2021**, *11*, 11818. [CrossRef] [PubMed]
42. Menon, P.G.; Teslovich, N.; Chen, C.Y.; Undar, A.; Pekkan, K. Characterization of neonatal aortic cannula jet flow regimes for improved cardiopulmonary bypass. *J. Biomech.* **2013**, *46*, 362–372. [CrossRef]
43. Mancini, V.; Bergersen, A.W.; Valen-Sendstad, K.; Segers, P. Computed poststenotic flow instabilities correlate phenotypically with vibrations measured using laser Doppler vibrometry: Perspectives for a promising in vivo device for early detection of moderate and severe carotid stenosis. *J. Biomech. Eng.* **2020**, *142*, 091007. [CrossRef]
44. Ozturk Papavassiliou, D.V.; O'Rear, E.A. An approach for assessing turbulent flow damage to blood in medical devices. *J. Biomech. Eng.* **2017**, *139*, 011008.
45. Sonntag, S.J.; Lipinski, E.; Neidlin, M.; Hugenroth, K.; Benkowski, R.; Motomura, T.; Kaufmann, T.A.S. Virtual fitting and hemodynamic simulation of the EVAHEART 2 left ventricular assist device and double-cuff tipless inflow cannula. *ASAIO J.* **2019**, *65*, 698–706. [CrossRef]
46. Kopernik, M. Modeling of blood thrombosis at microscopic and mesoscopic scales. *Comput. Assist. Methods Eng. Sci.* **2019**, *25*, 21–45.

symmetry

MDPI

Article

Local-Energy-Conservation-Based Decomposition Method for Wall Friction and Heat Flux

Mingzhi Tang [1,2], Wenfeng Zhou [1,2], Yanchao Hu [1,2,*], Gang Wang [1,2] and Yanguang Yang [2,3,*]

[1] Hypervelocity Aerodynamics Institute, China Aerodynamics Research and Development Center, Mianyang 621000, China; tangmingzhi@cardc.cn (M.T.); zhouwenfeng@cardc.cn (W.Z.); wanggang@cardc.cn (G.W.)

[2] National Key Laboratory of Aerospace Physics in Fluids, Mianyang 621000, China

[3] China Aerodynamics Research and Development Center, Mianyang 621000, China

* Correspondence: huyanchao@cardc.cn (Y.H.); yangyanguang@cardc.cn (Y.Y.)

Abstract: A novel decomposition method that adheres to both local time translation symmetry and spatial rotational symmetry is proposed in this study, thereby extending the limitations of existing methods, which are typically restricted to quasi-two-dimensional configurations. Grounded in the FIK and RD identities, this method provides a clear physical and reliable interpretation suitable for arbitrary-curvature profiles. Utilizing this method, an analysis of the aerothermodynamic characteristics of the bistable states of curved compression ramp flows was conducted. The results reveal that the generation of undisturbed and peak C_f is dominated by viscous dissipation. Specifically, flow separation happens when all of the energy input from the work exerted by the adverse pressure gradient (APG) is insufficient to be entirely converted into local viscous dissipation and kinetic energy. Furthermore, the propensity for flow separation at higher wall temperatures is firstly elucidated quantitatively from the perspective of the work by the APG. The peak heat flux is predominantly triggered by the work of viscous stress, with the secondary contribution from energy transport playing a more significant role in the generation of the peak heat flux of the separation state than that of the attachment state.

Keywords: time translation symmetry; decomposition method; generation mechanism; wall friction; heat flux

1. Introduction

Shock-wave/boundary-layer interaction (SBLI) frequently occurs on the deflected control surface of a vehicle during its supersonic/hypersonic flight. SBLI may cause significant flow separation and severe peak heat flux, resulting in a substantial reduction in control effectiveness and even a loss of control [1–4]. During flight, wall friction can contribute as much as 50 percent to the total drag [5], and the peak heat flux induced by SBLI and shock/shock interactions (SSIs) can reach $10 \sim 100$ times that without interactions [6]. Therefore, clarifying the generation mechanism of wall friction and peak heat flux is essential for further control intention.

To elucidate the individual contributions of different physical processes to the frictional drag for further guidance on wall friction control, Fukagata, Iwamoto and Kasagi [7] proposed a relationship (referred to as the FIK identity) by integrating the momentum equation. The FIK identity provides a quantitative analysis method for researchers to understand the generation and control mechanisms of wall friction and has been widely applied. By generalizing the FIK identity to compressible flows, Gomez et al. [8] analyzed the compressibility impact on the generation of mean wall friction drag, but their analysis did not quite focus on the impact of compressibility on the contribution terms. However, the FIK identity is generally thought to lack a simple physical interpretation involved in the derivation. Therefore, Renard and Deck [9] proposed a decomposition method (called the

RD identity) based on the Galileo transform and streamwise kinetic energy equation, which has a clearer physical interpretation. This method was extended to compressible flows by Li et al. [10]. Combined with empirical mode decomposition or spectral form equations, the FIK and RD identities help researchers investigate the roles of different flow structures in the generation of wall friction [11–13]. Another obvious benefit of decomposition methods is that they clarify the direction of each part of the energy [14–16], which is helpful in exploring a more efficient flow control method.

The accurate prediction of heat flux is very important for the thermal protection design of high-speed vehicles, and there exist some classical pressure–heat flux scaling relations [1,2,17–20]. However, the generation mechanism of heat flux is still vague. To understand this mechanism, the FIK and RD identities have also been extended to heat flux decomposition, such as in the studies by Zhang and Xia [21,22] and Sun et al. [23]. The heat flux decomposition method was also further derived to investigate the heat flux generation mechanism of turbulent flows in a high-enthalpy scenario by Li et al. [24]. In addition, Zhang and Xia [22] analyzed the results of different forms of heat flux decomposition by the FIK and RD identities and concluded that the twofold repeated integration of the FIK identity had a better physical interpretation. And Wenzel, Gibis and Kloker [25] hold the same viewpoint. However, they did not provide the physical image. In fact, some arbitrariness still exists in the selection of the order of integrals in the FIK identity or the transformation speed in the RD identity. In addition, different understandings of the physical processes of the contributions in the FIK and RD identities also lead to significant differences in understanding the influence of different contributions [26].

The present decomposition analyses mainly focus on the equilibrium flow in quasi-two-dimensional flows, such as channel flow and flat-plate flow. Sagaut and Peet [27] and Bannier, Garnier and Sagaut [28] extended the FIK identity to surfaces with spanwise complex geometries like riblets, enabling the quantitative investigation of such wall friction drag reduction control methods. However, these methods do not account for the streamwise geometry changes due to insufficient consideration of time translation symmetry (energy conservation). Therefore, there has been little discussion on the flow separation, wall friction change and severe heat flux caused by the SBLI induced by flow turning, such as compression ramp flows.

Due to the difference in initial flow states, maneuvering of the aircraft may result in multistable states of the flow, which is ubiquitous in aerospace flow systems [29]. Bistable states of SSIs between regular and Mach reflections are observed when varying the incident shock angle [30–34]. The underlying mechanism for this bistable behavior has been elucidated by Hu et al. through minimal viscous dissipation theory [35]. Recently, Hu et al. numerically observed bistable states of SBLI between separation and attachment in curved compression ramp (CCR) flows by varying the attack angle [29] and mathematically demonstrated their existence [36]. Multistable states can lead to quite different aerothermodynamic performance results, such as the lift and drag coefficients, wall parameters and the start margin of the inlet [37–42]. The mechanism of separation hysteresis and the aerothermodynamics associated with bistable states in CCR flows were thoroughly examined by Zhou et al. [43] and Tang et al. [44]. However, the origin of the disparities in the aerothermodynamic characteristics across multistable states needs further investigation.

The aim of this paper is to address the fact that the existing decomposition methods are limited in their applicability to complex geometries by proposing a decomposition method that satisfies local time translation symmetry and spatial translation symmetry. Moreover, this method can offer a unified framework for analyzing the generation mechanism of wall friction and heat flux for further guidance in researching their simultaneous control. Before deriving the decomposition method, the details of the numerical simulations are described in Section 2. Based on the local time translation symmetry and spatial rotation symmetry, a physically clear form of the wall friction and heat flux decomposition method, which can apply to a surface with a streamwise arbitrary curvature, as well as the verification of the method, is proposed in Section 3. The flow fields, the aerothermodynamic characteristics of bistable states in CCR flows, and the mechanism of wall friction and heat flux generation are analyzed in Sections 4–6. The conclusions are drawn in Section 7.

2. Numerical Simulation

2.1. Governing Equations and Numerical Methods

Direct numerical simulations (DNSs) of laminar flows were carried out for the present analysis and implemented by the in-house code OPENCFD-SC, which has been successfully validated and applied to many cases [29,45–47]. The governing equations, formulated in a non-dimensionalized conservative form and expressed within the framework of curvilinear coordinates, are as follows:

$$\frac{\partial U}{\partial t}+\frac{\partial F}{\partial \xi}+\frac{\partial G}{\partial \eta}=0 \tag{1}$$

with $U = J\{\rho,\rho u,\rho v,\rho e\}$ as the conservative vector flux, ρ as the density, (u,v) as the streamwise and vertical components of velocity, and $e = c_v T$ as the internal energy per volume. J is the Jacobian matrix transforming Cartesian coordinates (x,y) into a computational space defined by curvilinear coordinates (ξ,η). The flux term F in the ξ direction is composed of the convective term F_c and the viscous term F_v:

$$F = F_c + F_v = Jr_\xi\begin{bmatrix}\rho u^* \\ \rho u u^* + p s_x \\ \rho v u^* + p s_y \\ (\rho e + p)u^*\end{bmatrix} - Jr_\xi\begin{bmatrix}0 \\ s_x\sigma_{xx}+s_y\sigma_{xy} \\ s_x\sigma_{yx}+s_y\sigma_{yy} \\ s_x\tau_x + s_y\tau_y\end{bmatrix} \tag{2}$$

where

$$\begin{aligned}
&u^* = us_x + vs_y, \quad s_x = \xi_x/r_\xi, \quad \tau_x = u\sigma_{xx}+v \\
&r_\xi = \sqrt{\xi_x^2+\xi_y^2}, \quad s_y = \xi_y/r_\xi, \quad \tau_y = u\sigma_{xy}+v\sigma_{yy}-q_y, \\
&\sigma_{ij} = 2\mu\left[\frac{1}{2}\left(\frac{\partial u_i}{\partial x_j}+\frac{\partial u_j}{\partial x_i}\right)-\frac{1}{3}\frac{\partial u_k}{\partial x_k}\delta_{ij}\right], \\
&q_j = -\frac{\mu}{Pr(\gamma-1)Ma_\infty^2}\frac{\partial T}{\partial x_j}.
\end{aligned} \tag{3}$$

where $i,j = x,y$. G is the flux term in the η direction having similar forms to F [48]. The working fluid is an ideal gas, fulfilling the pressure condition $p = \rho T/\gamma Ma_\infty^2$, where $\gamma = 1.4$ is the ratio of specific heats, T the static temperature, and Ma the Mach number. The free-stream condition is denoted by the subscript "∞". The free-stream quantities are used to normalize the flow parameters ρ, u, p and T. The unit Reynolds number Re_∞ (m^{-1}) = 3×10^6 and Prandtl number $Pr = 0.7$. The viscosity μ fulfills Sutherland's law, $\mu = \frac{1}{Re_\infty}\frac{T^{3/2}(1+110.4/T_\infty)}{T+110.4/T_\infty}$, with $T_\infty = 108.1$ K as the free-stream static temperature.

2.2. Computational Domain and Flow Conditions

Curved compression ramps (CCRs), as shown in Figure 1, were selected for numerical simulation. The streamwise region x ranges from -82 mm to 108.9 mm ($x = -80$ mm is the start of the flat plate, and the extension line of the flat plate intersects the extension line of the tilted plate at $x = 0$ mm). The normalwise region y ranges from 0 mm to 40 mm. The flat plate and the tilted plate are linked by an arc wall with the curvature radius $R = \frac{L}{\sin\frac{\phi}{2}}$, where $L = 25$ mm, with the curved wall starting at $x = -L$ and with ϕ as the turning angle.

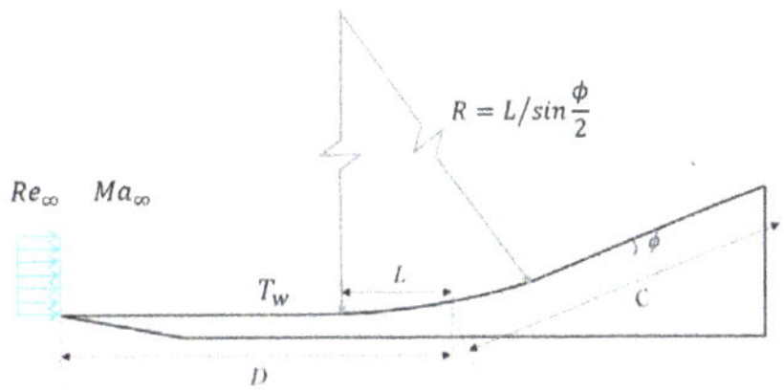

Figure 1. Sketch of curved compression ramp.

The algebraically generated computational mesh consists of 3401 × 401 nodes in the streamwise and normalwise directions, respectively. The streamwise grids consist of a uniformly spaced region with $\Delta x = 0.05$ mm and a buffer region containing 160 nodes with a stretch factor of 1.015. The normalwise grids are clustered in the near-wall region, with the first grid height $\Delta y = 0.01$ mm. The laminar boundary layer thickness $\delta|_{x=-50\text{mm}} \approx 0.5$ mm results in 41 points inside the boundary layer, ensuring good resolution.

The wall boundary conditions are set as no-slip and isothermal, with the wall temperature ratio given by $T_w = \hat{T}_w/T_\infty$, where $\hat{T}_w$ is the wall temperature and the subscript "w" represents the wall condition. A uniform inlet flow condition is applied at $x = -82$ mm. A non-reflecting boundary condition is applied at the outlet. Bistable states of CCR flows are obtained by altering ϕ and T_w, and the two series of DNSs conducted are summarized in Table 1.

Table 1. Flow and geometric conditions for the present simulations. (For each case, the attachment state and separation state, denoted by "Att" and "Sep", respectively, may exist with the same boundary condition and ramp angle but with different initial conditions.)

Ma_∞	T_w	ϕ (°)
6.0	1.5	17 (Att), 18 (Att/Sep), 19 (Att/Sep), 20 (Att/Sep), 21 (Att/Sep), 22 (Att/Sep), 23 (Att/Sep), 24 (Sep)
6.0	1.25 (Att), 1.5 (Att/Sep), 1.75 (Att/Sep), 2.0 (Att/Sep), 2.25 (Sep)	18

2.3. Convergence Study and Validation

A flat-plate ramp with a turning angle $\phi = 24°$ was selected for mesh convergence examination. The flow condition is $Ma_\infty = 6.0$ and $T_w = 1.5$ (denoted by IBC1 in the following). Time convergence will be discussed in Section 4. Four grid scales were selected, including 1201 × 141 (grid A), 1701 × 201 (grid B), 2401 × 281 (grid C) and 3401 × 401 (grid D). Surface pressure coefficient C_p distributions are shown in Figure 2, where $C_p = \frac{(p_w - p_\infty)}{\frac{1}{2}\rho_\infty u_\infty^2}$ and p_w is the wall pressure. The pressure plateau p_{pla} and pressure peak p_{pk} collapse well with that predicted by the minimal viscous dissipation (MVD) theorem [46]. As shown in Figure 2b, the separation point (the black circle) moves upstream, and the pressure rises with a denser grid. And, both the distributions of C_p and locations of the separation points of grids C and D collapse well, indicating that the mesh size of grid D is deemed adequate for the present analysis.

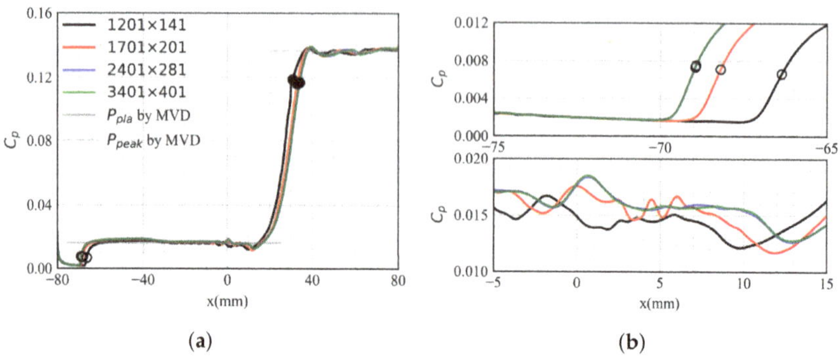

Figure 2. Four grid scales for mesh convergence examination by the case IBC1, $\phi = 24°$. (**a**) C_p distributions of four grid scales. The gray solid and dashed lines are the pressure plateau and pressure peak predicted by the MVD theorem [46], respectively. (**b**) Close-up plots of the separation and vertex regions.

The distributions of non-dimensionalized u and T in the normalwise direction inside the boundary layer are validated by comparison against theoretical Blasius solutions with compressible correction [49]. Good agreement validates the DNS results, as shown in Figure 3. Further validation with published experimental and numerical results can be found in Appendix A.

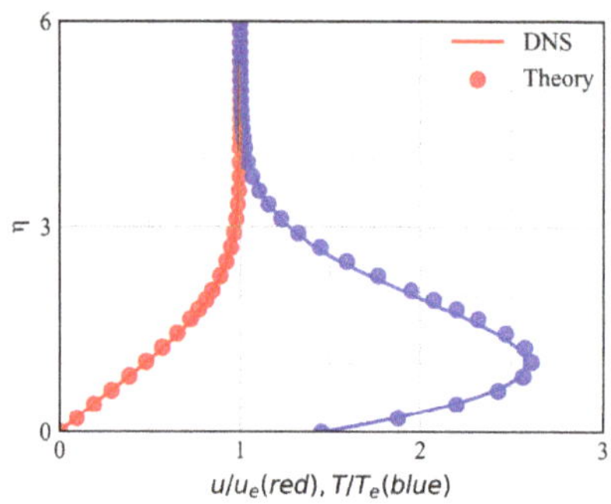

Figure 3. Comparison of u and T from DNS with theoretical predictions at $x = -43$ mm of the case IBC1, $\phi = 18°$, with the subscript "e" representing the external boundary layer flow.

3. Decomposition Method

Existing studies involving the simultaneous decompositions of wall friction and heat flux frequently employ a consistent decomposition methodology [25,50], and the most common one is the twofold repeated integration of the momentum and energy equations based on the the FIK identity [7]. By employing this method, individual physical mechanisms can be distinctly decoupled for analysis. However, the connection between wall friction and heat flux generation is not clear enough, as they are manifested through the distinct forms of momentum and energy, respectively. Therefore, it finds difficulty in guiding the efficient control of friction and heat flow simultaneously. Since all contributions in heat flux decomposition are in the form of energy, and the RD identity [9] is based on the work–energy conversion process between different mechanisms, the RD and FIK identities are adopted for wall friction and heat flux decompositions, respectively. In this way, the two decompositions are related from the perspective of energy. Noether's theorem [51–53] posits a profound relationship between conservation laws and the corresponding symmetries. Critical to the application of decomposition methods in fluid dynamics from the perspective of energy is their adherence to the principle of time translation symmetry. This adherence ensures that the decomposition methods uphold energy conservation, thereby providing a robust physical basis for its application to flows across diverse surfaces. In addition, the clear energy conversion path [14,16] obtained by the method may help improve flow control efficiency. In the following, the decomposition methods of wall friction and heat flux for an arbitrary surface are introduced, and the unique and disambiguated selection of the integral order in the FIK identity and the Galileo transformation speed in the RD identity are clarified from the perspective of local time translation symmetry.

3.1. Derivation of Wall Friction Decomposition

Renard and Deck analyzed the work–energy conversion process in differential form in their paper (Equation (2.5) in Ref. [9]). On the basis of their idea, the present paper will further discuss this process from the perspective of energy conversion between the wall and flow and demonstrate the disambiguation of Galileo transformation velocity u_b, since the arbitrariness of u_b may lead to non-compliance with local energy conservation, which will be explained in detail in the following. The following derivation is based on the rotated equations of the coordinate system, which will be discussed in Section 3.3. Specifically, the Reynolds-averaged streamwise momentum equation is expressed as

$$\frac{\partial \overline{\rho u}}{\partial t}+\frac{\partial \overline{\rho u u_j}}{\partial x_j}=\frac{\partial \overline{\tau_{x,j}}}{\partial x_j}-\frac{\partial \overline{p}}{\partial x} \tag{4}$$

where "$-$", ρ, u_j, $\tau_{x,j}$ and p represent the Reynolds average operator, density, velocity component, shear stress and pressure, respectively. This equation represents streamwise momentum conservation and satisfies spatial translation symmetry. So, we can follow the process of the RD identity [9,10] and adopt Galileo transformation with a speed of $-u_b$ to transform the reference frame from a stationary wall (the initial reference frame) to a moving wall (absolute reference frame). Then, the parameters in the absolute reference frame satisfy

$$t_a=t, \rho_a=\rho, x_a=x-u_b t, y_a=y, u_a=u-u_b, v_a=v \tag{5}$$

where the subscript "a" represents the variables in the absolute frame. The streamwise momentum equation in this reference frame can be obtained by substituting Equation (5) into (4):

$$\frac{\partial \overline{\rho u_a}}{\partial t_a}+\frac{\partial \overline{\rho_a u_a u_{aj}}}{\partial x_{aj}}=\frac{\partial \overline{\tau_{x,j}}}{\partial x_{aj}}-\frac{\partial \overline{p_a}}{\partial x_a} \tag{6}$$

Then, the energy budget equation of the averaged streamwise kinetic energy $K_a = 1/2\overline{\rho_a u_a u_a}$ in the absolute reference frame can be obtained by multiplying both sides of Equation (6) by $\tilde{u}_a$ and is expressed as

$$\frac{\partial \overline{K_a}}{\partial t_a}+\frac{\partial \overline{K_a u_{aj}}}{\partial x_{aj}}=\tilde{u}_a\frac{\partial \overline{\tau_{x,j}}}{\partial x_{aj}}-\tilde{u}_a\frac{\partial \overline{p_a}}{\partial x_a} \tag{7}$$

where the superscript "~" represents the Favre average operator. Integrating Equation (7) by parts over y from 0 to infinity and adopting the steady-state condition and non-slip condition on the wall, we can obtain the intermediate result as

$$u_b\bar{\tau}_w=\int_0^{\delta}\overline{\tau_{xy}}\frac{\partial \tilde{u}}{\partial y}dy+\int_0^{\delta}\frac{\partial \overline{K_a u_{aj}}}{\partial x_j}dy+\int_0^{\delta}\tilde{u}_a\left(-\frac{\partial \overline{\tau_{xx}}}{\partial x_a}\right)dy+\int_0^{\delta}\tilde{u}_a\frac{\partial \overline{p_a}}{\partial x_a}dy \tag{8}$$

and

$$u_b\bar{\tau}_w = \int_0^\delta \overline{\tau_{xy}}\frac{\partial \tilde{u}}{\partial y}dy + \int_0^\delta (\tilde{u} - u_b)[\frac{\partial \overline{\rho u u}}{\partial x} + \frac{\partial \overline{\rho u v}}{\partial y}]dy$$
$$+ \int_0^\delta (\tilde{u} - u_b)(-\frac{\partial \overline{\tau_{xx}}}{\partial x})dy + \int_0^\delta (\tilde{u} - u_b)\frac{\partial \overline{p}}{\partial x}dy \quad (9)$$

In the absolute reference frame with the moving wall and still flow after Galileo transformation, the boundary layer is generated by the moving wall because of the viscosity and non-slip condition. The left-hand sides of Equations (8) and (9) represent the work input from the wall to the fluid to maintain the boundary layer, and the work input should be equal to the work required for the wall of a vehicle to overcome frictional drag to ensure local energy conservation at any point T on the wall of a vehicle, as shown in Figure 4. The work by wall frictional drag at any point T in Figure 4 is equal to the product of the wall shear stress and velocity component in the direction of the force:

$$SW = \vec{\tau} \cdot \vec{u}_\infty = \bar{\tau}_w u_\infty \cos\varphi \quad (10)$$

where SW is the wall shear stress work by wall friction, $\tau_w = \mu_w \frac{\partial u}{\partial y_n}|_{y_n=0}$ the wall shear stress, and φ the angle between the tangential direction of the local wall surface and $\vec{u}_\infty$. As discussed above, the left-hand side of Equation (9) should be equal to SW at point T to ensure local energy conservation. It should be emphasized that the above derivation and analysis focus on point T locally. And, the Galileo transformation is also implemented locally along the wall surface, which is different from the global transformation in the traditional RD identity [9,10]. The benefit of this approach is that local time translation symmetry is guaranteed. Therefore, this method has a clear physical interpretation in evaluating frictional drag reduction research for vehicles with complex configurations. According to the analysis of Equation (9), the Galileo transformation velocity u_b is unique and disambiguated as

$$u_b = u_\infty \cos\varphi \quad (11)$$

rather than arbitrary.

Figure 4. Schematic diagram of Galileo transformation on curved wall with boundary layer profile on it.

Combining Equations (9) and (11), as well as the definition of wall friction as $C_f = \frac{\overline{\tau_w}}{1/2\rho_\infty u_\infty^2}$, the final form of wall friction decomposition can be obtained as

$$C_f = \underbrace{\frac{2}{\rho_\infty u_\infty^3 \cos\varphi}\int_0^\delta \overline{\tau_{xy}}\frac{\partial \tilde{u}}{\partial y}dy}_{C_f^L} + \underbrace{\frac{2}{\rho_\infty u_\infty^3 \cos\varphi}\int_0^\delta (\tilde{u} - u_\infty\cos\varphi)[\frac{\partial \overline{\rho u u}}{\partial x} + \frac{\partial \overline{\rho u v}}{\partial y}]dy}_{C_f^{MT}}$$
$$+ \underbrace{\frac{2}{\rho_\infty u_\infty^3 \cos\varphi}\int_0^\delta (\tilde{u} - u_\infty\cos\varphi)(-\frac{\partial \overline{\tau_{xx}}}{\partial x})dy}_{C_f^S} + \underbrace{\frac{2}{\rho_\infty u_\infty^3 \cos\varphi}\int_0^\delta (\tilde{u} - u_\infty\cos\varphi)\frac{\partial \overline{p}}{\partial x}dy}_{C_f^P} \quad (12)$$

where the contribution terms for wall friction generation are the viscous dissipation C_f^L, the kinetic energy transport C_f^{MT}, the streamwise distortion C_f^S and the work performed by the pressure gradient C_f^P. In the above derivations, φ is a local quantity to reflect the local wall shape. In addition, as previously mentioned, $\cos\varphi$ is a typical parameter to reflect the real local work performed by wall shear stress on a curved wall, which is quite different from the case of flat-plate flows. After non-dimensionalization, C_f represents the local wall shear work, which is converted into C_f^L, C_f^{MT} and C_f^S, while C_f^P can usually serve as a source (or sink) term, since C_f^P originates from the pressure gradient, which is generally induced by compression or expansion walls, or incident shocks. For an aircraft, C_f and C_f^P correspond to friction and wave resistance, respectively, and the conversion between C_f and C_f^P is achieved indirectly by C_f^L, C_f^{MT} and C_f^S. After decomposition, the generation mechanism of wall friction is decoupled into these quantified terms, which is crucial for understanding the relative importance of each effect and for designing targeted flow control strategies. Moreover, decomposition by the RD identity inherently satisfies energy conservation, providing a robust basis for analyzing the work–energy conversion process. Similar to Refs. [14,16], the whole energy conversion process can be depicted in Figure 5. With respect to this physical description, the wall friction decomposition method for arbitrary profiles is obtained.

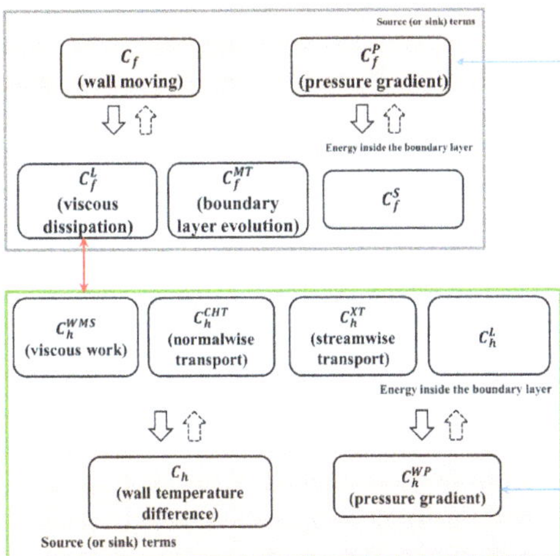

Figure 5. A schematic to illustrate the energy conversion pathways in the flow. The pathways in the gray and green boxes are based on the wall friction and heat flux decomposition methods, respectively. The red and blue arrows represent the links between the terms.

3.2. Derivation of Heat Flux Decomposition

For heat flux decomposition, we agree with the viewpoint of Zhang and Xia [21,22] and Wenzel et al. [25] that the twofold integration of the FIK identity has better interpretability. The following derivation is consistent with that of Zhang and Xia [21,22], but we will propose an explicit physical depiction and physical interpretation from the perspective of the redistribution of temperature (enthalpy) in the boundary layer. The equation for averaged static enthalpy is

$$\frac{\partial \overline{\rho e}}{\partial t} + \frac{\partial \overline{\rho h u_j}}{\partial x_j} = \frac{\partial \overline{q_j}}{\partial x_j} + \overline{\tau_{i,j}\frac{\partial u_i}{\partial x_j}} + \overline{u_j\frac{\partial p}{\partial x_j}} \tag{13}$$

where the static enthalpy $h = e + p/\rho = c_pT$, and the heat flux $\overline{q_j} = \overline{\kappa\partial T/\partial x_j}$.

For steady flows, integrating Equation (13) from the wall ($y|_w = 0$) to a distance y in the wall-normal direction and applying the no-slip condition $u_w = v_w = 0$ on the wall yields

$$q_w = q_y - \overline{\rho h v}\Big|_y + \int_0^y \overline{\tau_{i,j}\frac{\partial u_i}{\partial x_j}} dy_1 + \int_0^y u_j \frac{\partial p}{\partial x_j} dy_1 + \int_0^y -\frac{\partial \overline{\rho h u}}{\partial x} dy_1 \tag{14}$$

where q_w is the wall heat flux, and the terms on the right-hand side of Equation (14) are energy conduction at y, enthalpy transfer along the normal direction (convective heat transport, red arrow in Figure 6), the work by viscous stress, the work by the pressure gradient and enthalpy transport along the streamwise direction, respectively. This equation expresses the local equilibrium relation on the wall but also gives the essence of heat flux generation on the wall, namely, the redistribution of temperature (enthalpy) in the boundary layer along the normal direction. This process includes the local generation of enthalpy in the boundary layer and the transport along the streamwise and normal directions. In this case, the convective heat transport term $\overline{\rho h v}\Big|_y$ is the most obvious way to enhance the redistribution of the boundary-layer temperature along the normal direction, except q_y. And, the total normally convected enthalpy from the outer boundary layer to the wall (or in reverse) can be obtained by integrating this term from 0 to δ, with δ as the boundary layer thickness.

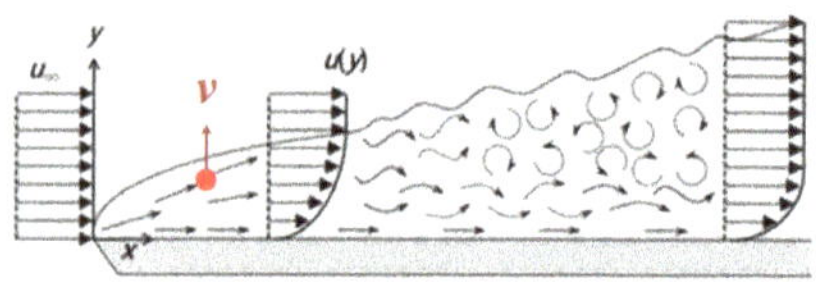

Figure 6. Normalwise enthalpy transfer.

Similar to the FIK identity and Zhang and Xia's derivation, by integrating Equation (14) by parts from 0 to δ along the wall-normal direction, we can obtain the final form of heat flux decomposition.

$$\begin{aligned} C_h = & \underbrace{\frac{1}{\rho_\infty u_\infty (h_{aw} - h_w)\delta} \int_0^\delta q_y dy}_{C_h^L} + \underbrace{\frac{1}{\rho_\infty u_\infty (h_{aw} - h_w)\delta} \int_0^\delta -\overline{\rho h v} dy}_{C_h^{\mathrm{CHT}}} \\ & + \underbrace{\frac{1}{\rho_\infty u_\infty (h_{aw} - h_w)\delta} \int_0^\delta (\delta - y) \overline{\tau_{i,j}\frac{\partial u_i}{\partial x_j}} dy}_{C_h^{\mathrm{WMS}}} + \underbrace{\frac{1}{\rho_\infty u_\infty (h_{aw} - h_w)\delta} \int_0^\delta (\delta - y) u_j \frac{\partial p}{\partial x_j} dy}_{C_h^{WP}} \\ & + \underbrace{\frac{1}{\rho_\infty u_\infty (h_{aw} - h_w)\delta} \int_0^\delta (\delta - y)\left(-\frac{\partial \overline{\rho h u}}{\partial x}\right) dy}_{C_h^{XT}} \end{aligned} \tag{15}$$

where the Stanton number $C_h = \frac{q_w}{\rho_\infty u_\infty (h_\infty + r u_\infty^2/2 - h_w)}$, and $r = \sqrt{Pr}$. This relation decouples the generation mechanism of the heat flux at the wall into five physical processes, including the heat transfer C_h^L, representing the energy transfer process achieved through molecular motion due to the temperature gradient; the convective heat transport C_h^{CHT}, representing the transport of enthalpy along the normal direction brought by the flow; the work from viscous stress C_h^{WMS}, representing the energy corresponding to the deformation of fluid microclusters and viscous dissipation caused by viscous stresses, where viscous dissipation leads to the loss of flow kinetic energy, which is converted into internal energy, resulting in an increase in local temperature; the pressure gradient C_h^{WP}, representing the energy corresponding to the deformation of fluid microclusters and viscous work by pressure gradients induced by shock waves; and streamwise enthalpy transport C_h^{XT}, representing the net increase in local enthalpy caused by the flow along the streamwise direction. Focusing on a certain location within the boundary layer, a significant portion of C_h^{WMS} is dissipated and converted into internal energy, leading to an increase in local temperature. According to the Reynolds analogy for an undisturbed boundary layer [54], changes in temperature within the boundary layer are closely related to changes in velocity, which, in turn, can cause variations in the local streamwise and normalwise velocities. The increased temperature will be transported by the altered velocities in both the streamwise and normalwise directions by C_h^{XT} and C_h^{CHT}. Specifically, C_h^{CHT} will further lead to a redistribution of the temperature profile within the boundary layer. This redistribution results in changes in the near-wall temperature gradient, thereby altering the local wall heat flux. In the strongly perturbed flows caused by SBLI, this process still holds. The work by APG injects much higher energy, and its impact on this process is more significant. The decomposition method used in this paper can obtain a comprehensive effect of this process through integration.

Comparing Equations (12) and (15), we can find that the integrated terms in $C_f^L \backslash C_h^{WMS}$ and $C_f^P \backslash C_h^{WP}$ are quite similar and represent the same energy generation mechanism. Therefore, the combined decomposition method links the generation of C_f and C_h from the perspective of energy conversion, as shown in Figure 5. And, one benefit of the method is that it provides a helpful evaluation tool for the simultaneous flow control of wall friction and heat flux.

3.3. Mesh-Independent Data Transformation Method for Arbitrary-Curvature Surface

According to the definition, the integrated terms in the wall friction and heat flux decompositions should be integrated along the wall-normal direction to ensure local time translation symmetry. The existing wall friction and heat flux decomposition analyses are usually established on quasi-two-dimensional profiles, such as flows on flat plates and in channels, and the integration along the wall-normal direction can be directly carried out along the grid lines, while the transformation of the flow field is needed for arbitrary-curvature profiles, since the vertical grid lines are not perpendicular to the wall. The transformation methods include using coordinate scaling, applying coordinate rotation, or directly dealing with the equations in curved surface coordinates.

- Coordinate scaling can be carried out by Jacobi transformation, i.e., transforming the flow field of any shape into a square, and then integration along grid lines with the Jacobian matrix is feasible, as the grid lines have been transformed to the wall-normal component.
- If we directly deal with the equations in the curved coordinate system, the equations are clear (Equation (16)). However, it is somewhat difficult to sort out the format of the grids for integrating these equations.

$$\rho\left[\frac{u_s}{h}\frac{\partial u_s}{\partial s}+u_n\frac{\partial u_s}{\partial n}+\frac{\kappa u_s u_n}{h}\right]=-\frac{1}{h}\frac{dp}{dx}+\frac{\partial \tau}{\partial n} \tag{16}$$

- The rotation of the computational domain with the given angle of the local slope satisfies spatial rotational symmetry and is also a feasible method. That is, the flow field is rotated to the concerned direction without changing the essence of the inertial system (Figure 7). This method is selected for the present analysis.

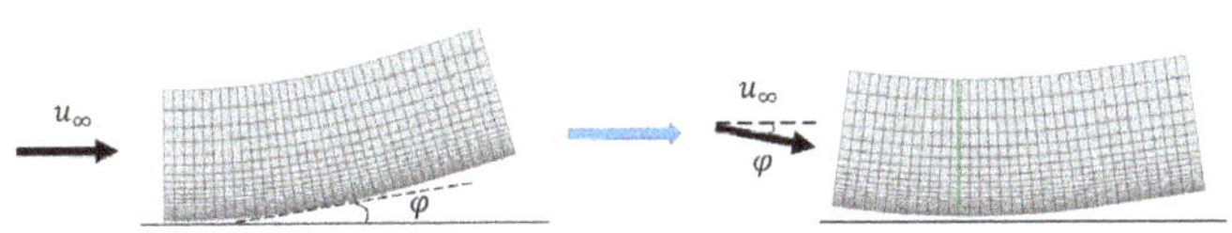

Figure 7. Coordinate rotating diagram.

The selection of the integral path direction is discussed below. Except for the definition, although the information on grid lines can transform to the normal position of the wall, it represents the upstream or downstream information of the actual position. In addition, the integral results will be associated with a grid set and may not be unique. Therefore, integration along the wall-normal direction may be a better choice to obtain the real wall friction and heat flux decomposition results. Specifically, grid interpolation can be performed; that is, a local normal grid is set (green line in Figure 7), along which the flow field is interpolated to obtain flow parameters along the normal direction.

3.4. Decomposition Method Verification

The wall friction C_f and heat flux C_h are determined through their definitions or Equations (12) and (15), which reflect the wall quantity and the overall effects inside the boundary layer, respectively. Figure 8 compares the results obtained by their definitions and Equations (12) and (15) (tagged with the superscript "d") with the conditions of IBC1, $\phi = 18°$, between the DNSs and decompositions. The relative errors (calculated by $error = |(C_m^d - C_m)/C_m| \times 100\%$ with $m = f, h$ representing the wall friction and heat flux, respectively) are within 3%. The small error indicates that the decomposition method offers a correct estimation of the overall effects inside the boundary layer on C_f and C_h. Further comparisons of the contribution terms with theoretical solutions are carried out and discussed in Sections 5 and 6.

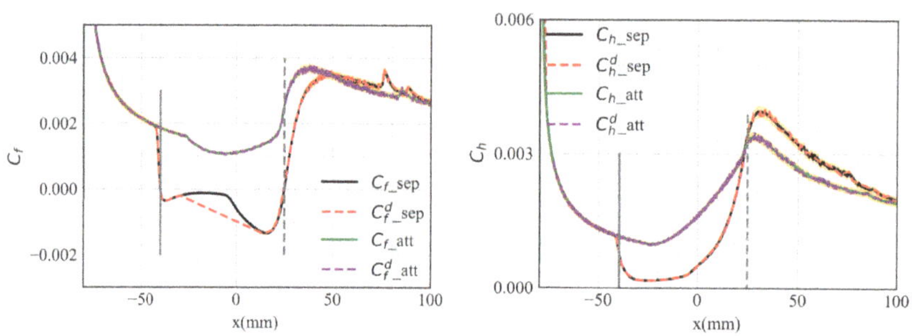

(a) (b)

Figure 8. A comparison of the results with the conditions of IBC1, $\phi = 18°$, between the numerical simulations and the decomposition (tagged with the superscript "d") of (**a**) wall friction and (**b**) heat flux (the orange area is the 3% error band, and the separation and attachment states are denoted by the subscripts "sep" and "att", respectively.) The vertical solid and dashed gray lines represent the streamwise locations of separation and reattachment points, respectively.

4. Bistable States of CCR Flows

The bistable states characterized by flow separation and attachment are achieved through variations in ϕ and T_w, as shown in Figure 9. For the ϕ-variation-induced bistable states, only the steady attachment flow state exists when $\phi = 17°$, as shown in Figure 9b. Increasing ϕ with $\Delta\phi = 1°$ after the flow reaches convergence and using it as the initial condition (state Att_$\phi = 17°$ to Att_$\phi = 23°$ in Figure 9a), the attachment state is maintained until $\phi = 23°$. However, a sudden flow separation occurs when $\phi = 24°$. In the opposite process of decreasing ϕ from 24° with the same $\Delta\phi$, the flow remains separated until $\phi = 17°$, at which point the attachment state suddenly appears again. The interval ($\phi_{att} = 17°$, $\phi_{sep} = 24°$) is the bistable state interval (BSI, as shown in the green square in Figure 9a) for the current inlet and boundary conditions (IBCs), in which the same IBCs may result in either separation or attachment flow states, contingent upon varying initial conditions.

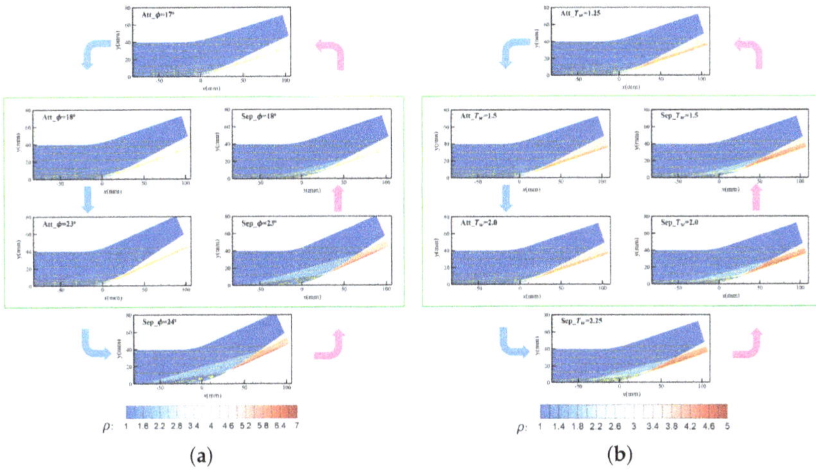

Figure 9. Bistable states of separation/attachment achieved through (**a**) ϕ variation with the condition IBC1 and (**b**) T_w variation with the conditions $Ma_\infty = 6.0$, $\phi = 18°$, colored by the local density. The streamlines are represented by golden lines with arrows. Inside the green square is the bistable state interval.

Bistable states induced by T_w variation are obtained in a similar process, as shown in Figure 9b. Starting with the attachment flow state of $T_w = 1.25$, the attachment state is maintained up to a threshold value of $T_w = 2.0$ when increasing T_w by $\Delta T_w = 0.25$. Beyond this threshold, specifically at $T_w = 2.25$, the flow state transitions to a distinct separation state. Decreasing T_w from $T_w = 2.25$ by the same ΔT_w, the separation state is maintained until $T_w = 1.5$. And, only the attachment state exists with $T_w = 1.25$. Therefore, the BSI is ($T_{w,att} = 1.25$, $T_{w,sep} = 2.25$) in this case.

The differences in C_f and C_h between bistable states are distinct, as shown in Figure 8. The separation state has a negative C_f and a higher peak C_h, and the reasons for these differences will be discussed in the following. More discussions about the differences in aerothermodynamic characteristics between bistable states can be found in Refs. [43,44].

The separation length L_{sep} is continuously observed to ascertain time convergence, where

$$L_{sep} = x_{reatt} - x_{sep} \tag{17}$$

with x_{reatt} and x_{sep} as the streamwise locations of reattachment and separation points, respectively. Figure 10 illustrates the evolution of separation lengths as the flow transitions from separation to attachment (ϕ18° _sep _ to _ ϕ17°, with "ϕ18° _sep" representing the separation state of CCR at $\phi = 18°$) and from attachment to separation (ϕ23° _att _ to _ ϕ24°). Starting from the attachment state, the flow separates after about 1000 τ (dimensionless time $1\tau = 1/(Ma_\infty\sqrt{\gamma R_g \hat{T}_{ref}})$), or equivalently, about five flow-through times on the ramp, but takes a long time to converge. The separation flow state is considered steady when the change in L_{sep} is below 0.01 mm over a time interval of 1000 τ. Starting from the separation state, L_{sep} drops to zero quickly after about 8000 τ and finally converges to zero. Therefore, the time convergence and the steady state are verified. For other cases, enough flow-through time is also carried out to ensure time convergence.

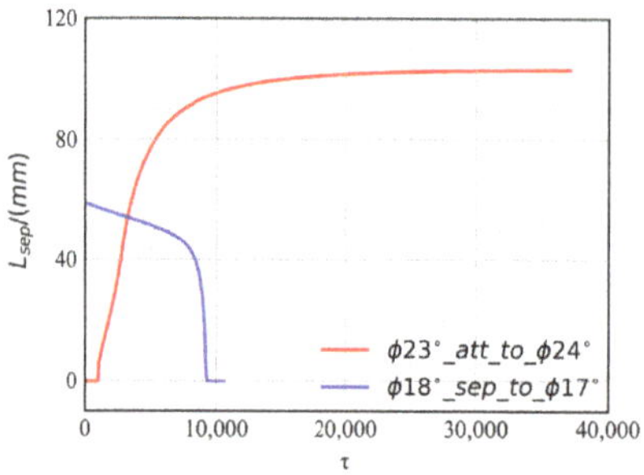

Figure 10. Time convergence process of separation length L_{sep} for IBC1. The suffixes "_sep" and "_att" denote flow states of separation and attachment, respectively.

5. Streamwise Evolution of C_f

The decomposition method and the data have been verified in the previous sections. In this section, the streamwise evolution of the contribution terms for C_f in Equation (12) at different locations (including the equilibrium flow region upstream of the origin interaction point, the separation region around the separation point, and the reattachment and peak wall friction region around the reattachment and peak wall friction points) are analyzed, as shown in Figure 11. It should be emphasized that for the wall friction decomposition, the following analysis takes place in the absolute reference frame by the Galilean transformation. That is, the boundary layer is generated by the moving wall, and the work from wall motion is a major source of the energy contained in the boundary layer, as described in Equations (10) and (12).

In the equilibrium flow region (Figure 11a), the flow remains undisturbed and attached. We further validate the decomposition results with theoretical Blasius solutions [49] (depicted by the scatters) in this region. The good agreement of each contribution term verifies the reliability of the decomposition analysis results. Moreover, we believe that the results of the separation and reattachment regions are reliable. From the distributions of the contribution terms, the dominant factor for wall friction generation is the viscous dissipation C_f^L, which indicates that the work performed by wall friction from the vehicle to the boundary layer is mainly dissipated by viscous stresses. The streamwise kinetic energy transport C_f^{MT} plays a secondary role, consistent with the streamwise progression of the boundary layer, as the boundary layer thickens and the kinetic energy contained in the boundary layer increases. But, the increase in kinetic energy also shows a gradual downward trend.

In the separation region (Figure 11b), the attachment state maintains the characteristics of the equilibrium flow region, since the flow is not disturbed. The wall friction of the separation state obviously decreases, which is the combined effect of C_f^{MT} and the work performed by the adverse pressure gradient (APG) C_f^P. Considering the energy conversion process shown in Figure 5, the positive and negative values of the contributions on the right-hand side of Equation (12) signify the direction of energy conversion, and a positive term indicates that the work by wall friction is converted into this contribution term, while a negative term is the opposite. Then, the negative C_f^P caused by the separation shock wave indicates that the work by the APG is injected into the boundary layer. Since the streamwise pressure gradient $\partial p/\partial x$ induced by shock waves is positive, and $\tilde{u} \leq u_\infty \cos\varphi$ inside the boundary layer, $C_f^P = \frac{2}{\rho_\infty u_\infty^3 \cos\varphi} \int_0^\delta (\tilde{u} - u_\infty \cos\varphi) \frac{\partial \overline{p}}{\partial x} dy$ around the separation and reattachment shock waves is always negative. Therefore, we refer to the absolute value of C_f^P when we talk about it in the following discussion. The APG will inevitably change the boundary-layer velocity profile, which is manifested as the obvious increase in kinetic energy in the boundary layer, as shown by the solid blue line in Figure 11b. The distortion of the velocity profile will also lead to local shear enhancement and result in higher C_f^L

(solid red line in Figure 11b). In this region, C_f^P and C_f^L are the two major contribution terms in the energy conversion process. The work performed by the APG leads to a significant decrease in the requirement for the work by the wall friction to maintain the boundary layer and even results in negative work by local wall friction, i.e., flow separation.

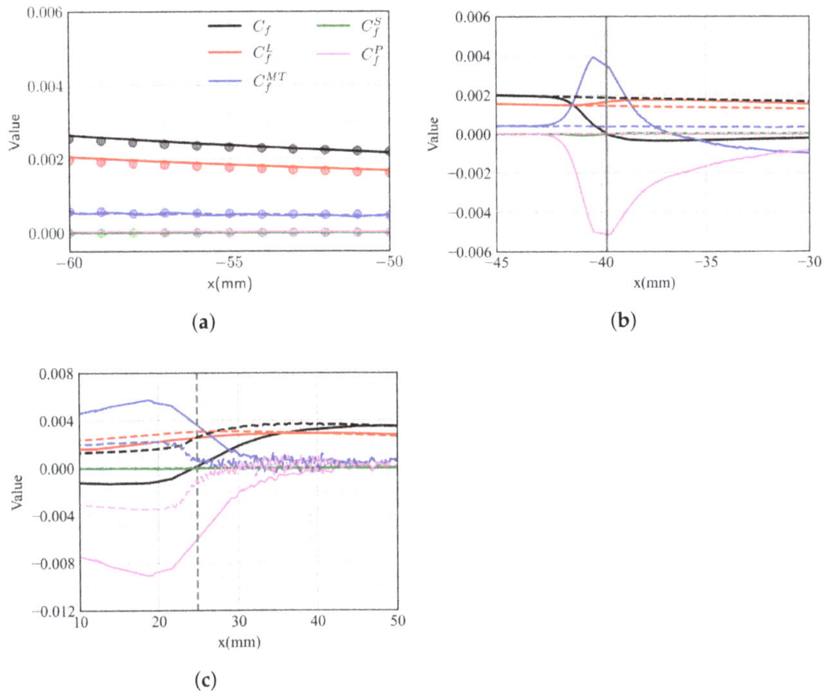

Figure 11. A comparison of different terms in the wall friction decomposition of separation/attachment states at different streamwise positions with the conditions of IBC1, $\phi = 18°$. The vertical solid and dashed gray lines indicate the streamwise separation and reattachment locations, respectively. The other solid and dashed lines indicate separation and attachment states, respectively. (**a**) The equilibrium flow region, with the scatters representing the results obtained theoretically; (**b**) the separation region; (**c**) the reattachment and peak wall friction region.

The relations of different contribution terms in the reattachment region (Figure 11c) are similar to those in the separation region, where C_f^P induced by the reattachment shock wave plays the dominant role. Specifically, the stronger intensity of the reattachment shock leads to a higher C_f^P.

As delineated in Equation (12), C_f^P is the coupling between the APG and the boundary-layer velocity profile, which will be affected by the wall temperature, thereby indirectly affecting C_f^P and the distribution of C_f. To elucidate the impact of the boundary-layer velocity profile on these parameters, it is advantageous to juxtapose the results from attachment states that share identical ramp angles and inlet Mach numbers but exhibit distinct wall temperature ratios T_w, as shown in Figure 12. In the equilibrium flow region ($x \leq -25$ mm), where the pressure gradient is nearly 0, the terms of wall friction decomposition are not affected by T_w variation, indicating that C_f^L, C_f^{MT} and C_f^S are independent of T_w. On the curved wall, with the increase in T_w, the minimum of C_f obviously decreases, and the reduction is primarily attributed to the escalation of C_f^P. Since the pressure rise on the surface is under nearly isentropic compression [43], the pressure gradient distribution is basically the same and independent of the wall temperature in these cases. The increase in C_f^P with higher T_w reflects a larger velocity profile deficit, which is consistent with the higher shape factor discussed by Zhou et al. [43]. However, C_f^L is nearly unaffected by T_w variation. Therefore, it is more difficult to dissipate the energy injection by the APG for higher T_w, and the flow is more likely to separate.

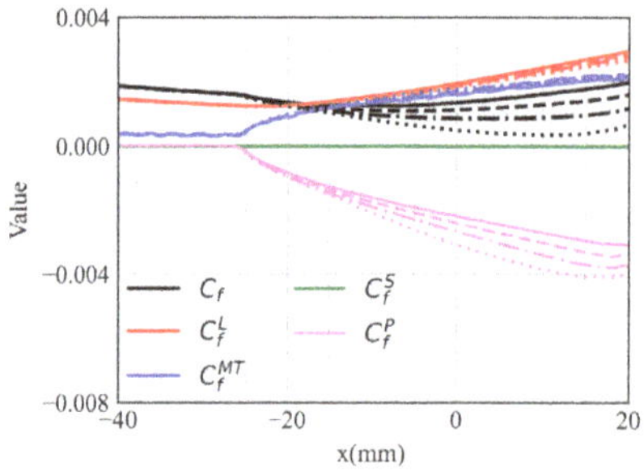

Figure 12. A comparison of the terms in the wall friction decomposition of attachment states with the conditions of $Ma_\infty = 6.0, \phi = 18°$, but different T_w (solid lines: $T_w = 1.25$; dashed lines: $T_w = 1.5$; dash-dot lines: $T_w = 1.75$; dotted lines: $T_w = 2.0$).

From Figures 11b,c and 12, C_f^{MT} and C_f^P are obviously highly correlated at the separation and reattachment regions of separation states or the curved walls of attachment states. The anti-correlation between them indicates that most of C_f^P is converted into C_f^{MT}, which is the conversion inside mechanical energy. Then, we add C_f^{MT} and C_f^P to obtain the mechanical energy transport term C_f^{ME}, i.e., $C_f^{ME} = C_f^{MT} + C_f^P$, to further investigate the energy conversion process, as shown in Figure 13. Similar to the previous discussion, the negative mechanical energy also means that the work performed by the APG supplies the mechanical energy input into the boundary layer.

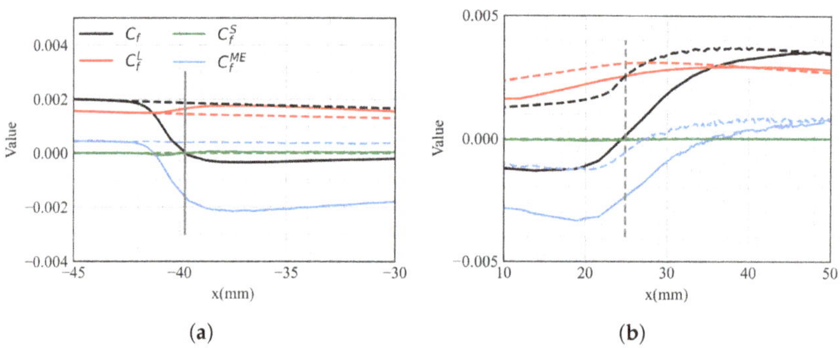

Figure 13. A comparison of the terms in wall friction decomposition in the mechanical energy form at different streamwise positions with the conditions of IBC1, $\phi = 18°$. The vertical solid and dashed gray lines indicate the streamwise separation and reattachment locations, respectively. The other solid and dashed lines indicate separation and attachment states, respectively. (**a**) The separation region; (**b**) the reattachment and peak wall friction region.

Figure 11b illustrates that the reduction in wall friction prior to the separation point is mainly caused by C_f^P. Figure 13 shows more clearly that the viscous dissipation increases slightly; therefore, it is difficult to dissipate further mechanical energy input, which results in a decrease in C_f (non-dimensionalized SW) until the flow separates. So, a key to keeping the boundary layer attached is the dissipation of the mechanical energy input via viscous dissipation. In the reattachment region, viscous dissipation obviously increases, and the mechanical energy input starts to decrease after the reattachment shock wave, so an increase in wall friction is needed to maintain the boundary layer. The coupling of C_f^L and C_f^{ME} leads to the appearance of peak wall friction, and C_f^L is the principal contribution term in the generation of C_f after the reattachment point. The enhanced reattachment shock, triggered by an increased ramp angle ϕ, leads to reduced minimum wall friction within the separation bubble, as illustrated in Figure 14. This reduction is primarily attributed to the significant disparity in C_f^{ME}. However, it is noteworthy that the peak wall friction remains largely unaffected by variations in C_f^{ME}.

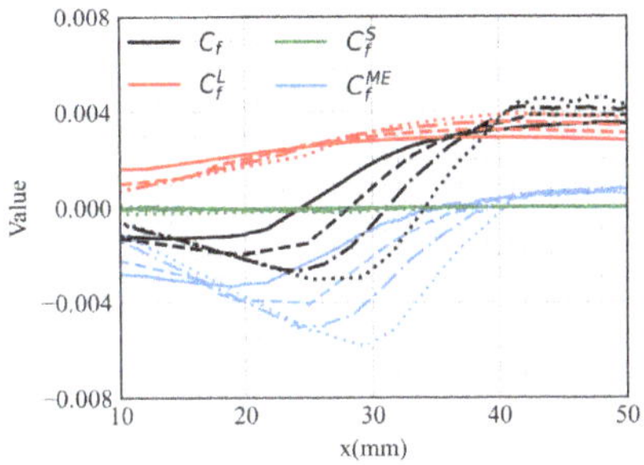

Figure 14. Effect of ramp angles on wall friction decomposition results of separation states at reattachment position (solid lines: $\phi = 18°$; dashed lines: $\phi = 20°$; dash-dot lines: $\phi = 22°$; dotted lines: $\phi = 24°$).

6. Streamwise Evolution of C_h

The analysis of wall heat flux decomposition takes place in the initial reference frame, different from the above analysis of C_f. The streamwise evolution of each contribution term is also analyzed first to elucidate the generation mechanism of heat flux C_h, as shown in Figure 15. For different regions, the heat transfer term C_h^L is very small and can be neglected, which represents a notable divergence from the results of Sun et al. [23]. They adopted the RD identity, resulting in the heat transfer weighted by the normal gradient of streamwise velocity, which will amplify the near-wall heat transfer. However, in the present analysis, C_h^L is the summation of heat transfer across the boundary layer. The difference may originate from these factors. In the equilibrium flow region (Figure 15a), we also validate the decomposition results with theoretical Blasius solutions [49] (depicted by the scatters). Good agreement of each contribution term is also obtained, verifying the reliability of the heat flux decomposition analysis results. From the distributions of the contribution terms, the convective heat transport term C_h^{CHT}, the viscous stress work term C_h^{WMS} and streamwise enthalpy transport term C_h^{XT} gradually decrease as the boundary layer develops. Although C_h^{CHT} convects the enthalpy outward from the wall to the outer flow, C_h^{XT} will supplement these parts. In addition, C_h^{WMS} is higher than C_h, so it can be inferred that the enthalpy convection along the normal direction by C_h^{CHT} is higher than the streamwise accumulation. In other words, C_h^{CHT} exerts a more pronounced influence on the generation of C_h.

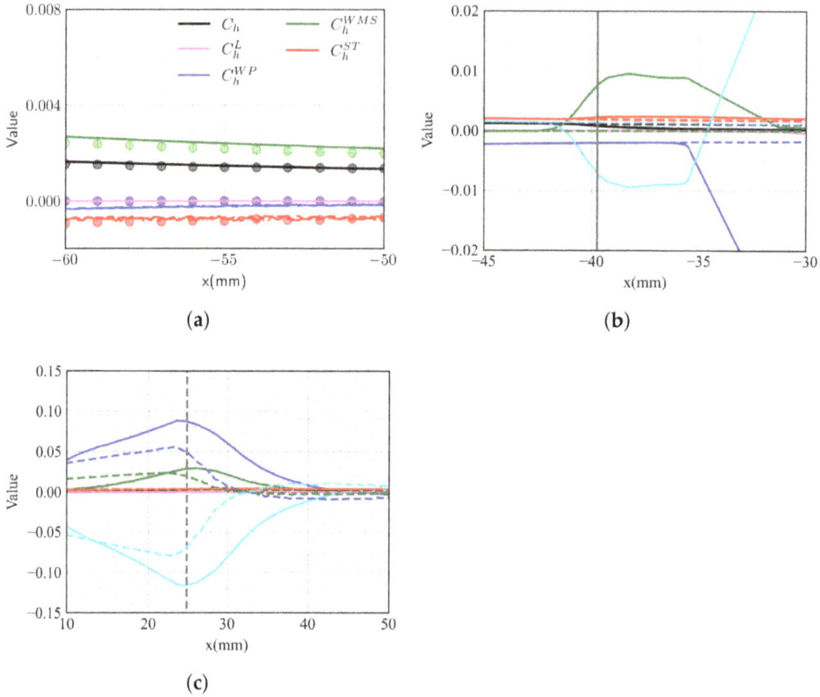

Figure 15. A comparison of different terms in the heat flux decomposition of separation/attachment states at different streamwise positions with the conditions of IBC1, $\phi = 18°$. The vertical solid and dashed gray lines indicate the streamwise separation and reattachment locations, respectively. The other solid and dashed lines indicate separation and attachment states, respectively. (**a**) The equilibrium flow region, with the scatters representing the results obtained theoretically; (**b**) the separation region; (**c**) the reattachment and peak heat flux region.

In the separation region (Figure 15b), similar to the previous analysis of C_f, prominent work by the APG C_h^{WP} originating from the separation shock wave can be found. Different from the results of wall friction decomposition, C_h^{WP} in heat flux decomposition presents as positive work and directly inputs energy into the boundary layer. Within the interval from the origin interaction point, located at approximately $x = -41$ mm, to the region where the boundary layer begins to noticeably deflect, situated at approximately $x = -36$ mm, the energy input associated with C_h^{WP} is transported downstream along the flow direction. This transport predominantly occurs through the streamwise enthalpy transport C_h^{XT}, which accounts for a larger proportion of the energy transfer. Consequently, this leads to a reduction in the wall heat flux. Furthermore, at $x > -36$ mm, an escalation in the normal velocity within the boundary layer contributes to a significant decrease in C_h^{CHT}. A very interesting phenomenon in this area is that C_h^{CHT} and C_h^{XT} together lead to an energy reduction, rather than anti-correlation between them in the equilibrium flow region. The energy input by C_h^{WP} accelerates the enthalpy transport, while the change in C_h^{WMS} is small, which is consistent with the results of the wall friction decomposition analysis.

In the reattachment and peak heat flux region (Figure 15c), C_h^{CHT} plays a positive role in the increase in wall heat flux as a consequence of the shear layer's impingement during the reattachment process of the separation state, and its value is significantly larger than those of C_h and C_h^{WP}. However, most of this energy is transported downstream by C_h^{XT}. For the attachment state, C_h^{CHT} also directs energy to the wall as the flow deflection falls behind the wall shape, but the amount is significantly lower than that of the separation state.

Comparing the values and distributions of C_h^{CHT}, C_h^{XT} and C_h^{WP}, we can find that these terms always exhibit quasi-balance characteristics. In addition, the sum of C_h^{CHT} and C_h^{XT} is the local enthalpy variation. Here, we sum the three terms as the energy transport term C_h^{Trans}, i.e., $C_h^{Trans} = C_h^{CHT} + C_h^{XT} + C_h^{WP}$. And, the distributions are shown in Figure 16.

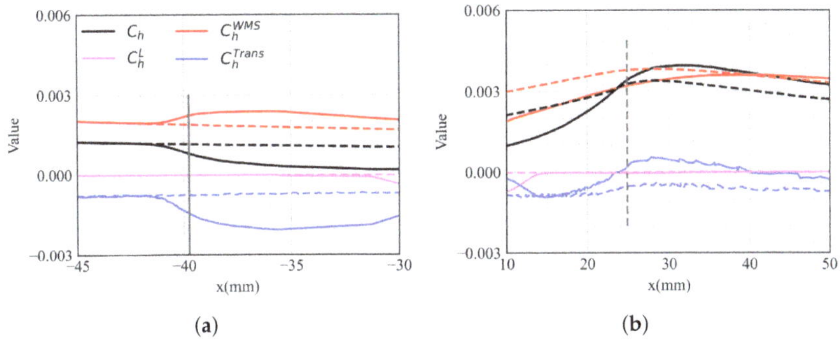

Figure 16. A comparison of different terms in wall heat flux decomposition in the energy transport form at different streamwise positions with the conditions of IBC1, $\phi = 18°$. The vertical solid and dashed gray lines indicate the streamwise separation and reattachment locations, respectively. The other solid and dashed lines indicate separation and attachment states, respectively. (**a**) The separation region; (**b**) the reattachment and peak heat flux region.

In the separation region, although C_h^{WMS} increases to offer more enthalpy, C_h continues to decrease, which aligns with the results of Sun et al. [23]. The reason is that the additionally generated enthalpy is transported by C_h^{Trans}, as shown in Figure 16a. In the peak heat flux region (Figure 16b), C_h^{WMS} dominates the wall heat flux generation. This result indicates that, although the enthalpy brought by the impinging flow is large and the work induced by the APG is also considerable, most of them are transported downstream, and little is left locally. For the attachment state, C_h^{Trans} remains nearly constant along the streamwise direction. As a result, the streamwise positions of peak wall heat flux and peak C_h^{WMS} are nearly the same as those of the attachment state, while the position of the peak heat flux of the separation state is affected by C_h^{Trans}. In addition, the difference in the peak heat flux between the separation and attachment states is mainly attributed to C_h^{Trans}. A larger C_h^{Trans} results in higher peak heat flux in the separation state, although the C_h^{WMS} associated with the attachment state within the vicinity of the peak heat flux is a little higher than that of the separation state.

The influences of the ramp angle and wall temperature ratio on the contribution terms to wall heat flux generation are illustrated in Figure 17a and Figure 17b, respectively. With the increase in the ramp angle, C_h^{WMS} increases slightly. In contrast, the increase in C_h^{Trans} is more obvious, and the increase in peak heat flux is primarily caused by C_h^{Trans}. However, the difference in wall temperature has less influence on C_h^{WMS} and C_h^{Trans}. The results indicate that the increase in C_h^{WMS} is mainly caused by the enhanced distortion of the velocity profile induced by a stronger APG, while the influence of the changed molecular viscosity caused by different wall temperature ratios is small.

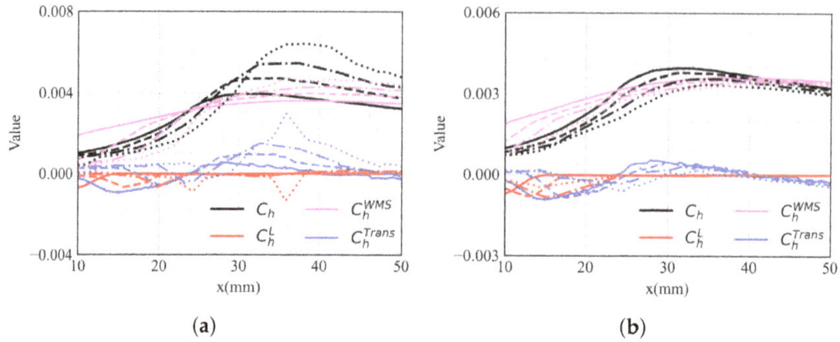

Figure 17. A comparison of decomposition in energy transport forms of separation states in peak heat flux region. (**a**) The influence of ramp angle (solid lines: $\phi = 18°$; dashed lines: $\phi = 20°$; dash-dot lines: $\phi = 22°$; dotted lines: $\phi = 24°$); (**b**) the influence of T_w (solid lines: $T_w = 1.5$; dashed lines: $T_w = 1.75$; dash-dot lines: $T_w = 2.0$; dotted lines: $T_w = 2.25$).

Compared with the results in Section 5, viscous dissipation/work dominates the generation of peak wall friction and peak heat flux. Therefore, one control method to simultaneously decrease peak wall friction and peak heat flux may be altering the matching of the velocity profile and stress distribution there to reduce the work by viscous stresses.

7. Conclusions

This study presents a significant advancement in the analysis of wall friction and heat flux in complex configurations, overcoming the resistance of existing decomposition methods that are suitable only for quasi-two-dimensional configurations. This decomposition method, which incorporates local time translation symmetry and spatial rotational symmetry, allows for a more accurate and comprehensive understanding of the generation mechanisms of wall friction and heat flux in complex flow configurations. This approach incorporates a specific Galilean transformation parameter that considers the variation in the wall slope. Based on the proposed decomposition method, the generation mechanism of the wall friction and heat flux of bistable states of curved compression ramp flows is investigated.

The generation of wall friction is dominated by viscous dissipation and the work by the adverse pressure gradient (APG). Moreover, these two contribution terms are correlated. The separation process is depicted from the perspective of energy conversion. In the absolute reference frame where the flow is still, the work by the APG will input energy into the boundary layer, a large proportion of which is converted into kinetic energy, while the remaining portion should be dissipated by viscous stresses; otherwise, the flow will be separated. The generation of peak wall friction is dominated by viscous dissipation.

The peak heat flux is primarily induced by the work of viscous stress. In the separation–reattachment process caused by shock-wave/boundary-layer interactions, although the energy brought by shear-layer impinging is high, most of the energy is transported downstream. However, the difference in peak heat flux between the separation and attachment states is mainly caused by energy transport. In addition, the impact of energy transport on the generation of peak heat flux increases with the increase in the ramp angle, i.e., a more intense APG, while viscous dissipation varies much less during this process.

The inherent symmetry within this analytical approach facilitates its straightforward extension to the evaluation of any three-dimensional engineering configuration, with a more accurate and reasonable interpretation of wall friction and heat flux generation. A key innovation of this work is the potential for the simultaneous control of wall friction and heat flux, originating from the unified framework that links these phenomena through energy conversion within the method, in contrast to existing methods, which often focus on isolated aspects of the flow, such as skin friction or heat flux independently. Therefore, a further research endeavor involves conducting flow control studies aimed at reducing wall friction and heat flux simultaneously. Given that both peak wall friction and heat flux are primarily driven by viscous dissipation, diminishing this dissipative effect emerges as a promising strategy for achieving concurrent reductions in these parameters. However, it is crucial to recognize that a reduction in viscous dissipation could potentially compromise the boundary layer's capacity to resist the APG. Therefore, this method may serve as a valuable tool in exploring more efficient control methodologies that balance the reduction in wall friction and heat flux with the maintenance of adequate boundary-layer resilience against the APG.

Author Contributions: Conceptualization, M.T. and Y.H.; methodology, M.T., W.Z. and Y.H.; software, M.T. and Y.H.; validation, M.T., W.Z. and Y.H.; formal analysis, M.T. and W.Z.; investigation, M.T. and W.Z.; resources, G.W. and Y.Y.; data curation, M.T. and W.Z.; writing—original draft preparation, M.T.; writing—review and editing, Y.H., G.W. and Y.Y.; visualization, M.T., W.Z. and Y.H.; supervision, Y.H., G.W. and Y.Y.; project administration, G.W. and Y.Y.; funding acquisition, Y.Y. All authors have read and agreed to the published version of the manuscript.

Funding: This work was supported by the National Key R & D Program of China (Grant No. 2019YFA0405300) and the National Natural Science Foundation of China (Grant Nos. 12102449 and 12302309).

Data Availability Statement: The data that support the findings of this study are available from the corresponding author upon reasonable request.

Acknowledgments: We are grateful to Xi Chen of Beihang University and Dong Sun for their helpful discussion.

Conflicts of Interest: The authors declare no conflicts of interest.

Abbreviations

The following abbreviations are used in this manuscript:

SBLI	Shock-wave/boundary-layer interaction
CCR	Curved compression ramp
FIK	Fukagata, Iwamoto and Kasagi
RD	Renard and Deck
DNS	Direct numerical simulation
BSI	Bistable state interval
APG	Adverse pressure gradient

Appendix A. Further Validation of the Numerical Simulation

The selected compression ramp configuration with $\phi = 15^\circ$ is consistent with the Aachen shock tube TH2 experiment [55]. The flat plate and the inclined plate both exhibit a length dimension of $L = 100$ mm. The Reynolds number $Re_\infty(m^{-1}) = 4.2 \times 10^6$. Ma_∞ and Pr are 7.7 and 0.7, respectively. The wall boundary condition is no-slip and isothermal with a wall temperature ratio $T_w = 2.344$. Uniform inflow at a 0° attack angle is specified as the inlet condition. The computational mesh of the case is 1021×251.

The Mach number distributions in this case are shown in Figure A1. The separation angle θ_s and separation shock angle β_s are all quite consistent with the results predicted by the MVD theorem [46].

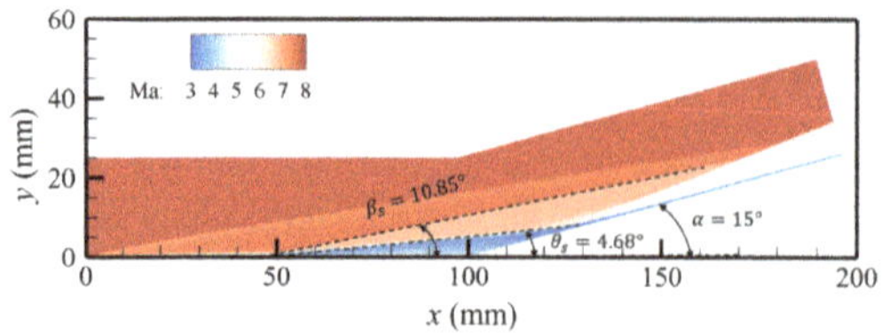

Figure A1. Mach number distributions with labeled separation angle θ_s and separation shock angle β_s predicted by MVD theorem.

Figure A2 compares the distributions of the pressure and Stanton number with experimental [55] and numerical results [56]. The distributions are in good agreement, including the separation length, the pressure plateau and pressure peak, as well as heat flux peak.

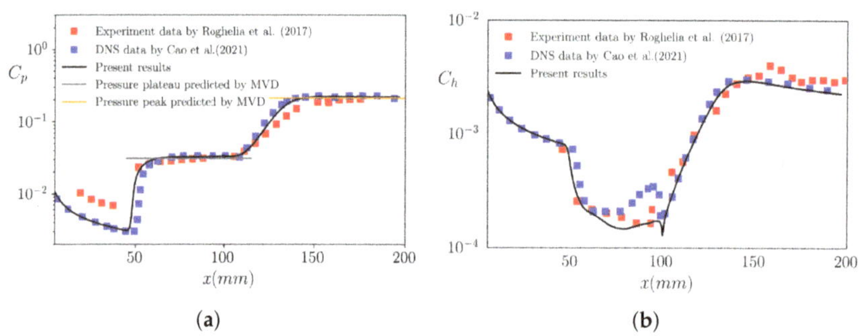

Figure A2. A comparison of the numerical simulations with published experimental [55] and numerical results [56]. (**a**) Distributions of the pressure coefficient; (**b**) distributions of the Stanton number.

References

1. Simeonides, G.; Haase, W.; Manna, M. Experimental, Analytical, and Computational Methods Applied to Hypersonic Compression Ramp Flows. *AIAA J.* **1994**, *32*, 301–310. [CrossRef]
2. Simeonides, G.; Haase, W. Experimental and Computational Investigations of Hypersonic Flow about Compression Ramps. *J. Fluid Mech.* **1995**, *283*, 17–42. [CrossRef]
3. Babinsky, H.; Harvey, J. (Eds.) *Shock Wave-Boundary-Layer Interactions*; Cambridge Aerospace Series; Cambridge University Press: Cambridge, UK; New York, NY, USA, 2011.
4. Zhang, K. *Hypersonic Curved Compression Inlet and Its Inverse Design*; Advanced Topics in Science and Technology in China; Springer: Singapore, 2020; Volume 56. [CrossRef]
5. Silvester, T.B.; Morgan, R.G. Skin-Friction Measurements and Flow Establishment within a Long Duct at Superorbital Speeds. *AIAA J.* **2008**, *46*, 527–536. [CrossRef]
6. Dolling, D.S. Fifty Years of Shock-Wave/Boundary-Layer Interaction Research: What Next? *AIAA J.* **2001**, *39*, 1517–1531. [CrossRef]
7. Fukagata, K.; Iwamoto, K.; Kasagi, N. Contribution of Reynolds Stress Distribution to the Skin Friction in Wall-Bounded Flows. *Phys. Fluids* **2002**, *14*, 5. [CrossRef]
8. Gomez, T.; Flutet, V.; Sagaut, P. Contribution of Reynolds Stress Distribution to the Skin Friction in Compressible Turbulent Channel Flows. *Phys. Rev. E* **2009**, *79*, 035301. [CrossRef] [PubMed]
9. Renard, N.; Deck, S. A Theoretical Decomposition of Mean Skin Friction Generation into Physical Phenomena across the Boundary Layer. *J. Fluid Mech.* **2016**, *790*, 339–367. [CrossRef]
10. Li, W.; Fan, Y.; Modesti, D.; Cheng, C. Decomposition of the Mean Skin-Friction Drag in Compressible Turbulent Channel Flows. *J. Fluid Mech.* **2019**, *875*, 101–123. [CrossRef]
11. de Giovanetti, M.; Hwang, Y.; Choi, H. Skin-friction generation by attached eddies in turbulent channel flow. *J. Fluid Mech.* **2016**, *808*, 511–538. [CrossRef]
12. Shi, Y.; Xia, Z.; Chen, S. A new identification method in sampled quadrant analysis for wall-bounded turbulence. *Phys. Fluids* **2016**, *28*, 061702. [CrossRef]
13. Chan, C.; Örlü, R.; Schlatter, P.; Chin, R. Large-scale and small-scale contribution to the skin friction reduction in a modified turbulent boundary layer by a large-eddy break-up device. *Phys. Rev. Fluids* **2022**, *7*, 034601. [CrossRef]

14. Gatti, D.; Cimarelli, A.; Hasegawa, Y.; Frohnapfel, B.; Quadrio, M. Global Energy Fluxes in Turbulent Channels with Flow Control. *J. Fluid Mech.* **2018**, *857*, 345–373. [CrossRef]
15. Ji, Y.; Yao, J.; Hussain, F.; Chen, X. Vorticity Transports in Turbulent Channels under Large-Scale Control via Spanwise Wall Jet Forcing. *Phys. Fluids* **2021**, *33*, 095112. [CrossRef]
16. Chen, X.; Yao, J.; Hussain, F. Theoretical framework for energy flux analysis of channels under drag control. *Phys. Rev. Fluids* **2021**, *6*, 013902. [CrossRef]
17. Holden, M. Shock Wave-Turbulent Boundary Layer Interaction in Hypersonic Flow. In Proceedings of the 10th Aerospace Sciences Meeting, San Diego, CA, USA, 17–19 January 1972. [CrossRef]
18. Holden, M. A Study of Flow Separation in Regions of Shock Wave-Boundary Layer Interaction in Hypersonic Flow. In Proceedings of the 11th Fluid and Plasma Dynamics Conference, Seattle, WA, USA, 10–12 July 1978. [CrossRef]
19. Hung, F. Interference Heating Due to Shock Wave Impingement on Laminar Boundary Layers. In Proceedings of the 6th Fluid and PlasmaDynamics Conference, Palm Springs, CA, USA, 16–18 July 1973. [CrossRef]
20. Chang, E.W.K.; Chan, W.Y.; McIntyre, T.J.; Veeraragavan, A. Hypersonic Shock Impingement Studies on a Flat Plate: Flow Separation of Laminar Boundary Layers. *J. Fluid Mech.* **2022**, *951*, A19. [CrossRef]
21. Zhang, P.; Xia, Z. Contribution of Viscous Stress Work to Wall Heat Flux in Compressible Turbulent Channel Flows. *Phys. Rev. E* **2020**, *102*, 043107. [CrossRef]
22. Zhang, P.; Song, Y.; Xia, Z. Exact Mathematical Formulas for Wall-Heat Flux in Compressible Turbulent Channel Flows. *Acta Mech. Sin.* **2022**, *38*, 321403. [CrossRef]
23. Sun, D.; Guo, Q.; Yuan, X.; Zhang, H.; Li, C.; Liu, P. A Decomposition Formula for the Wall Heat Flux of a Compressible Boundary Layer. *Adv. Aerodyn.* **2021**, *3*, 33. [CrossRef]
24. Li, J.; Yu, M.; Sun, D.; Liu, P.; Yuan, X. Wall Heat Transfer in High-Enthalpy Hypersonic Turbulent Boundary Layers. *Phys. Fluids* **2022**, *34*, 085102. [CrossRef]
25. Wenzel, C.; Gibis, T.; Kloker, M. About the Influences of Compressibility, Heat Transfer and Pressure Gradients in Compressible Turbulent Boundary Layers. *J. Fluid Mech.* **2022**, *930*, A1. [CrossRef]
26. Agostini, L.; Leschziner, M. The Connection between the Spectrum of Turbulent Scales and the Skin-Friction Statistics in Channel Flow At. *J. Fluid Mech.* **2019**, *871*, 22–51. [CrossRef]
27. Sagaut, P.; Peet, Y. Theoretical Prediction of Turbulent Skin Friction on Geometrically Complex Surfaces. In *Progress in Wall Turbulence: Understanding and Modeling*; Stanislas, M., Jimenez, J., Marusic, I., Eds.; Springer: Dordrecht, The Netherlands, 2011; Volume 14, pp. 39–49. [CrossRef]
28. Bannier, A.; Garnier, É.; Sagaut, P. Riblet Flow Model Based on an Extended FIK Identity. *Flow Turbul. Combust.* **2015**, *95*, 351–376. [CrossRef]
29. Hu, Y.C.; Zhou, W.F.; Wang, G.; Yang, Y.G.; Tang, Z.G. Bistable States and Separation Hysteresis in Curved Compression Ramp Flows. *Phys. Fluids* **2020**, *32*, 113601. [CrossRef]
30. Hornung, H.G.; Oertel, H.; Sandeman, R.J. Transition to Mach reflexion of shock waves in steady and pseudosteady flow with and without relaxation. *J. Fluid Mech.* **1979**, *90*, 541–560. [CrossRef]
31. Chpoun, A.; Ben-Dor, G. Numerical confirmation of the hysteresis phenomenon in the regular to the Mach reflection transition in steady flows. *Shock Waves* **1995**, *5*, 199–203. [CrossRef]
32. Chpoun, A.; Passerel, D.; Li, H.; Ben-Dor, G. Reconsideration of Oblique Shock Wave Reflections in Steady Flows. Part 1. Experimental Investigation. *J. Fluid Mech.* **1995**, *301*, 19–35. [CrossRef]
33. Vuillon, J.; Zeitoun, D.; Ben-Dor, G. Reconsideration of Oblique Shock Wave Reflections in Steady Flows. Part 2. Numerical Investigation. *J. Fluid Mech.* **1995**, *301*, 37–50. [CrossRef]
34. Ivanov, M.S.; Ben-Dor, G.; Elperin, T. Flow-Mach-Number-Variation- Induced Hysteresis in Steady Shock Wave Re Ections. *AIAA J.* **2001**, *39*, 972–974. [CrossRef]
35. Hu, Y.C.; Zhou, W.F.; Tang, Z.G.; Yang, Y.G.; Qin, Z.H. Mechanism of Hysteresis in Shock Wave Reflection. *Phys. Rev. E* **2021**, *103*, 023103. [CrossRef]
36. Hu, Y.C.; Wang, G.; Zhou, W.F.; Tang, M.Z.; Yang, Y.G.; Tang, Z.G. The Bistability of Curved Compression Ramp Flows. *arXiv*, **2023**, arXiv:2304.03429.
37. Yang, Z.; Igarashi, H.; Martin, M.; Hu, H. An Experimental Investigation on Aerodynamic Hysteresis of a Low-Reynolds Number Airfoil. In Proceedings of the 46th AIAA Aerospace Sciences Meeting and Exhibit, Reno, Nevada, 7–10 January 2008. [CrossRef]
38. McCroskey, W.J. Unsteady airfoils. *Annu. Rev. Fluid Mech.* **1982**, *14*, 285–311. [CrossRef]
39. Mueller, T.J. The influence of laminar separation and transition on low Reynolds number airfoil hysteresis. *J. Aircr.* **1985**, *22*, 763–770.
40. Biber, K.; Zumwalt, G.W. Hysteresis Effects on Wind Tunnel Measurements of a Two-Element Airfoil. *AIAA J.* **1993**, *31*, 326–330.
41. Mittal, S.; Saxena, P. Prediction of Hysteresis Associated with the Static Stall of an Airfoil. *AIAA J.* **2000**, *38*, 933–935. [CrossRef]
42. Jin, Y.; Sun, S.; Tan, H.; Zhang, Y.; Huang, H. Flow Response Hysteresis of Throat Regulation Process of a Two-Dimensional Mixed-Compression Supersonic Inlet. *Chin. J. Aeronaut.* **2022**, *35*, 112–127. [CrossRef]
43. Zhou, W.F.; Hu, Y.C.; Tang, M.Z.; Wang, G.; Fang, M.; Yang, Y.G. Mechanism of Separation Hysteresis in Curved Compression Ramp. *Phys. Fluids* **2021**, *33*, 106108. [CrossRef]

44. Tang, M.Z.; Wang, G.; Xie, Z.X.; Zhou, W.F.; Hu, Y.C.; Yang, Y.G. Aerothermodynamic Characteristics of Hypersonic Curved Compression Ramp Flows with Bistable States. *Phys. Fluids* **2021**, *33*, 126106. [CrossRef]
45. Li, X.; Fu, D.; Ma, Y.; Liang, X. Direct numerical simulation of shock/turbulent boundary layer interaction in a supersonic compression ramp. *Sci. China Phys. Mech. Astron.* **2010**, *53*, 1651–1658.
46. Hu, Y.C.; Zhou, W.F.; Yang, Y.G.; Tang, Z.G. Prediction of plateau and peak of pressure in a compression ramp flow with large separation. *Phys. Fluids* **2020**, *32*, 101702.
47. Zhang, Z.; Tong, F.; Duan, J.; Li, X. Direct numerical simulation of supersonic turbulent expansion corner with shock impingement. *Phys. Fluids* **2021**, *33*, 105104.
48. Versteeg, H.K. *An Introduction to Computational Fluid Dynamics the Finite Volume Method, 2/E*; Pearson Education: London, UK, 2007.
49. Schlichting, H.; Gersten, K. *Boundary-Layer Theory*; Springer: Berlin/Heidelberg, Germany, 2017.
50. Xu, D.; Wang, J.; Chen, S. Skin-Friction and Heat-Transfer Decompositions in Hypersonic Transitional and Turbulent Boundary Layers. *J. Fluid Mech.* **2022**, *941*, A4. [CrossRef]
51. Noether, E. Invariante variationsprobleme. In *Gesammelte Abhandlungen-Collected Papers*; Springer: Berlin/Heidelberg, Germany, 1983; pp. 231–239.
52. Hanc, J.; Tuleja, S.; Hancova, M. Symmetries and conservation laws: Consequences of Noether's theorem. *Am. J. Phys.* **2004**, *72*, 428–435.
53. Halder, A.K.; Paliathanasis, A.; Leach, P.G. Noether's theorem and symmetry. *Symmetry* **2018**, *10*, 744. [CrossRef]
54. Duan, L.; Martin, M. Direct numerical simulation of hypersonic turbulent boundary layers. Part 4. Effect of high enthalpy. *J. Fluid Mech.* **2011**, *684*, 25–59.
55. Roghelia, A.; Olivier, H.; Egorov, I.; Chuvakhov, P. Experimental investigation of Görtler vortices in hypersonic ramp flows. *Exp. Fluids* **2017**, *58*, 1–15.
56. Cao, S.; Hao, J.; Klioutchnikov, I.; Olivier, H.; Wen, C.Y. Unsteady effects in a hypersonic compression ramp flow with laminar separation. *J. Fluid Mech.* **2021**, *912*, A3.

symmetry

MDPI

Article

The Influence of Low-Frequency Oscillations on Trailing-Edge Tonal Noise with Symmetry Spanwise Source Regions

Zhangchen Song [1], Peiqing Liu [1], Hao Guo [1,*], Yifeng Sun [2] and Shujie Jiang [3]

1 School of Aeronautic Science and Engineering, Beihang University, Beijing 100191, China; songdbs_work@buaa.edu.cn (Z.S.); lpq@buaa.edu.cn (P.L.)
2 COMAC Shanghai Aircraft Design and Research Institute, Shanghai 201315, China; sunyifeng@comac.cc
3 Laboratory of Aerodynamic Noise Control, China Aerodynamics Research and Development Center, Mianyang 621000, China; anjsj@126.com
* Correspondence: guohao@buaa.edu.cn; Tel.: +86-135-1106-0706

Abstract: For noise reduction at a low-to-moderate Reynolds number, airfoil trailing-edge tonal noise has multiple prominent tones. Among these tones, secondary tones are greatly influenced by external disturbances such as oscillations commonly in the environment. In previous experiments, the spatial movement of sources was found to be related to an inherent high-frequency oscillation. Therefore, the spatial influence of external low-frequency oscillations was investigated in this study. By using tripping tapes to construct different symmetry source regions on the pressure side with side secondary tones, a transient spatial analysis of an NACA0012 airfoil at 2 degrees was performed by microphone arrays when a 10 Hz pressure oscillation was significant at 24 m/s. Temporally, this 10 Hz periodic strength change became more intense at a broader frequency bandwidth for a longer source region. Furthermore, a substantial time delay, significantly larger than the sound propagating time difference between microphones, was observed exclusively along the spanwise direction. This delay led to a periodic directivity pattern, particularly when two 0.2 m source regions were separated by a 0.2 m or 0.4 m tripping region. This low-frequency oscillation introduces an asymmetric transient switching pattern for symmetric spanwise source regions. Consequently, the response of airfoils to external oscillations in field tests should be considered.

Keywords: aeroacoustics; microphone array; transient analysis; low-frequency oscillation

1. Introduction

Aerodynamic noise has been an unignorable factor for engineering applications such as unmanned aerial vehicles and small wind turbines at a low-to-moderate Reynolds number [1–3]. Among these noises, airfoil noise is of great importance and is directly related to aerodynamic performance, which is critical in quiet airfoil designs [4–6]. Among these airfoil self-noises, airfoil trailing-edge tonal noise is so prominent that scientists all over the world have spent decades conducting experiments and simulations to understand the noise and noise reduction mechanism [7–29]. Generally, at a moderate Reynolds number between $Re_c = 2 \times 10^4$ and $Re_c = 2 \times 10^6$, this noise has one or several narrowband tones and some broadband humps in the acoustic spectra [20]. Due to the influence of the acoustic feedback loop (AFL) and the flow instability growth in noise amplification, the frequency of the multiple tones has a piecewise ladder structure with jumps at certain velocity ranges [16]. The dominant tone is related to the steady periodic symmetric shedding vortical structures along the spanwise direction, while the secondary tones are attributed to the intermittent unsymmetric changes in the frequency, time and spatial domains [23]. As a result, interesting transient patterns related to secondary tones have been found, such as switching tones [19,24], side secondary tones [21] and the periodic spanwise movement of sources [22]. Meanwhile, the difference between experiments and simulations [19] has attracted researchers to analyze the occurrence of transient patterns in spatial distributions.

Hence, it is of great significance to investigate trailing-edge tonal noise in the frequency, time and spatial domains.

Basically, according to Longhouse [10], trailing-edge tonal noise began as an initial disturbance on the transitional region, which grew in the boundary layer towards the trailing edge, scattering as shedding vortical structures around the edge to release acoustic waves backward, causing new initial disturbances that propagate forward and maintain a phase-match state. In this process, AFL dominated the selection of amplificated tones on the broadband hump from growing instabilities along the airfoil, which produced corresponding flow coherent structures. In experiments based on the transient correlation analysis between velocity signals from particle image velocity (PIV) data and acoustic signals from a single microphone, Pröbsting [15] confirmed that the vortical structures shed at the same frequency as the dominant tone frequency. And the growing instabilities were the result of Tollmien–Schlichting (T-S) instability along the boundary layer [15] and Kelvin–Helmholtz (K-H) instability related to the laminar separation bubbles [17]. In simulations, Ricciardi [23] found that although the AFL confirmed that only a dominant tone existed during the initial cycles, secondary tones would gradually emerge due to the unsteadiness of growing instabilities after several cycles. In addition, scientists pointed out that the strength of noise sources in the time domain was receptive to the pressure oscillations [13] and dynamics of flow coherent structures [23,25]. Furthermore, experiments confirmed that trailing-edge tonal noise was sensitive to external disturbances from a surface-mounted plasma actuator [30], acoustic excitation [31] and serrations [32]. Therefore, as an inevitable disturbance in the open test section from the wind tunnel buffeting [33], the low-frequency pressure oscillation should be considered in the transient source localization of trailing-edge noise measurement.

In the literature, this low-frequency pressure oscillation generally exists in the open test section of low-speed wind tunnels [33,34]. It usually has a frequency lower than 20 Hz accompanied by velocity fluctuations [35,36]. Called wind tunnel buffeting, pumping or swaying in the literature [37–40], this phenomenon is generally attributed to a feedback loop between the shear layers around the inlet nozzle and flows around the collector through the coherent vortex structures [34]. Consequently, since trailing-edge tonal noise is receptive to the flow of coherent structures [41], it is possible for the low-frequency pressure fluctuation in the flow to influence the trailing-edge tonal noise in the frequency, time and spatial domains.

According to the amplitude modulation theory [42], a periodic strength change can produce side tones around the dominant tone in equal frequency intervals. In experiments, Pröbsting [15] confirmed that this periodic modulation of the fluctuation amplitude in the vortical structures on one side of the airfoil could produce secondary tones. In simulations, Ricciardi [23] concluded that the different periodic changes in spanwise coherence led to different secondary tones. Specifically, for airfoil trailing-edge tonal noise, when the coherent flow structure passes the airfoil edge, a spatial–temporal unsteady pattern influenced the amplitude of the dominant tone and produced secondary tones. Hence, based on the results of experiments and simulations by the transient wavelet analysis between acoustic fields and velocity fields, the emergence of secondary tones is a result of the periodic and intermittent coherent flow structures [15,17,23,24]. Correspondingly, according to Yang's PIV measurement experiments with proper orthogonal decomposition (POD) [21], low-frequency oscillations beneath the incoming flow produced the side secondary tones at a frequency gap lower than 20 Hz. In addition, similar patterns could be found due to forced disturbances in the literature [30–32]. Therefore, being influenced by the low-frequency pressure oscillation, the source region should have a spatial transient source distribution pattern, which awaits further experiments.

In source localization problems, the microphone array technique has been widely used in experiments [43–46]. It utilizes the difference of time in propagating between microphones to localize sources. In recent years, thanks to the development of wavelet-based beamforming methods [47–49], it is possible for airfoil experiments to investigate transient source localization. It was found by Yu [22] that the major noise sources moved along the spanwise direction at a frequency according to the amplitude modulation frequency. Hence, if the low-frequency oscillations changed the dominant tone's strength in a similar way, there could be an obvious transient spatial movement pattern due to the long period time of the low frequency. Since the deformations of airfoils and unsteady flow oscillations might exist at a low frequency for UAVs and wind turbines in the environment, this response to the external oscillations from the environment should be considered in airfoil design if the results show significant patterns.

In this study, aimed at investigating the influence of low-frequency oscillation on trailing-edge tonal noise in the frequency, time and spatial domains, two microphone arrays were introduced to localize sources on different specially designed spanwise source regions of an NACA0012 airfoil model. This airfoil had a chord length of 0.1 m and a span of 1.6 m on the open test section of the Beihang D5 aeroacoustics wind tunnel [50]. The reason to choose an NACA0012 profile is due to the existing large number of trailing-edge tonal noise investigations based on NACA0012 with different chord lengths and spans [4,7,11,12,15,19,22]. NACA0012 shows a steady time-average pattern within a considerable velocity range, which makes it credible to highlight the influence of the flow oscillations. For the currently open test section of a D5 aeroacoustics wind tunnel, a 10 Hz strong oscillation due to wind tunnel buffeting could be found at 24 m/s according to the background noise measurement, which was able to produce a strong periodic strength change in the time domain. To further highlight the role of this low-frequency pressure oscillation on the noise signals, the tripping tapes on the airfoil suction side diminished the phase modulation between flow structures from two sides of the airfoil [13], and cases with high-frequency intervals for primary secondary tones were excluded. Therefore, different configurations for symmetry spanwise distributions of source regions could be created on the pressure side, where only one dominant tone with only side secondary tones due to the 10 Hz oscillation was present.

Inspired by the previous experiments in localizing transient sources [22], microphone arrays were employed with wavelet-based beamforming methods. Based on the 64-channel spiraled microphone array, source strength distribution maps in a high spatial resolution were generated to validate the time-average consistent pattern along the spanwise direction. Based on the 21-channel cross-shaped microphone array, simultaneous acoustic measurement along the spanwise direction and horizontal direction could be achieved to analyze the different transient patterns along the spanwise and horizontal directions directly. By using continuous wavelet analysis and a corresponding wavelet-based beamforming method, periodic strength changes in the frequency, time and spatial domains were detected. According to the transient source strength distribution maps, the symmetry source region along the spanwise direction produced a switching pattern between consecutive line sources and separated sources in the time domain. During this periodic switching process, unsymmetric strength distributions were found along the spanwise direction for each microphone as a periodic directivity pattern. It was found that a significant large time delay between microphones only along the spanwise direction was responsible for the periodic directivity pattern. Hence, the response to the low-frequency oscillations in the flow field became a significant characteristic. For future airfoil designs involved with trailing-edge tonal noise, the noise measurement experiments should consider the airfoil's response to environmental disturbances.

The structure of this paper is as follows: Section 2 will briefly introduce the flow facility, experiment setup and analysis tools. The configurations with different spanwise source regions will be illustrated in diagrams, and beamforming methods will be discussed. Section 3 is concerned with the results of the time-average, transient and spatial–temporal analyses. The time-average characteristics, such as the ladder structure of tonal frequencies and spatial distribution of sound sources, will be illustrated in plots and contour maps. The strong 10 Hz oscillation at 24 m/s was first proposed in the background noise measurement. Then came the time-average analysis for the trailing-edge noise validation in the frequency and spatial domains. Transient analysis by continuous wavelet transform was performed to evaluate the difference between different spanwise distributions of source regions. For cases with a 0.2 m tripping region inside the symmetry 0.6 m source region, it was found that the 10 Hz oscillation had a significant time delay between different microphones only along the spanwise directions, which contributed to a switching pattern between consecutive strong line sources and weak separated sources. In contrast, the microphones along the horizontal direction showed consistent results in the time delay. As a result, by decreasing the horizontal microphones' weight in the calculation, nearly equal strength of sources for two types of source strength distribution could be achieved at the cost of a low horizontal resolution.

2. Materials and Methods

2.1. Experiment Setup

The experiment for the trailing-edge tonal noise investigation was conducted at the Beihang D5 aeroacoustics wind tunnel [50] at Beihang University. The D5 wind tunnel is a small-scale, closed-circuit wind tunnel with a cross-section of 1 m × 1 m (height × width). The length of the open test section is 2 m from the nozzle to the collector.

The airfoil is an aluminum 0.1 m chord length NACA0012 with a sharp trailing edge (around 0.2 mm at height). It has a span of 1.6 m, which is long enough to cover the shear layer length and leave enough spanwise source region within the uniform flow from the nozzle. The airfoil is placed vertically with a sweep angle of less than 0.5 degrees according to a laser level. The suction side of the airfoil is fully tripped by coarse tapes (the tape has a height of 0.5 mm). In this study, the angle of attack is fixed at 2 degrees and the incoming flow velocity ranges from 10 m/s to 60 m/s.

For noise measurement, a 64-channel spiraled microphone array and a 21-channel cross-shaped microphone array were placed at a distance of 1.5 m toward the pressure side of the airfoil in two separate experiments with the same model setup. As shown in Figure 1, general views in experiments are presented for these two microphone arrays with the airfoil. Specifically, a geometry placement between the airfoil and arrays is shown in Figure 2. Both the centers of these arrays were placed towards the symmetry line of the trailing edge. For the cross-shaped microphone array, 11 microphones are in the same horizontal line, and 9 microphones are in the same spanwise line vertically.

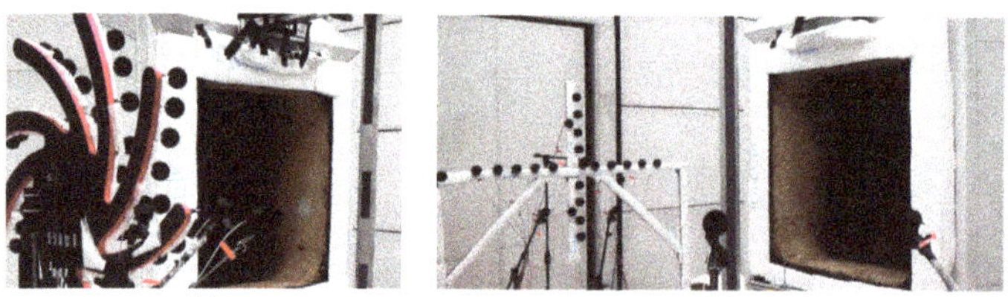

(a) (b)

Figure 1. The experiment setups of (**a**) the 64-channel spiraled microphone array in a view from pressure side of the airfoil, and (**b**) the 21-channel cross-shaped microphone array in a view from the suction side of the airfoil. The suction side surface of the airfoil is fully tripped.

GRAS 40PH (GRAS Sound & Vibration from Holte, Denmark) and Brüel & Kjær 4954A (Hottinger Brüel & Kjær from Hertfordshire, UK) microphones with a sampling rate of 51,200 Hz for a duration of 30 s were applied for the 64-channel spiraled array and the 21-channel cross-shaped array, respectively. They are both high-accuracy free-field microphones and calibrated under the same GRAS 42AA pistonphone (GRAS Sound & Vibration from Holte, Denmark). An average periodogram method was used to generate estimates of the power spectrum density (PSD) and sound pressure level (SPL). For the general time-average analysis, the number of samples per window was 12,800, resulting in a frequency resolution of 4 Hz. For the low-frequency oscillation analysis and auto-spectrum of wavelet power coefficients, each window block consisted of 102,400 samples, which contributed to a frequency resolution of 0.5 Hz. The Hanning window and an overlap of 50% were applied in the analysis.

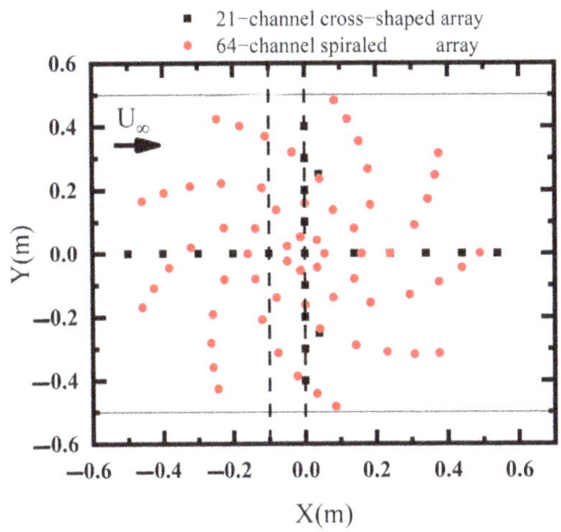

Figure 2. The geometry position of arrays with microphones and airfoils in the X-Y plane. The dashed black lines indicate the leading edge and trailing edge of the airfoil on the left and on the right, respectively. The black squares refer to the BK 4189A microphones for the 21-channel cross-shaped array, while the red circles refer to the GRAS 40PH microphones for the 64-channel spiraled array. The incoming flow direction is highlighted by the arrow.

As previously shown in Figure 1, the suction side was fully tripped, while the pressure side was tripped partially to create different source regions. In this study, the main configurations are shown in Figure 3 with different spanwise conditions on the pressure side. A set of terms, such as *L*0, *L*2 and *L*8*T*4, are defined according to the lengths of the clean region and tripped region. The number after *L* refers to the length of clean regions in a unit of 0.1 m, and the number after *T* stands for the length of the tripping region within the clean region in a unit of 0.1 m. Three dotted square lines are used to further distinguish configurations in different clusters. For configurations within region (a) and region (b), a symmetry placement of clean regions is present. Configurations within region (c) are aimed at evaluating the time-average consistency along the spanwise direction. Hence, an additional microphone was positioned on the pressure side towards the center of the 0.2 m clean region at a distance of 1.2 m as the reference microphone. For other cases, the reference microphone from the microphone arrays was positioned on the pressure side towards the center of the airfoil trailing edge within the uniform flow region at a distance of 1.5 m.

2.2. *Microphone Array Methods*

In comparison with the velocity measurement from invasive hotwires and short-time-period particle image velocity (PIV) in the flow field, the microphone array method is non-invasive and can obtain long-time-period data [43]. By using wavelet-based beamforming methods [22], it is possible to describe the high-correlated flow-induced noise sources in the time and spatial domains, especially when the spanwise source strength distribution is inconsistent due to tripping devices.

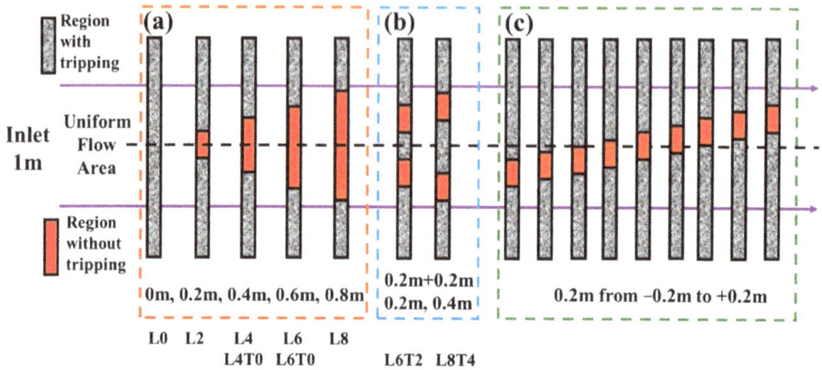

Figure 3. The surface conditions of the pressure side of airfoils for different configurations. The red filled area indicates the source region without tripping. The grey shadow area indicates the region with tripping. (**a**) Fully tripped region or source regions in consistent symmetrical distributions. (**b**) A disruption of tripped region between source regions. (**c**) Different locations of the 0.2 m clean region. The incoming flow has a spanwise length of 1 m. Without *T*, *L* stands for the length of symmetry clean region without tripping. *T* refers to the length of symmetric tripping region inside clean regions from *L*.

For source localizations by microphone array, a brief introduction of the beamforming method and the wavelet beamforming method are illustrated below. A detailed discussion of different microphone array methods can be found in a review [51]. According to the delay-and-sum approach, the estimated spatial strength distributions of sound sources are deduced by the microphones' time signals and the spatial distributions between scanning points and microphones. The sound pressure at a scanning position $\vec{x}_b$ is deduced in Equations (1) and (2) in the time and frequency domain, respectively.

$$bf\left(\vec{x}_b, t\right) = \frac{1}{M}\sum_{m=1}^{M} w_m A_m\left(\vec{x}_b, \vec{x}_m\right) p_m\left(t - \frac{\left|\vec{x}_b - \vec{x}_m\right|}{c}\right) \tag{1}$$

$$bf\left(\vec{x}_b, \omega\right) = \frac{1}{M}\sum_{m=1}^{M} w_m A_m\left(\vec{x}_b, \vec{x}_m\right) p_m(\omega) e^{j\omega\frac{\left|\vec{x}_b - \vec{x}_m\right|}{c}} \tag{2}$$

$$p_m(\omega) = \frac{1}{t_2 - t_1}\int_{t_1}^{t_2} p_m(t) e^{j\omega t} dt \tag{3}$$

In these equations, the following apply:

- t is global time and $\omega(= 2\pi f)$ is the angular frequency for the Fourier transform;
- p_m is the sound pressure signal for the number m microphone. In the time domain, $p_m(t)$ is the transient sound pressure according to the global time. In the frequency domain, $p_m(\omega)$ represents the estimated complex strength and phase in a frequency for a time period ranging from t_1 to t_2. By confining the value of $t_2 - t_1$, a relatively transient estimation of sound source is achieved in the frequency domain;
- $\vec{x}_m$ is the position of the number m microphone;
- w_m is a weight function to determine the influence of the number m microphone;
- $A_m\left(\vec{x}_b, \vec{x}_m\right)$ is the propagating scale factor to determine the decaying of signal strength according to the distance between the microphone and the sound source. For the monopole point source and Green function, $A_m\left(\vec{x}_b, \vec{x}_m\right) = \frac{1}{4\pi\left|\vec{x}_b - \vec{x}_m\right|}$.

By using the short-term Fourier transform and defining the length of the time period, an estimation of source positions in different frequencies is produced in a relatively large time resolution. And by averaging over the time periods, a typical time-average conventional frequency domain beamforming result can be produced. In the current experiments, the maximum time delay between each microphone should be below 0.002 s, which is much less than the 10 Hz time period of 0.1 s.

Analogously, by substituting continuous wavelet transform from the Fourier transform in Equation (3), a high-time-resolution time domain microphone array method can be produced as the wavelet-based beamforming method. In the literature [22,47–49], the wavelet-based beamforming methods showed extraordinary abilities in detecting the movement of sources, identifying the relative strength shift between sources and increasing the signal-to-noise ratio to perform better cross—correlation analysis. In this paper, complex Morlet wavelets are applied to acquire the temporal phase information. The selection of an appropriate non-dimensional frequency and other parameters has been undertaken to better describe the spatial and time characteristics. And a source integration method [52,53] is introduced to calculate summed SPL over different source regions. For the spanwise source regions, the calculated horizontal regions cover a whole chord length (0.1 m) from the center of the airfoil toward the trailing edge.

3. Results

3.1. Time-Average Analysis

As has been mentioned in the introduction, low-frequency pressure oscillations commonly exist in wind tunnel experiments [34]. For the open test section of the Beihang D5 aeroacoustics wind tunnel, strong pressure oscillations can be detected through background noise measurement. As shown in Figure 4, a contour map of PSD for the background noise is produced within a velocity range from 10 m/s to 60 m/s. The constant tone at 4 Hz is related to the inherent natural frequency of the anechoic chamber. The tone changing at different velocities is a result of the unsteady coherent vortices in the shear layer. For example, a dashed line indicates a velocity-dependent frequency law between the wind tunnel's inner frequency and incoming flow velocity as $f = 0.2U + 5.2$ in standard science units. Among these tones, the significantly strong tone is a result of amplification due to resonance between the nozzle and collector, which contributes to the well-known wind tunnel buffeting phenomenon [35]. The dashed line and dotted line in Figure 4 indicate a strong 10 Hz oscillation around 24 m/s, which is focused on in the present study as it is the strong low-frequency pressure oscillation that influences the trailing-edge tonal noise.

In order to validate the airfoil trailing-edge noise and check if the tripping tapes succeed in having only pressure side sources, a series of acoustic measurements has been performed for configuration $L2$ and $L0$ at different incoming flow velocities. The angle of attack is fixed at 2 degrees, so the upper surface is the suction side and the down surface is the pressure side towards the reference microphones. The time-average PSD by the reference microphone from the 64-channel microphone array is presented in contour maps and plots in Figure 5. In Figure 5a, ladder-type structures are indicated by black dashed lines with jumps. The dominant tones are highlighted by red cycles whose frequencies follow a 0.85th power law of velocity on ladders with a jump between different ladders. During a velocity range from 18 m/s to 28 m/s, no primary secondary tones are found in a high frequency interval around the dominant tones. Specifically, for cases ranging from 20 m/s to 28 m/s at an interval of 2 m/s, the dominant tones in Figure 5b are significantly stronger than the fully tripped broadband noise. In general, the time-average noise power spectra have validated the airfoil trailing-edge tonal noise experiments with the suction side being fully tripped among all the other experiments in the literature [7,11,16]. In addition, the 10 Hz oscillation has been found in Figure 5c at 24 m/s and 26 m/s for configuration $L2$ and $L0$, which will be focused on in the following transient analysis.

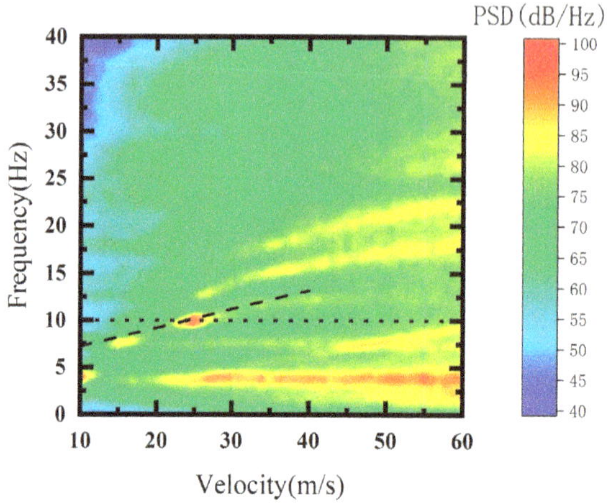

Figure 4. A contour map of acoustic spectra of PSD for the background noise. Without the airfoil within the flow regions, the background noise is measured by the reference microphone at a distance of 1.5 m towards the centerline of the open test section. The dashed line is an approximate line as $f = 0.2U + 5.2$ for low-frequency oscillations with a 10 Hz peak at 24 m/s. The dotted line indicates the 10 Hz oscillation.

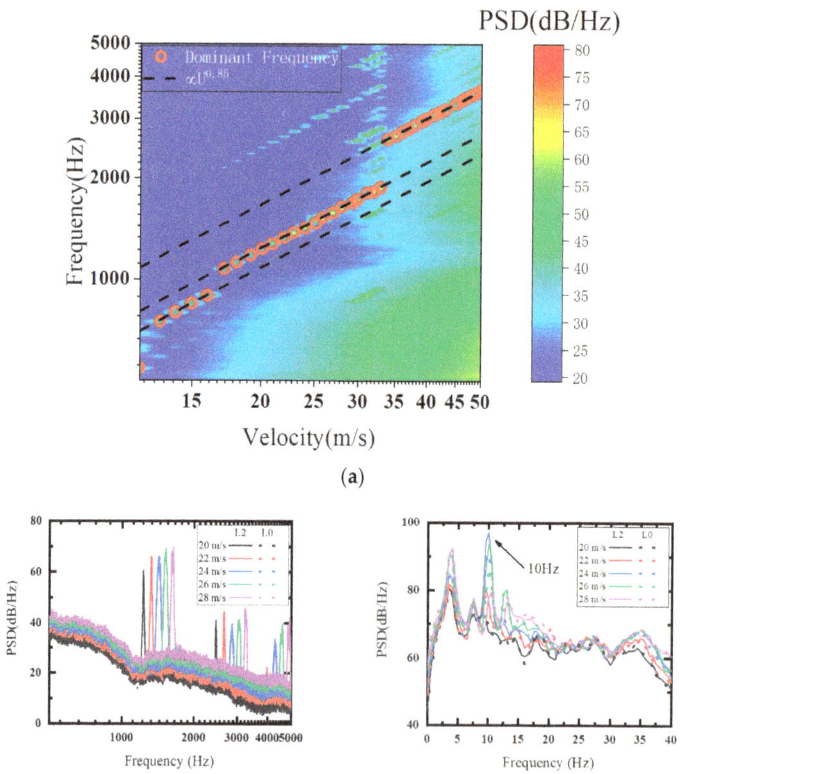

(a)

(b) (c)

Figure 5. The time-average noise characteristics according to the reference microphone from the 64-channel spiraled array. (**a**) A contour map of acoustic spectra of PSD for the configuration *L*2 with incoming flow velocities ranging from 10 m/s to 50 m/s; (**b**) plots of PSD ranging from 20 m/s to 28 m/s for the configuration *L*2 and *L*0. *L*2 is the straight line and *L*0 is the dotted line; (**c**) zoomed plots of (**b**) in a low frequency range. The 10 Hz oscillations at 24 m/s and 26 m/s are highlighted by an arrow.

In order to exclude the inherent influence on different spanwise configurations, an evaluation of the spanwise consistency was performed by moving the 0.2 m source region in a spanwise range from $z = -0.2$ m to $z = 0.2$ m in Figure 3c. A moving microphone towards the center of the 0.2 m source region and the 64−channel spiraled microphone array were used to compare the time-average characteristic in the frequency–spatial domains. As shown in Figure 6, contour maps of PSD for the moving microphone towards different spanwise locations of the source region were present at a velocity range from 20 m/s to 28 m/s. Due to the velocity fluctuations, a difference in dominant tone frequency occurs, but a generally similar pattern was confirmed along the spanwise directions. As shown in Figure 7, time-average source localization of the 0.2 source region at different spanwise locations validated that the tripping tapes confined the sound sources within the 0.2 m clean region and there was a similar spatial pattern along the spanwise direction. SPL was summed at a 1/3 octave around the center frequency of 1424 Hz.

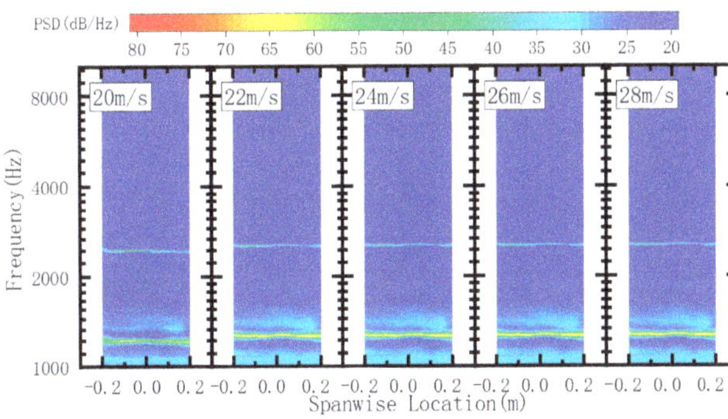

Figure 6. The time-average PSD for the moving 0.2 m clean region along the spanwise location at different velocities according to the moving microphone. The moving microphone is placed towards the center of 0.2 m clean region at a fixed distance of 1.2 m.

Based on the results of source localization and strength measurement for the 0.2 m source region, a symmetric source distribution should be present for a symmetric source region in source localization. As shown in Figure 8, the time-average source localization maps at 24 m/s for different spanwise configurations show a generally symmetric source strength distribution, especially for the configuration *L*6. SPL is summed at a 1/3 octave around the center frequency of 1424 Hz, which is enough to consider the frequency deviation along the spanwise direction. In addition, a spurious source was found in the tripping region for the configuration *L*6*T*2 in comparison with *L*6. According to the spatial resolution of beamforming methods established from the Rayleigh limit [43], the spurious source is attributed to the insufficient distance between two symmetric 0.2 m source regions. Proof is given in Figure 9 as source localization maps, where two 0.2 m source regions are measured independently with a sum of the strength of the two source localization maps. The source strength distribution map in Figure 8e is similar to the one in Figure 9c. Therefore, it could be assumed that the impact of the low-frequency pressure oscillation on the time-average result of the summed SPL over a wide frequency range is insignificant. However, in the following transient analysis with a better frequency-time resolution, the influence of the low-frequency oscillation becomes notable.

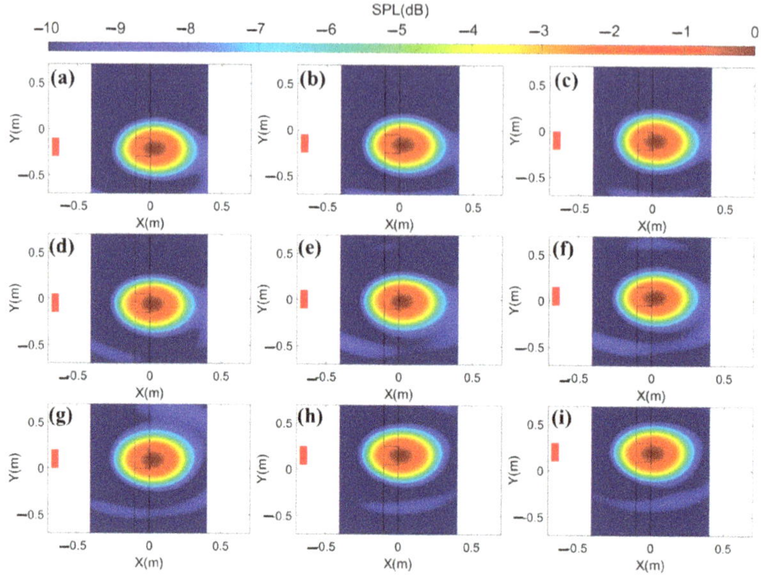

Figure 7. The time-average source localization for the moving 0.2 m clean region along the spanwise location at 24 m/s according to the 64-channel spiraled microphone array. (**a–i**) refer to the spanwise location from $z = -0.2$ m and $z = 0.2$ m in Figure 6. The black X−direction lines refer to the leading edge and trailing edge, while the black Y-direction lines indicate the red source regions on the left.

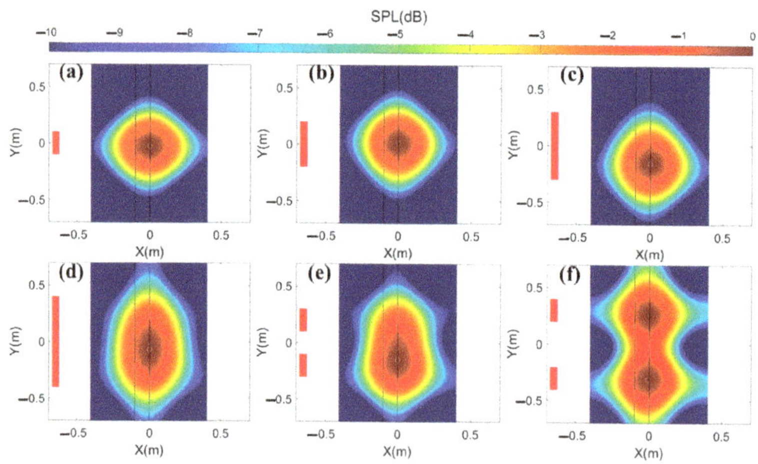

Figure 8. The time-average source localization of different source regions along the spanwise location at 24 m/s according to the 21-channel cross-shaped microphone array. (**a**) *L*2; (**b**) *L*4; (**c**) *L*6; (**d**) *L*8; (**e**) *L*6*T*2; (**f**) *L*8*T*4. The black lines refer to the leading edge and trailing edge. The unfilled white cycles indicate the positions of maximum SPL, which is located around the trailing edge.

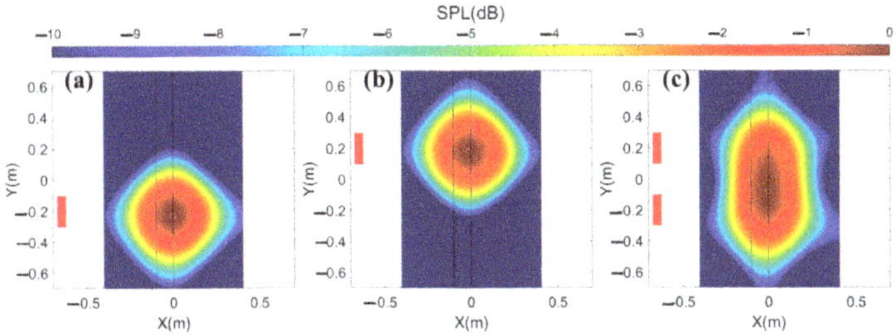

Figure 9. The time-average source localization of different source regions along the spanwise location at 24 m/s according to the 21-channel cross-shaped microphone array. (**a**) The 0.2 m source region in the down part; (**b**) 0.2 m source region in the upper part; (**c**) a sum of (**a**,**b**) to produce $L6T2$. The black lines refer to the leading edge and trailing edge. The unfilled white cycles indicate the positions of maximum SPL, which is located around the trailing edge.

3.2. Transient Analysis

By using continuous wavelet analysis, the transient characteristics of noise signals can be revealed. As a time-frequency-dependent parameter, the wavelet coefficient W stands for the transient efficient amplitude at a certain frequency. W_n^2 is defined as a non-dimensional normalization of the square absolute values of the wavelet coefficient W according to the maximum values in the contour maps. The reference microphone at the symmetrical line towards the airfoil trailing edge is chosen to compare the transient results between different spanwise configurations at 24 m/s.

As shown in Figure 10, a periodic 10 Hz strength change around the dominant tone could be found in the frequency-time domain, especially for configurations $L2$ and $L4$. Therefore, according to amplitude modulation, side secondary tones should be present in the time-average acoustic spectra, which are present in Figure 11. The black square and red cycle symbols indicated a 10 Hz interval, which was similar to the side secondary tones' pattern in the literature [21]. Meanwhile, as shown in Figure 10, a larger spanwise length of the source region (from $L4$ to $L8$) can greatly increase the unsteadiness towards a higher value in the time domain. For $L6T2$ and $L8T4$, a decrease in clean regions by adding tripping tapes inside the total spanwise clean region tends to control the periodic strength change in a much more regular and steady way.

Moreover, by using an auto-spectrum of the total strength around the center frequency (1424 Hz at 24 m/s) in a 1/3 octave way, the 10 Hz oscillations were prominent as peaks for all configurations in Figure 12. From $L2$ to $L8$, an increase in the total spanwise source regions led to a larger sound strength in a more broadband-like way. While $L2$ and $L4$ mainly show the 10 Hz peak and 20 Hz harmonic peak, a tendency of larger frequency oscillation can be observed in $L6$ and $L8$. When tripping regions were added, $L6T2$ and $L8T4$ were degenerations of $L6$ and $L8$ in strength and regularized in the time domain. In addition, the normalization of the auto-spectrum strength as $E(f)/E(0)$ clearly exhibited this similar normalization pattern at 10 Hz. The tripping region within the clean region increased the quality factors of the dominant tone and decreased the lower frequency broadband components.

Similar to previous experiments in the literature [21], by transient analysis, a periodic strength change in the dominant tone was confirmed with the 10 Hz oscillation to produce side secondary tones. Therefore, it is feasible to carry on spatial–temporal analysis for the influence of low-frequency oscillation on trailing-edge tonal noise. Based on the above transient results for the single reference microphone, the procedure to compare different microphones in the horizontal line and spanwise line can be performed, and transient source localization by the wavelet beamforming method could be undertaken.

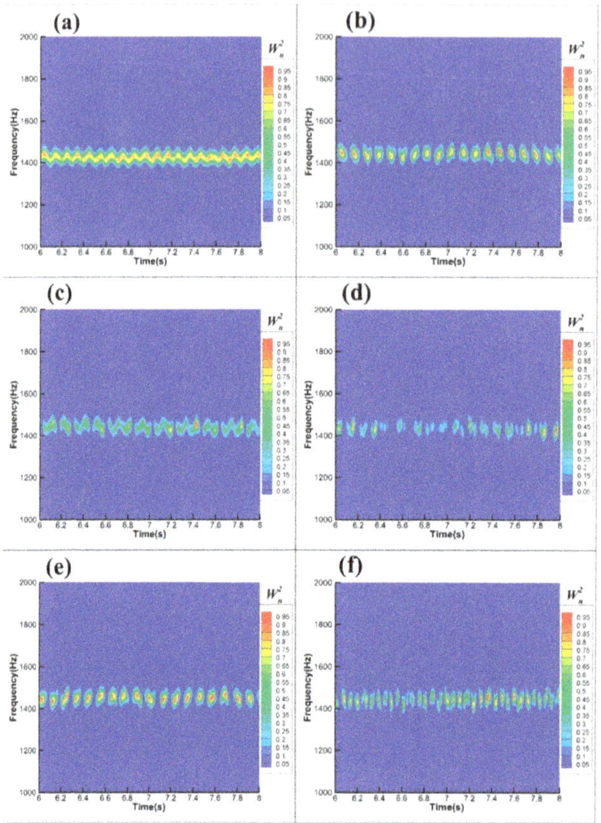

Figure 10. The normalized wavelet coefficients in contour maps for different spanwise configurations. The incoming velocity is 24 m/s. (**a**) *L*2; (**b**) *L*4; (**c**) *L*6; (**d**) *L*8; (**e**) *L*6*T*2; (**f**) *L*8*T*4.

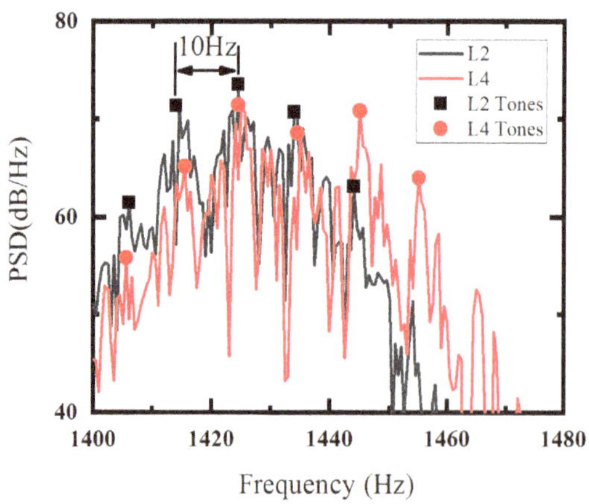

Figure 11. The time-average PSD according to the reference microphone from the 21-channel cross-shaped array for *L*2 and *L*4 at 24 m/s. A sampling time from 6 s to 8 s is applied as in Figure 9. A 10 Hz interval could be found between tones. The square black symbols refer to tones for *L*2, while the red cycle symbols refer to tones for *L*4.

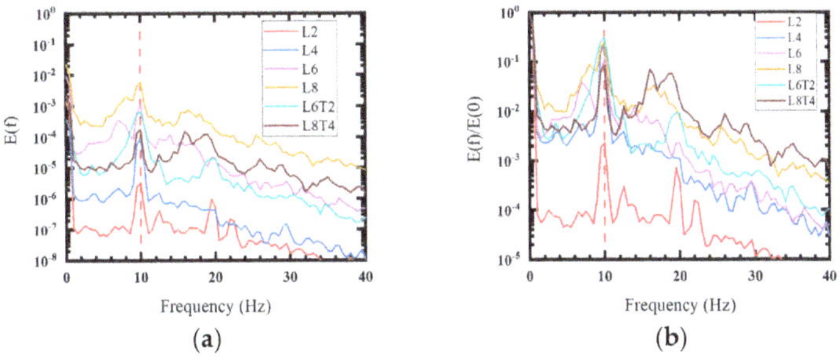

Figure 12. (**a**) Auto-spectrum of the summed wavelet coefficients around the dominant tone frequency in a 1/3 octave for different spanwise configurations at 24 m/s; (**b**) normalization of the auto-spectrum from $E(f)$ to $E(f)/E(0)$. Dashed red lines indicate 10 Hz peaks.

3.3. Spatial–Temporal Analysis

To address the influence of intermittency in the time and spatial domains, the transient results of microphones at different locations are compared. According to the spatial distribution of the 21-channel cross-shaped microphone array in Figure 2, microphones from the 21-channel microphone array can be divided into two parts along the horizontal and spanwise locations, respectively. By using the same wavelet analysis process in Section 3.2, the SPL is summed from the center frequency (1424 Hz at 24 m/s) in a 1/3 octave frequency range. The results of the SPL for microphones along horizontal and spanwise directions are shown in Figures 13 and 14, respectively.

As shown in Figure 13, the microphones at horizontal locations reveal a consistent periodic strength change pattern in the time domains for all configurations from *L*2 to *L*8. Although an increased length generates a more complicated pattern in the time domain, the strengths along the horizontal line remain identical in the time domain. In contrast, in Figure 14, the microphones along the spanwise direction indicate an unsteady, uneven time delay for the low-frequency oscillation. As has been mentioned in Section 2.2 about the microphone array methods, the maximum time delay between microphones in current experiments should not exceed 0.002 s, which is much less than the delay time period in Figure 14 between horizontal microphones for *L*6*T*2 and *L*8*T*4. In comparison, a single 0.2 m clean region (*L*2) shows an almost constant time of arrival for the 10 Hz oscillation, which indicated that a longer length of source region was more sensitive to the strength change in the time domain. Therefore, it can be assumed that these periodic strength changes in microphones along the spanwise direction produce a periodic directivity pattern in the time domain, which occurs in spanwise configurations with long enough source regions. In addition, the tripping regions inside *L*6 and *L*8 produce *L*6*T*2 and *L*8*T*4 with a more uniform source strength distribution and a stronger periodic directivity pattern along the spanwise direction.

As a result, this periodic directivity pattern for the low-frequency oscillation contributes to periodic source localization within a 10 Hz period time. By selecting a range of 0.2 s for the two period times at 10 Hz, the wavelet-based beamforming method produces the transient source localization for different spanwise configurations in Figure 15. For a compact single 0.2 m clean source region, *L*2 shows a consistent pattern with the single microphone, resulting in a periodic strength change without spatial change. For an increased length of source region from *L*4 to *L*8, an unsteady change along the spanwise direction can be found more and more clearly. For *L*6*T*2 and *L*8*T*4, transient maps indicate sound sources in the tripping regions. Although it should be partially due to the Rayleigh limit, which is mentioned in previous time-average discussions, the periodic switch between two types of source distribution should be greatly influenced by the periodic directivity pattern along the spanwise direction. The significant time delays along the spanwise direction directly change the transient strength from different microphones in Equation (1) to produce source strength distribution maps.

Furthermore, by using source power integration methods [52,53], transient total strength as the summed SPL along the spanwise direction was produced, which highlighted the spanwise difference in the time domains. As shown in Figure 16, the consecutive spanwise configuration from *L*2 to *L*8 tends to show a consecutive change along the spanwise direction, while for *L*6*T*2 and *L*8*T*4, an interruption between two 0.2 sources region was found, which was unlike the time-average results in Figure 8. It should be noted that although the unsteady intermittent patterns prevail in source regions with a large spanwise length, periodic strength change at 10 Hz always exists in Figures 12 and 16. Since the strong low-frequency oscillation is inherent due to the wind tunnel buffeting, it can be assumed that a periodic vortical structure is strong enough to dominate the acoustic feedback loop, which contributes to the current spanwise periodic directivity pattern in Figure 14.

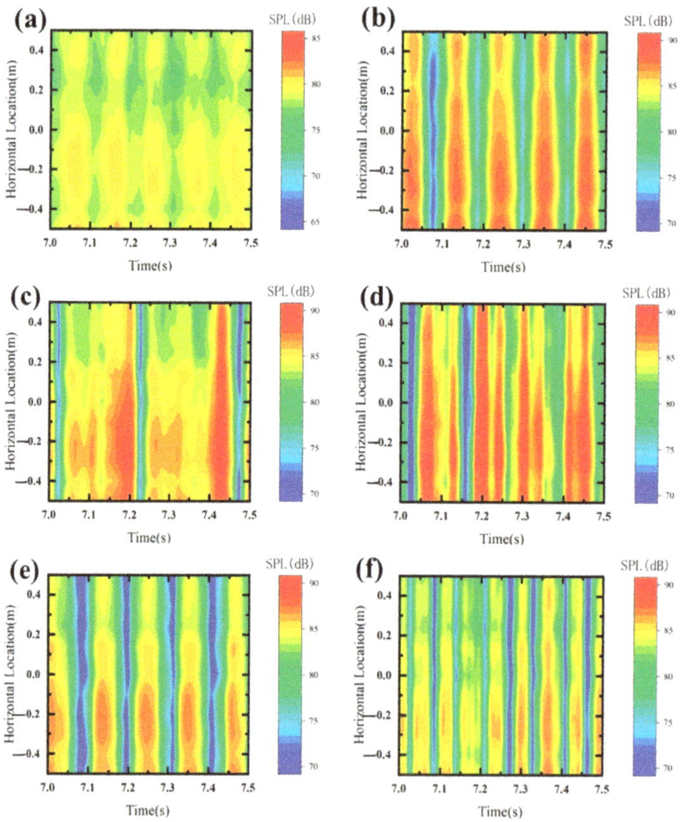

Figure 13. The contour maps of sound pressure level (SPL) for each microphone in the horizontal line in time domain for different spanwise configurations. SPL is summed around the tonal frequency in a 1/3 octave. The incoming velocity is 24 m/s. (**a**) *L*2; (**b**) *L*4; (**c**) *L*6; (**d**) *L*8; (**e**) *L*6*T*2; (**f**) *L*8*T*4.

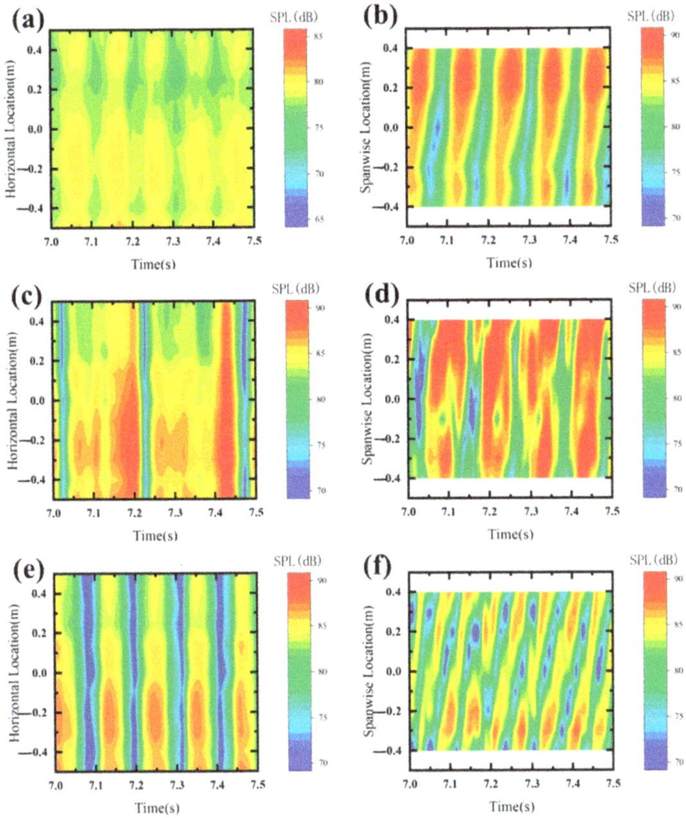

Figure 14. The contour maps of sound pressure level (SPL) for each microphone in the spanwise line in time domain for different spanwise configurations. SPL is summed around the tonal frequency in a 1/3 octave. The incoming velocity is 24 m/s. (**a**) *L*2; (**b**) *L*4; (**c**) *L*6; (**d**) *L*8; (**e**) *L*6*T*2; (**f**) *L*8*T*4.

Based on the above results, the strength influence can be changed to add weights in Equation (1) for different microphones. Considering the periodic directivity pattern in the time domain along the spanwise direction, weight coefficients are added to decrease the influence of consistent horizontal microphones. As a result, the transient source localization in Figure 17 for *L*6*T*2 and *L*8*T*4 reduces the strength change between the consecutive source and separated sources. This conclusion can also be applied to time-average source localization. As shown in Figure 18, the weighted source localization tends to increase the spanwise distribution along the spanwise direction between two source regions. Therefore, it can be concluded that the 10 Hz oscillation is critical in changing the roles of different microphones in source localizations, which should be paid attention in experiments.

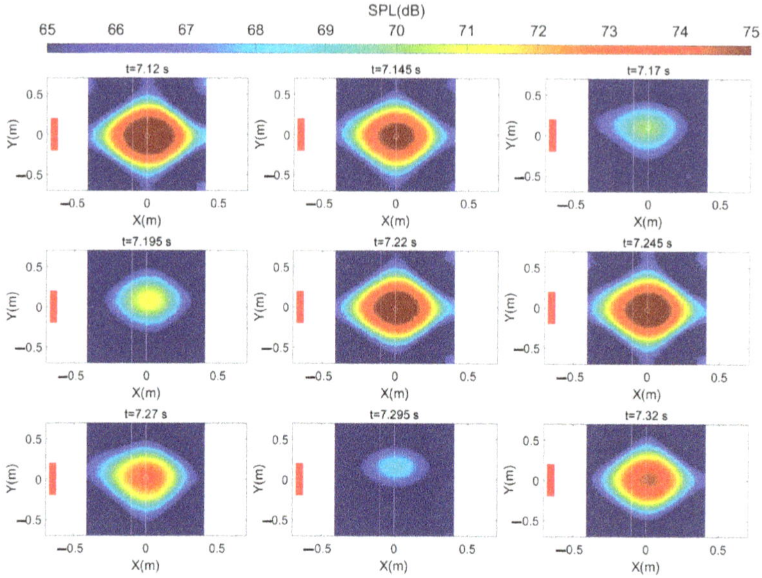

(a)

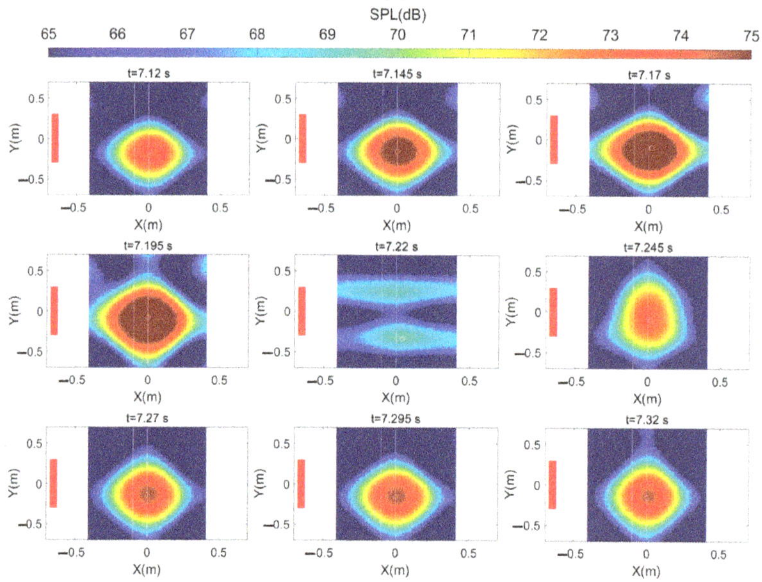

(b)

Figure 15. *Cont.*

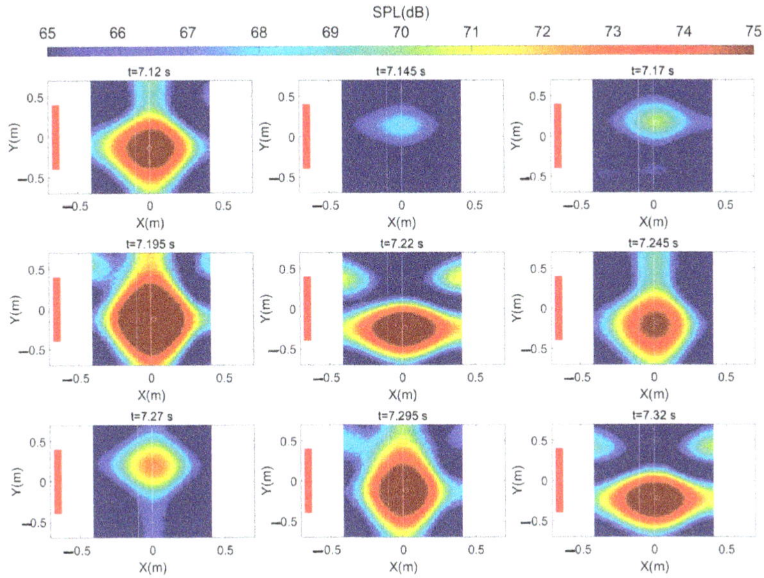

(**c**)

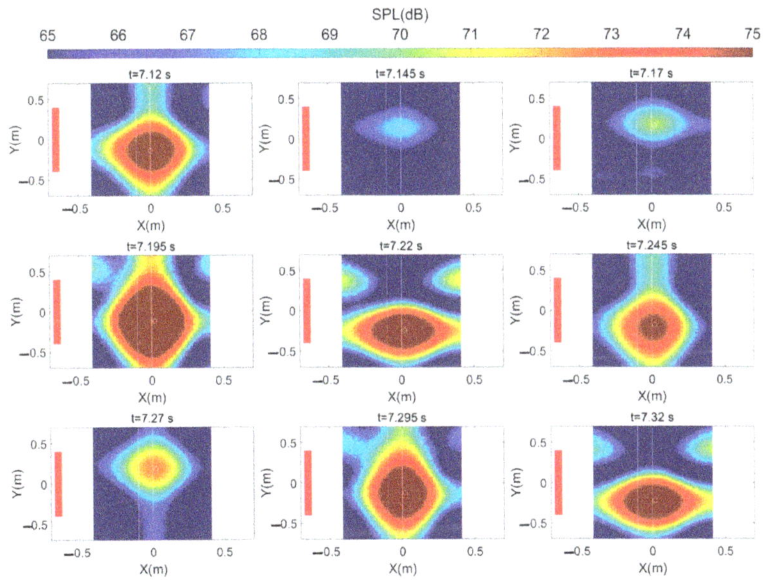

(**d**)

Figure 15. *Cont.*

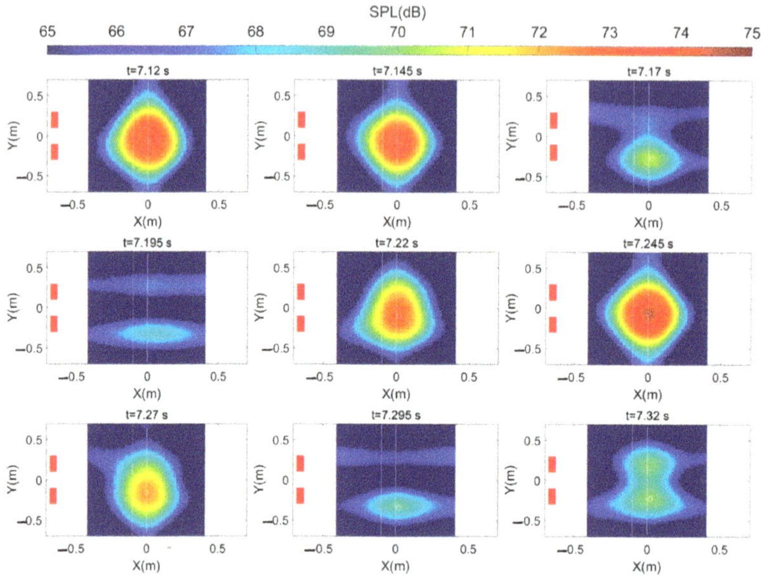

(e)

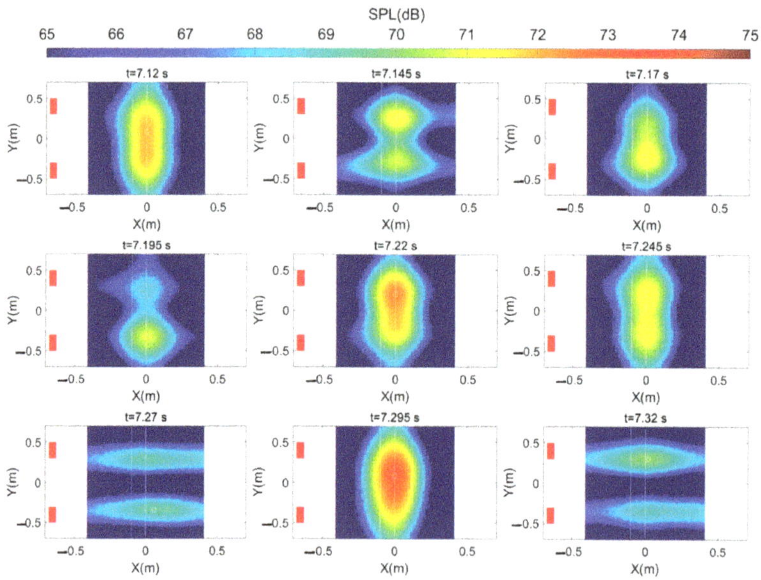

(f)

Figure 15. The transient wavelet-based results at 24 m/s for different spanwise configurations. The columns from the left and rows from the bottom are the transient wavelet-based imaging results from 7.12 s to 7.32 s, which indicate two periods at 10 Hz. (**a**) *L*2; (**b**) *L*4; (**c**) *L*6; (**d**) *L*8; (**e**) *L*6*T*2; (**f**) *L*8*T*4. The unfilled white cycles indicate the positions of maximum values.

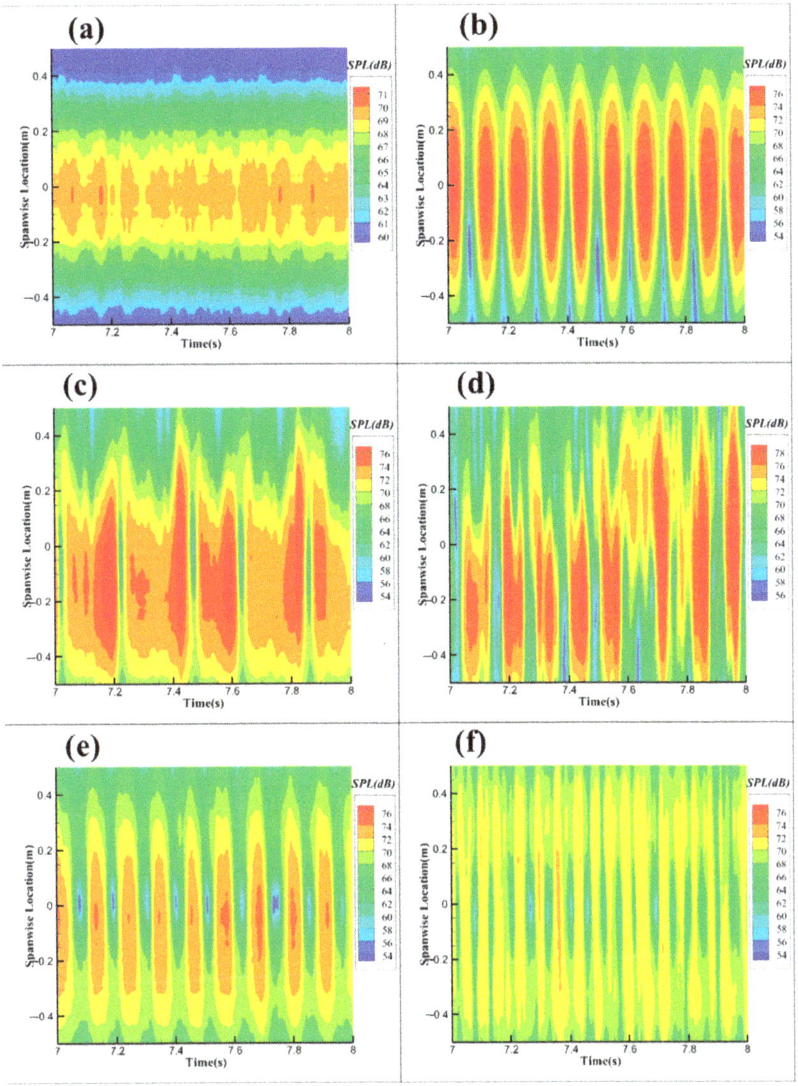

Figure 16. The summed SPL of wavelet-based beamforming results at different spanwise locations near the trailing edge in the time and spatial domain. (**a**) $L2$; (**b**) $L4$; (**c**) $L6$; (**d**) $L8$; (**e**) $L6T2$; (**f**) $L8T4$.

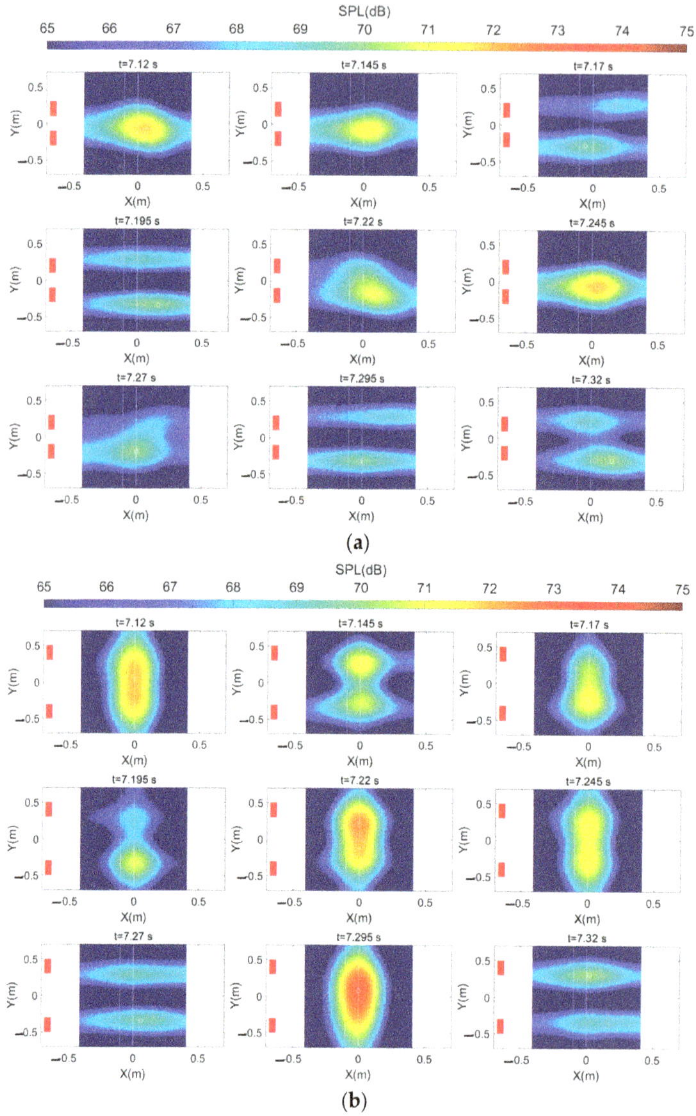

Figure 17. The weighted transient wavelet-based results at 24 m/s for different spanwise configurations. The columns from the left and rows from the bottom are the transient wavelet-based imaging results from 7.12 s to 7.32 s, which indicate two periods at 10 Hz. (**a**) *L6T2*; (**b**) *L8T4*. The unfilled white cycles indicate the positions of maximum values.

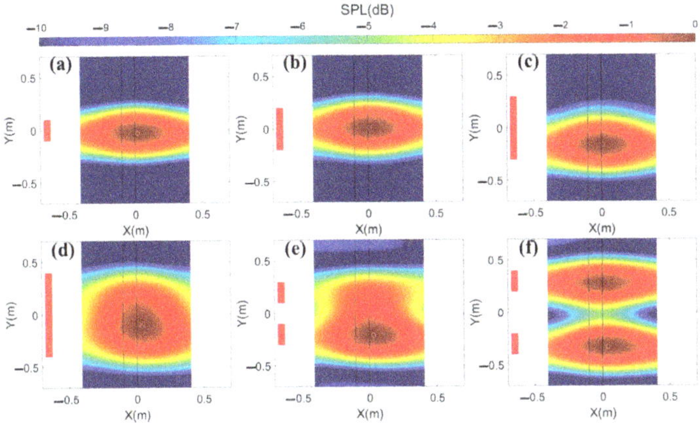

Figure 18. The weighted time-average source localization of different source regions along the spanwise location at 24 m/s according to the 21-channel cross-shaped microphone array. (**a**) *L*2; (**b**) *L*4; (**c**) *L*6; (**d**) *L*8; (**e**) *L*6*T*2; (**f**) *L*8*T*4. The unfilled white cycles indicate the positions of maximum values.

4. Conclusions

In this paper, an investigation of the low-frequency oscillation on the trailing-edge tonal noise was carried out. Inspired by previous experiments for high-frequency oscillations from unsteady coherent flows, the present study aimed to figure out the influence of low-frequency oscillations from the wind tunnels. By using tripping tapes to diminish the local sources, configurations for symmetry spatial distributions of source regions were generated to check the noise characteristics. Periodic strength change was focused upon by continuous wavelet analysis to reveal the spatial difference between microphones at different locations.

In wind tunnel experiments, the NACA0012 airfoil trailing-edge tonal noise was measured by two microphone arrays that were used to analyze the time-average and transient characteristics. For the time-average patterns, a consistent source strength distribution in a 0.2 m source region along the spanwise direction was validated by a 64-channel spiraled microphone array. And a symmetric source strength distribution was found for symmetric spanwise configurations by a 21-channel cross-shaped array. The impact of a low-frequency oscillation on the time-average results was generally insignificant, except for on the side secondary tones due to amplitude modulation.

In contrast, for the transient analysis by continuous wavelet methods, a periodic strength change pattern around the dominant tone emerges from each microphone's result. The increasing length of source regions without tripping from *L*2 to *L*8 produced more intermittent broadband patterns in the time domain. Meanwhile, based on the simultaneous transient wavelet coefficients, a significantly large delay of time for the 10 Hz oscillation occurs between microphones along the spanwise direction rather than the horizontal direction. The delay time is almost 0.1 s, which is much larger than the maximum propagating delay time between microphones of 0.02 s. Therefore, a significant periodic directivity pattern was found along the spanwise direction. As a result, the transient source localization caused by the wavelet-based beamforming method contributes to a switching pattern between the two types of results. One of the results is similar to the time-average results as a consecutive line source strength distribution, while the other separates the sources in different source regions better than the Rayleigh limit. Hence, low-frequency oscillations play a role in changing the transient spatial directivity of sources, which was quite interesting and few discussions had been undertaken before.

Moreover, for microphone arrays in transient source localization, the time delay as a shift of phase in the frequency-time domains greatly influences the results of wavelet-based beamforming methods, which produces a switching pattern between different spatial distributions. Therefore, based on the weighted results, to decrease the influence of horizontal microphones in the symmetrical line, it can be assumed that a low-frequency oscillation increases the spanwise complexity in transient source localization even for symmetric spanwise source regions. However, since the low-frequency oscillation is inherent due to the wind tunnel buffeting, a periodic pattern should exist, which contributes to the current results in the spatial domain. Since this pattern was greatly involved with low-frequency oscillations, the response from different airfoils or even different regions within a three-dimensional wing model might be considered in experiments. Furthermore, the same NACA0012 airfoil in different wind tunnels might possess different transient noise spectra and these spectra were highly related to the location of sources due to this periodic directivity pattern. However, since only NACA0012 as a 0.1 m chord length symmetric airfoil is involved in the current study with different spanwise configurations, there might be other patterns for unsymmetric airfoils. Future investigations should be conducted for other different airfoils. In addition, artificial low-frequency oscillations could be generated to further check the effects of the oscillations at different frequencies.

Considering that all these complicated patterns exist with only side secondary tones being involved in this experiment, the primary secondary tones with a much higher frequency interval should be handled with more caution. For further investigations of trailing-edge tonal noise, the current study indicates that transient analysis could be greatly influenced by external disturbances in the wind tunnel, which partially explains the difference between experiments and simulations. Noise measurement of different airfoils should be performed with multiple microphones along the spanwise direction to check if the periodic directivity pattern exists. Meanwhile, since it was confirmed that the source's directivity might change dramatically in the time domain due to wind tunnel buffeting, it should be similar for other external disturbances in a low frequency in the real environment. As a result, the environment in applications should be considered in the design of airfoils. Supplementary experiments should be conducted to investigate the potential influence of low-frequency oscillations.

Author Contributions: Conceptualization, Z.S. and H.G.; data curation, Z.S.; formal analysis, Z.S.; funding acquisition, H.G., Y.S. and S.J.; investigation, Z.S.; methodology, Z.S.; project administration, H.G.; resources, P.L., H.G., Y.S. and S.J.; software, Z.S.; supervision, P.L.; validation, H.G.; visualization, Z.S.; writing—original draft, Z.S.; writing—review and editing, Z.S., P.L. and H.G. All authors have read and agreed to the published version of the manuscript.

Funding: This research was funded by the National Natural Science Foundation of China (No. 12072016), National Key R&D Program of China (No. 2022YFB2602000), Research Fund of the Key Laboratory of Aerodynamic Noise Control (No. ANCL20220303), and the Fundamental Research Funds for the Central Universities.

Data Availability Statement: Some or all of the data, models, or codes that support the findings of this study are available from the corresponding author upon reasonable request.

Acknowledgments: Thanks to all team members for participating in the experiments and analyzing process.

Conflicts of Interest: The authors declare no conflicts of interest. The funders had no role in the design of the study; in the collection, analyses or interpretation of data; in the writing of the manuscript; or in the decision to publish the results.

References

1. Dobrzynski, W. Almost 40 Years of Airframe Noise Research: What Did We Achieve? *J. Aircr.* **2010**, *47*, 353–367. [CrossRef]
2. Grande, E.; Ragni, D.; Avallone, F.; Casalino, D. Laminar Separation Bubble Noise on a Propeller Operating at Low Reynolds Numbers. *AIAA J.* **2022**, *60*, 5324–5335. [CrossRef]
3. Zhang, Y.; Avallone, F.; Watson, S. Wind Turbine Blade Trailing Edge Crack Detection Based on Airfoil Aerodynamic Noise: An Experimental Study. *Appl. Acoust.* **2022**, *191*, 108668. [CrossRef]
4. Brooks, F.; Stuart, D.; Marcolini, A. *Airfoil Self-Noise and Prediction*; NASA Reference Publication 1218; NASA: Washington, DC, USA, 1989.

5. Lee, S. The Effect of Airfoil Shape on Trailing Edge Noise. *J. Theor. Comp. Acout.* **2019**, *27*, 1850020. [CrossRef]
6. Liu, C.; Lee, S. Parametric Airfoil Design and Sensitivity Analysis for Turbulent Boundary-Layer Trailing-Edge Noise Reduction. *AIAA J.* **2022**, *60*, 2324–2341. [CrossRef]
7. Paterson, R.W.; Vogt, P.G.; Fink, M.R.; Munch, C.L. Vortex Noise of Isolated Airfoils. *J. Aircr.* **1973**, *10*, 296–302. [CrossRef]
8. Tam, C.K.W. Discrete Tones of Isolated Airfoils. *J. Acoust. Soc. Am.* **1974**, 55. [CrossRef]
9. Fink, M.R. Prediction of Airfoil Tone Frequencies. *J. Aircr.* **1975**, *12*, 118–120. [CrossRef]
10. Longhouse, R.E. Vortex Shedding Noise of Low Tip Speed, Axial Flow Fans. *J. Sound Vib.* **1977**, *53*, 25–46. [CrossRef]
11. Arbey, H.; Bataille, J. Noise Generated by Airfoil Profiles Placed in a Uniform Laminar Flow. *J. Fluid Mech.* **1983**, *134*, 33–47. [CrossRef]
12. Nash, E.C.; Lowson, M.V.; McALPINE, A. Boundary-Layer Instability Noise on Aerofoils. *J. Fluid Mech.* **1999**, *382*, 27–61. [CrossRef]
13. Desquesnes, G.; Terracol, M.; Sagaut, P. Numerical Investigation of the Tone Noise Mechanism over Laminar Airfoils. *J. Fluid Mech.* **2007**, *591*, 155–182. [CrossRef]
14. Arcondoulis, E.J.G.; Doolan, C.J.; Zander, A.C.; Brooks, L.A. A Review of Trailing Edge Noise Generated by Airfoils at Low To Moderate Reynolds Number. *Acoust. Aust.* **2010**, *3*, 135–139.
15. Pröbsting, S.; Serpieri, J.; Scarano, F. Experimental Investigation of Aerofoil Tonal Noise Generation. *J. Fluid Mech.* **2014**, *747*, 656–687. [CrossRef]
16. Pröbsting, S.; Scarano, F.; Morris, S.C. Regimes of Tonal Noise on an Airfoil at Moderate Reynolds Number. *J. Fluid Mech.* **2015**, *780*, 407–438. [CrossRef]
17. Pröbsting, S.; Yarusevych, S. Laminar Separation Bubble Development on an Airfoil Emitting Tonal Noise. *J. Fluid Mech.* **2015**, *780*, 167–191. [CrossRef]
18. Padois, T.; Laffay, P.; Idier, A.; Moreau, S. Tonal Noise of a Controlled-Diffusion Airfoil at Low Angle of Attack and Reynolds Number. *J. Acoust. Soc. Am.* **2016**, *140*, EL113–EL118. [CrossRef]
19. Yakhina, G.; Roger, M.; Moreau, S.; Nguyen, L.; Golubev, V. Experimental and Analytical Investigation of the Tonal Trailing-Edge Noise Radiated by Low Reynolds Number Aerofoils. *Acoustics* **2020**, *2*, 293–329. [CrossRef]
20. Golubev, V. Recent Advances in Acoustics of Transitional Airfoils with Feedback-Loop Interactions: A Review. *Appl. Sci.* **2021**, *11*, 1057. [CrossRef]
21. Yang, Y.; Pröbsting, S.; Li, P.; Liu, Y.; Li, Y. A Secondary Modulation Mechanism for Aerofoil Tonal Self-Noise Generation. *J. Fluid Mech.* **2022**, *943*, A13. [CrossRef]
22. Yu, J.; Chen, W.; Zhou, T.; Lee, C.; Huang, X. Transient Analysis of Trailing Edge Noise Assisted by Wavelet-Based Beamforming and Flow Visualisation. *J. Sound Vib.* **2022**, *526*, 116751. [CrossRef]
23. Ricciardi, T.R.; Wolf, W.R.; Taira, K. Transition, Intermittency and Phase Interference Effects in Airfoil Secondary Tones and Acoustic Feedback Loop. *J. Fluid Mech.* **2022**, *937*, A23. [CrossRef]
24. Ricciardi, T.R.; Wolf, W.R. Switch of Tonal Noise Generation Mechanisms in Airfoil Transitional Flows. *Phys. Rev. Fluids* **2022**, *7*, 084701. [CrossRef]
25. Sano, A.; Cavalieri, A.V.G.; Da Silva, A.F.C.; Wolf, W.R. On the Emergence of Secondary Tones in Airfoil Noise. *J. Fluid Mech.* **2023**, *966*, A8. [CrossRef]
26. Smith, T.A.; Klettner, C.A. Airfoil Trailing-Edge Noise and Drag Reduction at a Moderate Reynolds Number Using Wavy Geometries. *Phys. Fluids* **2022**, *34*, 117107. [CrossRef]
27. Chen, W.; Wang, X.; Xing, Y.; Wang, L.; Qiao, W. Effect of Wavy Leading Edges on Airfoil Tonal Noise. *Phys. Fluids* **2023**, *35*, 097136. [CrossRef]
28. Kojima, Y.; Skene, C.S.; Yeh, C.-A.; Taira, K.; Kameda, M. On the Origin of Quadrupole Sound from a Two-Dimensional Aerofoil Trailing Edge. *J. Fluid Mech.* **2023**, *958*, A3. [CrossRef]
29. Wang, Y.; Zuo, K.; Guo, P.; Zhao, K.; Kopiev, V.F. Experimental Investigation of Airfoil Instability Tonal Noise Reduction Using Structured Porous Trailing Edges. *Appl. Sci.* **2024**, *14*, 2992. [CrossRef]
30. Yarusevych, S.; Kotsonis, M. Steady and Transient Response of a Laminar Separation Bubble to Controlled Disturbances. *J. Fluid Mech.* **2017**, *813*, 955–990. [CrossRef]
31. Andan, A.D.; Lee, D.-J. Effect of External Acoustic Excitation on NACA0015 Discrete Tonal Noise. *Appl. Acoust.* **2018**, *141*, 374–381. [CrossRef]
32. Gelot, M.B.R.; Kim, J.W. Effect of Serrated Trailing Edges on Aerofoil Tonal Noise. *J. Fluid Mech.* **2020**, *904*, A30. [CrossRef]
33. Ahuja, K.; Massey, K.; D'Agostino, M.; Ahuja, K.; Massey, K.; D'Agostino, M. Flow/Acoustic Interactions in Open-Jet Wind Tunnels. In Proceedings of the 3rd AIAA/CEAS Aeroacoustics Conference, Atlanta, GA, USA, 12–14 May 1997; American Institute of Aeronautics and Astronautics: Reston, VA, USA, 1997.
34. Jin, L.; Deng, X.B.; Wang, X.; Gu, Y.; Liang, Y.; Lian, Z. Standing Waves in the Plenum of an Open Jet Wind Tunnel: Resonance and Self-Excited Oscillation. *AIP Adv.* **2022**, *12*, 025105. [CrossRef]
35. Jin, L.; Gu, Y.; Deng, X.B.; Sun, H.; Yue, T.; Zhang, J. Standing Wave and Its Impact on the Low-Frequency Pressure Fluctuation in an Open Jet Wind Tunnel. *J. Wind Eng. Ind. Aerodyn.* **2021**, *208*, 104413. [CrossRef]
36. Ma, R.; Li, H.; Han, S.; Wu, C.; Zhang, S.; Wang, X. Numerical Simulation of Low Frequency Oscillation in an Aeroacoustic Wind Tunnel. *J. Phys. Conf. Ser.* **2022**, *2329*, 012017. [CrossRef]

37. Sellers, W.L.; Applin, Z.T.; Molloy, J.K. *Effect of Jet Exit Vanes on Flow Pulsations in an Open-Jet Wind Tunnel*; Technical Report No. NASA-TM-86299; Langley Research Center: Hampton, VA, USA, 1985.
38. Martin, R.M.; Brooks, T.F.; Hoad, D.R. Reduction of Background Noise Induced by Wind Tunnel Jet Exit Vanes. *AIAA J.* **1985**, *23*, 1631–1632. [CrossRef]
39. von Heesen, W.; Höpfer, M. Suppression of Wind Tunnel Buffeting by Active Flow Control. *SAE Trans.* **2004**, *113*, 431–441.
40. Kudo, T.; Komatsu, Y.; Maeda, K.; Nishimura, M. Techniques for Reducing Low-Frequency Fluctuations in Aeroacoustic Wind Tunnels. *J. Environ. Eng.* **2009**, *4*, 289–301. [CrossRef]
41. Chen, W. Acoustic Receptivity in the Airfoil Boundary Layer: An Experimental Study in a Closed Wind Tunnel. *Phys. Fluids* **2023**, *35*, 043616. [CrossRef]
42. Roder, H. Amplitude, Phase, and Frequency Modulation. *Proc. IRE* **1931**, *19*, 2145–2176. [CrossRef]
43. Brandstein, M.; Ward, D. (Eds.) *Microphone Arrays: Signal Processing Techniques and Applications. Digital Signal Processing*; Springer: Berlin/Heidelberg, Germany, 2001; ISBN 978-3-642-07547-6.
44. Sanders, M.P.; Botero, L.; De Santana, L.D.; Venner, C. Airfoil Trailing Edge Noise Measurements in an Open-Jet, Hard-Wall and Kevlar-Wall Test Section: A Benchmark Exercise. In Proceedings of the AIAA AVIATION 2021 Forum, Virtual Event, 2–6 August 2021; American Institute of Aeronautics and Astronautics: Reston, VA, USA, 2021.
45. Herold, G.; Sarradj, E. Performance Analysis of Microphone Array Methods. *J. Sound Vib.* **2017**, *401*, 152–168. [CrossRef]
46. Zhou, H.; Liu, Z.; Luo, L.; Wang, M.; Song, X. An Improved Two-Stage Spherical Harmonic ESPRIT-Type Algorithm. *Symmetry* **2023**, *15*, 1607. [CrossRef]
47. Chen, W.; Huang, X. Wavelet-Based Beamforming for High-Speed Rotating Acoustic Source. *IEEE Access* **2018**, *6*, 10231–10239. [CrossRef]
48. Chen, W.; Peng, B.; Liem, R.P.; Huang, X. Experimental Study of Airfoil-Rotor Interaction Noise by Wavelet Beamforming. *J. Acoust. Soc. Am.* **2020**, *147*, 3248–3259. [CrossRef] [PubMed]
49. Chen, W.; Yang, Z.; Peng, B.; Huang, X. On Trailing Edge Noise from Propellers with Interactions to Shear Layers. *J. Sound Vib.* **2021**, *495*, 115901. [CrossRef]
50. Liu, P.; Xing, Y.; Guo, H.; Li, L. Design and Performance of a Small-Scale Aeroacoustic Wind Tunnel. *Appl. Acoust.* **2017**, *116*, 65–69. [CrossRef]
51. Merino-Martínez, R.; Sijtsma, P.; Snellen, M.; Ahlefeldt, T.; Antoni, J.; Bahr, C.J.; Blacodon, D.; Ernst, D.; Finez, A.; Funke, S.; et al. A Review of Acoustic Imaging Methods Using Phased Microphone Arrays: Part of the "Aircraft Noise Generation and Assessment" Special Issue. *CEAS Aeronaut. J.* **2019**, *10*, 197–230. [CrossRef]
52. Brooks, T.; Humphreys, W., Jr. Effect of Directional Array Size on the Measurement of Airframe Noise Components. In Proceedings of the 5th AIAA/CEAS Aeroacoustics Conference and Exhibit, Bellevue, WA, USA, 10–12 May 1999; American Institute of Aeronautics and Astronautics: Reston, VA, USA, 1999.
53. Merino-Martínez, R.; Sijtsma, P.; Carpio, A.R.; Zamponi, R.; Luesutthiviboon, S.; Malgoezar, A.M.; Snellen, M.; Schram, C.; Simons, D.G. Integration Methods for Distributed Sound Sources. *Int. J. Aeroacoustics* **2019**, *18*, 444–469. [CrossRef]

symmetry

MDPI

Article

Tomographic Background-Oriented Schlieren for Axisymmetric and Weakly Non-Axisymmetric Supersonic Jets

Tong Jia [1,†], Jiawei Li [1,†], Jie Wu [2] and Yuan Xiong [1,3,*]

1 Key Laboratory of Fluid Mechanics of Ministry of Education & Aircraft and Propulsion Laboratory, Ningbo Institute of Technology, Beihang University, Beijing 100191, China
2 School of Aerospace Engineering, Huazhong University of Science & Technology, Wuhan 430074, China
3 Tianmushan Laboratory, Xixi Octagon City, Yuhang District, Hangzhou 310023, China
* Correspondence: xiongyuan@buaa.edu.cn
† These authors contributed equally to this work.

Abstract: The Schlieren technique is widely adopted for visualizing supersonic jets owing to its non-invasiveness to the flow field. However, extending the classical Schlieren method for quantitative refractive index measurements is cumbersome, especially for three-dimensional supersonic flows. Background-oriented Schlieren has gained increasing popularity owing to its ease of implementation and calibration. This study utilizes multi-view-based tomographic background-oriented Schlieren (TBOS) to reconstruct axisymmetric and weakly non-axisymmetric supersonic jets, highlighting the impact of flow axisymmetry breaking on TBOS reconstructions. Several classical TBOS reconstruction algorithms, including FDK, SART, SIRT, and CGLS, are compared quantitatively regarding reconstruction quality. View spareness is identified to be the main cause of degraded reconstruction quality when the flow experiences axisymmetry breaking. The classic visual hull approach is explored to improve reconstruction quality. Together with the CGLS tomographic algorithm, we successfully reconstruct the weakly non-axisymmetric supersonic jet structures and confirm that increasing the nozzle bevel angle leads to wider jet spreads.

Keywords: tomographic background-oriented Schlieren; supersonic jet; visual hull; tomography algorithm

1. Introduction

Jet propulsion systems play a pivotal role in aviation and aerospace. These systems are based on the momentum theorem, and the core mechanism for generating propulsion is to accelerate the exhaust of high-pressure gas so that the aircraft or rocket obtains reacting force. The high-speed exhaust forms a supersonic jet and produces complex flow structures such as shock waves, expansion waves, and shear flow under the influence of the boundary conditions of the nozzle and the ambient atmosphere. These flow structures significantly impact the propulsion performance and jet noise emissions, thus requiring further in-depth research.

The state of the supersonic jet depends on the relationship between the total pressure at the nozzle outlet and the ambient pressure. It can be characterized by the nozzle pressure ratio (NPR) shown in the following equation:

$$NPR = \frac{P_t}{P_{amb}} = (1 + \frac{\gamma - 1}{2} Ma^2)^{\frac{\gamma}{\gamma-1}} \quad (1)$$

Inside P_t and P_{amb} are the stagnation pressure of the jet and the ambient pressure, respectively; Ma represents the jet Mach number, and γ is the specific heat ratio. The jet of aero and rocket engines is usually in an under-expansion state; that is, its Ma is greater than the designed Mach number of the nozzle Ma_d so that the jet can gain higher momentum.

Many studies have explored the impact of complex flow structures such as shock reflection, shock hysteresis, and nozzle flow separations on supersonic jet performance [1–4].

Supersonic jets are also the main noise sources of aircraft, which produce turbulent mixing noise, broadband shock-associated noise, and screech tones [5–7]. Some studies have pointed out that enhancing the mixing of jets with surrounding fluids can significantly alleviate turbulent mixing noise and shock-associated noise, and many methods for promoting low-speed jet mixing have been developed [8–11]. However, due to the significant compressibility effect, effectively mixing supersonic jets with ambient flows is not trivial. Appropriate modification of the nozzle trailing edge is a method that has been widely studied to enhance the mixing of supersonic jets. The extended perimeter of the trailing edge will facilitate the interaction between high-speed and low-speed flows [12–14]. Wu and New [15] studied single-beveled nozzles and observed changes in the shock cell structure caused by the presence of the nozzle notch, and the beveled nozzle shortened the potential core length, indicating better jet mixing capabilities. Rice and Raman [16] found that compared with the single-beveled nozzle, a double-beveled nozzle with a more symmetrical geometry can produce a more symmetrical jet and may provide better mixing performance. Raman [17] studied the shock cell structure and the noise of beveled rectangular jets, and they noted that jets produced by inclined notches generally produce weak screech tones over a limited range of Mach numbers.

Further improvements in the propulsion system require a detailed and accurate three-dimensional description of the supersonic jet flow structures. The compressibility of supersonic flow causes its density to change, and according to the Gladstone–Dale relation ($n - 1 = K\rho$), the gas index of refraction (IOR, n) is proportional to its density ρ (assuming that the G–D constant K of air is constant). Therefore, the flow structures can be characterized by the IOR distribution of the flow field. The classical Schlieren method uses this feature of the compressible flow to visualize supersonic flow fields. The background-oriented Schlieren (BOS) method was proposed around 2000 [18,19], and due to the advantages of simple optical design and low implementation cost, it has been widely used in the two-dimensional (2D) measurement of supersonic jets [20,21].

Tomographic background-oriented Schlieren (TBOS), which combines tomography with BOS measurement, makes it possible to measure three-dimensional (3D) non-uniform density fields. Abel inversion was first applied to the reconstruction of axisymmetric flow fields [22,23]. Later, Venkatakrishnan and Meier [24] proposed using the Filtered Back Projection (FBP) algorithm to reconstruct the axisymmetric density field to avoid the algorithm singularities and noise sensitivity of Abel inversion. Tan et al. [25] successfully implemented the density field measurement of axisymmetric supersonic jets using the adaptive Fourier–Hankel (AFH) methods. Xiong et al. [26] compared the onion peeling (OP), three-point Abel (TPA), FBP, and AFH methods and found that, in the presence of noise, the three-point Abel inversion yielded the best axisymmetric IOR reconstruction results. Facing the reconstruction of the more general 3D supersonic jet, multi-view BOS measurement is required. Kirby [27] reported the use of TBOS technology to measure supersonic jets generated by elliptical nozzles, which study used BOS observations of 40 views and the FBP-SART combined algorithm; although his results restored flow structures such as shocks to a certain extent, errors and artifacts are still evident. Nicolas et al. [28] used the least squares optimization algorithm and spatial mask constraints to achieve axisymmetric under expanded supersonic jet reconstruction with fewer than 20 viewing angles. However, achieving high-quality flow reconstruction for non-axisymmetric supersonic jets with sparse camera views remains challenging.

In this paper, we systematically investigated the application of TBOS reconstruction to under-expansion supersonic jets with/without axisymmetry breaking. The paper is organized as follows: Section 2 outlines the measurement principle of TBOS, the tomographic reconstruction algorithm, and virtual device settings used subsequently; Section 3 studies the reconstruction of the jet structures transitioning from axisymmetry to weak non-axisymmetry, focusing on the impact of algorithms, noise, and visual-hull-like mask methods on reconstruction, and the physics of jets are discussed; and Section 4 provides the main conclusions of this study.

2. Methodology

2.1. Principle of TBOS

The TBOS principle has been discussed in detail in previous studies [29,30], so it is only briefly introduced here. As shown in Figure 1, the **X** point on the background pattern is mapped to the **Y** point on the imaging plane after passing through the space of constant IOR; however, in the presence of a flow field with non-uniform IOR, the light ray is deflected and mapped on **Y**′. At this time, the optical path can be described by the following ray equation:

$$\frac{d}{ds}\left(n\frac{d\mathbf{r}}{ds}\right) = \nabla n \tag{2}$$

where **r** represents the coordinates of a certain point on the optical path, ds represents the differential length, and ∇n represents the local refractive index gradient. This equation can be rewritten as follows:

$$n\frac{d\mathbf{r}}{ds} = \mathbf{d}, \quad \frac{d\mathbf{d}}{ds} = \nabla n \tag{3}$$

where **d** represents the ray vector. For a gas environment with $n \approx 1$, **d** can be approximated as a unit vector. Therefore, the deflection $\vec{\varepsilon}$ of a light ray passing through the entire non-uniform IOR region is defined as the integral of ∇n along the optical path S:

$$\vec{\varepsilon} = \mathbf{d}_{out} - \mathbf{d}_{in} = \int_S \nabla n ds. \tag{4}$$

where $\mathbf{d}_{in}$ and $\mathbf{d}_{out}$ represent the direction vectors of a light ray entering and leaving the flow field, respectively. In the actual BOS experiment, the background pattern is first imaged to obtain the **Y** position under a uniform IOR condition. The imaging settings are kept unchanged and performed in the presence of non-uniform IOR. Based on this pair of images and the cross-correlation algorithm or the optical flow method [31], the imaging/pixel displacement $\vec{\Delta}$ is calculated, and $\hat{\mathbf{Y}'} = \mathbf{Y} + \vec{\Delta}$. Finally, through simple geometric optics calculations, the approximate $\mathbf{d}_{in}$ and $\mathbf{d}_{out}$ and deflection $\vec{\varepsilon}$ can be obtained. For a multi-camera TBOS measurement, through imaging with multiple view angles and further discretizing Equation (4), we can obtain the following:

$$\varepsilon_u = T \cdot (\nabla n)_u \tag{5}$$

The inside subscript u represents the directions x, y, z. The multi-camera imaging determines I rays, and the measurement volume contains a total of J discrete voxels. Then $\varepsilon_u \in \mathbb{R}^I$ represents the vector of all ray deflections; $(\nabla n)_u \in \mathbb{R}^J$ is a vector recording all IOR gradients; and $T \in \mathbb{R}^{I \times J}$ is the tomographic coefficient matrix, in which the elements marked by (i, j) record the contribution weight of ∇n of the j-th voxel to the deflection of the i-th ray. Obviously, Equation (5) forms a large linear system of equations in the form $b = Ax$. Solving x, that is, $(\nabla n)_u$ in Equation (5), is to solve the inverse problem of linear equations.

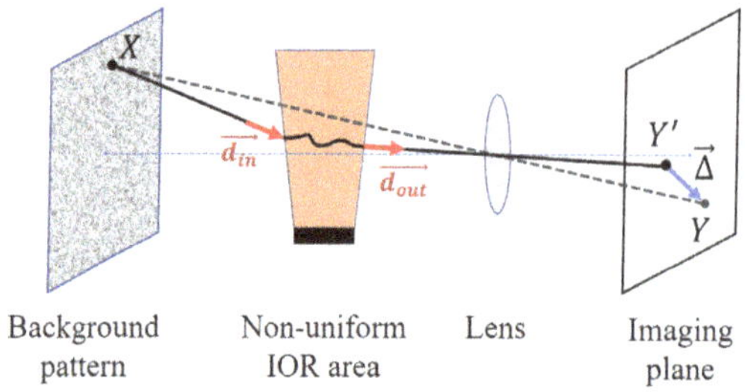

Figure 1. BOS imaging schematic diagram.

2.2. *Tomographic Reconstruction Algorithm*

Under a limited observation angle, the sampled I (number of effective rays) is usually much smaller than J. Furthermore, Equation (5) is seriously ill-posed due to measurement errors and noise. Therefore, the direct inversion method cannot solve the tomographic reconstruction problem. Still, some numerical solution techniques must be used to estimate a solution as close to the true value as possible.

Many algorithms have been used for tomographic reconstruction, driven by multiple disciplines, such as computed tomography and magnetic resonance imaging. Based on the principle of solving, tomography reconstruction algorithms are mainly divided into analytical and iterative types. The analytical algorithm is based on the mathematical model of the measurement system and is directly calculated analytically. Iterative algorithms are based on statistical models of measured data, using iterative optimization operations to estimate the best reconstructions. The two types of reconstruction algorithms have specific characteristics, so during TBOS reconstruction, they must be selected based on objective conditions and requirements.

When the flow field to be measured has the characteristics of axisymmetry, it can be considered that the BOS measurements performed around the symmetry axis are the same. Therefore, only the BOS measurement of one view is needed to carry out the tomographic reconstruction of any multiple view angles. Currently, the number of effective rays (I in Equation (5)) can be increased without limit, making the system of equations close to a positive definite state. In this case, analytical algorithms, such as Abel transform, AFH, and FBP, can be used for reconstruction. Multi-view measurements are required for the TBOS reconstruction of a non-axisymmetric flow field. However, the number of observation view angles (projection planes) is limited due to laboratory space and equipment costs. In this case, iterative algorithms show better reconstruction results in practice. For a detailed introduction to various volume tomography reconstruction algorithms, please refer to the review by Grauer et al. [32]. The following is a brief introduction to the tomographic reconstruction algorithm used in this paper:

2.2.1. Analytical Algorithm

FBP is currently the most widely used analytical reconstruction algorithm. Its basis is the Fourier central slice theorem [33]: If a 3D measured target x is imaged from different viewing angles, a series of 2D projections b can be obtained; the 1D Fourier transform of these projections b is equal to the slice of the 2D Fourier transform of x at the corresponding viewing angle. Therefore, when enough views of b are sampled, the slices can be combined into a complete 2D Fourier transform of x, and finally, the reconstruction of x can be completed through the inverse Fourier transform. Another necessary operation of FBP is performing convolution filtering on the Fourier transformed b to eliminate the measurement error and the noise diffusion. Therefore, the reconstruction process of FBP can be summarized as follows:

1. Perform a 1D Fourier transform on the projection data.
2. Select appropriate filters to perform convolutional filtering on the transformed projection.

3. Back-project the filtered projection data into the measurement volume.
4. All back projections are superimposed to obtain the reconstructed projection.

The traditional FBP algorithm can only process projection data under parallel beam and uniform distribution conditions, so it cannot be directly applied to TBOS reconstruction based on the camera imaging model (cone beam) and arbitrary viewing angle sampling. We use the improved FBP algorithm proposed by Feldkamp, Davis, and Kress, also known as the FDK algorithm [34]. It introduces a more complex reconstruction geometric model and overcomes the limitations of FBP through weighted correction in the operation. In addition, the Ram-Lak filter is used in the current FDK algorithm, which is equivalent to an ideal high-pass filter in the frequency domain, so it can remove low-frequency information and enhance high-frequency information, thereby highlighting the flow structure.

2.2.2. Iterative Algorithm

The algebraic reconstruction technique (ART) is the most commonly used iterative algorithm family. It was originally used to replace the FBP algorithm and outperforms FBP in the case of fewer projections. For inverse problems of the form $b = Ax$, the ART follows the following iteration structure:

$$x^{k+1} = x^k + \lambda VA^T W(b - Ax^k) \tag{6}$$

where λ is the relaxation coefficient used to control the convergent speed of iteration, and V and W are weight matrices based on the length of the light ray. The basic ART algorithm sequentially calculates A_i ($i-$th row of A, corresponding to one ray) and projection b_i to update x and completes an iteration after traversing all rays.

To improve computational efficiency and robustness, modified ART algorithms such as SART and SIRT [35] emerged: the SART algorithm updates x by projection planes one by one, while SIRT uses all projection planes to update. Generally, the more data are used in an update, the faster each iteration step will be, but the slower the convergence (more iterations) will be. The current study uses these two algorithms, and the relaxation coefficients are all set to 0.99^{k-1}.

Another type of iterative algorithm is performing optimization operations on Equation (7) under the least squares framework to obtain the reconstruction result with the highest probability. This method is derived from the statistical model based on the TBOS system and measurement data.

$$\hat{x} = \arg\min_x \|b - Ax\|_2^2 \tag{7}$$

This paper used the CGLS [36] algorithm for the above problem, which is a derivative iterative algorithm of Krylov methods: at the $k-$th iteration, the optimal solution x^k belonging to the Krylov subspace of increasing dimensions is found according to the optimality criterion defined by the specific Krylov solver. Krylov subspaces are linear combinations of matrices' first $k-1$ powers acting on vectors. In the tomographic reconstruction problem, we mainly focus on the subspace of $A^T A$ acting on $A^T b$, as shown in the following equation:

$$\mathcal{K}_k(A^T A, A^T b) = \text{span}\{A^T b, (A^T A)A^T b, ..., (A^T A)^{k-1} A^T b\}. \tag{8}$$

CGLS uses the conjugate gradient method to find the fastest descent direction. Therefore, the result of the k-th iteration is the least square solution under the k-order Krylov subspace:

$$x^k = \arg\min_{x \in \mathcal{K}_k} \|b - Ax\|_2^2 \tag{9}$$

which can make the residual statistics $\|b - Ax^k\|$ monotonically decrease. Algorithm 1 below shows the simplified CGLS process.

It is apparent that FDK, utilizing the Fourier central slice theorem, circumvents the cumbersome iterative process by directly deriving the reconstruction result from projection data. Consequently, FDK offers a significantly faster computational speed, which is widely acknowledged as one of its primary advantages. In contrast, iterative algorithms necessitate continuous projection and back-projection operations to iteratively minimize the final residual. As a result, they incur higher computational loads and usually require more time to accomplish the reconstruction process.

Algorithm 1 CGLS

Initialize: $x^{(0)}; r^{(0)} = b - Ax^{(0)}; p^{(0)} = s^{(0)} = A^T r^{(0)}; \gamma^{(0)} = ||p^{(0)}||^2$
for $i = 0, 1, 2, \ldots, k$ **do**
$q^{(i)} = Ap^{(i)}$
$\alpha^{(i)} = \gamma^{(i)} / ||q^{(i)}||^2$
$x^{(i+1)} = x^{(i)} + \alpha^{(i)} q^{(i)}$
$r^{(i+1)} = r^{(i)} - \alpha^{(i)} q^{(i)}$
$s^{(i+1)} = A^T r^{(i+1)}$
$\gamma^{(i+1)} = ||s^{(i+1)}||^2$
$\beta^{(i)} = \gamma^{(i+1)} / \gamma^{(i)}$
$p^{(i+1)} = s^{(i+1)} + \beta^{(i)} p^{(i)}$
end for
Output: x^{k+1}

2.3. Experimental Setting

In this paper, we adopted an experimental scheme of TBOS measurement based on CFD data and a synthetic BOS image generation platform [37]. Its process was as follows: the target flow fields were computed via CFD simulation, the flow field data were incorporated within the ray-tracing process of synthetic BOS image generation, and then the background pattern imaging was performed through ray tracing technology, and the synthetic BOS images were used for tomographic reconstruction studies. Because of the existence of CFD data as the ground truth, this scheme has the advantage of quantifiable algorithm accuracy.

This paper studied the under-expanded jets produced by three types of nozzles, as shown in Figure 2. These nozzles had the same parameters: the internal diameter ($D = 12.7$ mm), the total length (i.e., average of the shortest and longest axial lengths), and the design Mach number ($Ma_d = 1.45$). Three-dimensional Reynolds-averaged Navier-Stokes (RANS) simulations were conducted for these nozzles, assuming that the supersonic shock structure was quasi-steady. The ANSYS CFX solver was used to simulate arbitrary flow scenarios involving complex physical geometries under low- and high-speed conditions [38,39]. To ensure the stability of the numerical simulations, shock and expansion waves were captured via a high-resolution advection scheme. The jets of the three nozzles were all set to $NPR = 4.0$, and the corresponding Mach number was 1.56. For more detailed nozzle and CFD simulation settings, please refer to previous research [40].

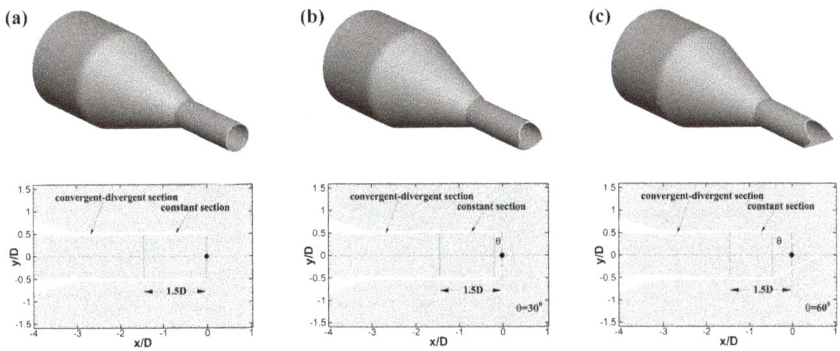

Figure 2. Geometrical details of the (**a**) non−beveled nozzle, (**b**) $\theta = 30°$ double−beveled nozzle, and (**c**) $\theta = 60°$ double−beveled nozzle [40].

Figure 3 shows the configuration for generating the synthetic BOS images. The cameras adopted an annular coplanar arrangement with the nozzle center as the axis, and the supersonic jet was located between the camera and the background pattern. Lang et al. [41] discussed the impact of the number and distribution of BOS camera views on reconstruction accuracy and concluded that odd-numbered views and coplanar annular arrangements can achieve optimal results. At the same time, arranging dozens of cameras in actual experiments is usually not feasible. Therefore, we referred to the studies of Atcheson et al. [42] and Nicolas et al. [28], and used 15 cameras/BOS in the present work. Under this measurement setting, we synthesized multi-view BOS images through ray tracing technology, and their resolution was 1280 × 2560 pixels. The data of pixel displacement $\vec{\Delta}$ were obtained through an in-house optical flow method [43]. The inquiry window size was set to 3 × 3 pixels, so the resolution of the $\vec{\Delta}$ field was 382 × 764 (Figure 4a). Further, the deflection ε_u required for the tomographic reconstruction operation, as shown in Figure 4b, can be obtained. Table 1 lists the detailed parameters of the TBOS device and data processing.

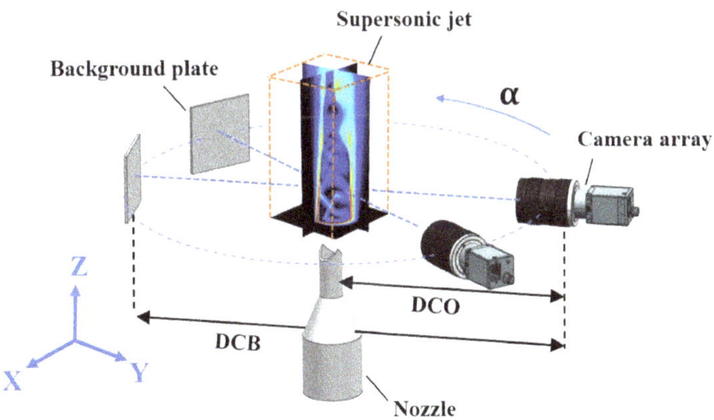

Figure 3. A schematic for the TBOS setup.

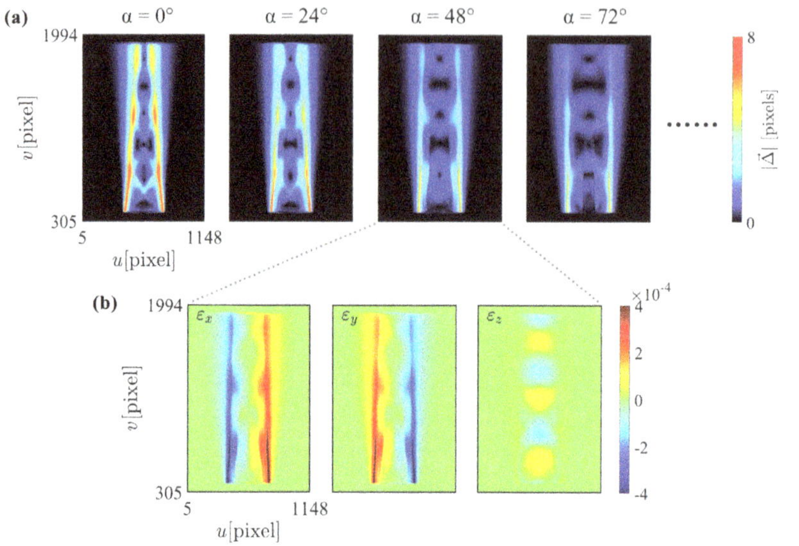

Figure 4. (**a**) Fields of displacement amplitude at selected views and (**b**) associated light deflections.

Table 1. Settings for the generation of synthetic BOS images.

Parameter	Setting
DCB	1500 mm
DCO	750 mm
Reconstruction resolution	250 × 250 × 500 voxels
Measurement size	25 × 25 × 50 mm^3
Lens focal length	50 mm
Pixel resolution	1280 × 2560 pixels
Pixel physical size	3.45 × 3.45 μm^2
$\vec{\Delta}$ resolution	382 × 764
Camera array	Coplanar
Camera number	15

3. Results and Discussion

3.1. Axisymmetric Jet Flow

We investigate the reconstruction of an under-expanded jet from the non-beveled nozzle. Considering its special axisymmetric structure, all projection planes are identical within the annular region. Therefore, only one projection plane was utilized to simulate the projection planes obtained from 180 viewing angles. Figure 5 illustrates the spatial distribution of $|\nabla n|$ obtained from CFD and reconstruction data. Observational analysis was performed on two selected cross-sections from the radial and flow direction positions. Figure 6 compares $|\nabla n|$ at two sampling positions marked in Figure 5. Note that the dashed line ② is not strictly linear but represents a series of points with maximum $|\nabla n|$ selected successively along the flow direction.

First, as depicted in Figure 5, it can be observed that the reconstruction results from four algorithms effectively identify the primary flow structures of the supersonic jet. Moreover, except for the SIRT algorithm, which captures comparatively lower peak values of $|\nabla n|$, the remaining algorithms reconstruct profiles matching the ones from CFD well.

Figure 6a illustrates that FDK attains superior reconstruction accuracy at the peaks of $|\nabla n|$, with specific peak errors delineated in Table 2. However, its capability to capture peak values in the downstream region might slightly lag behind that of SART and CGLS, as portrayed in Figure 6b. Besides peak errors, we also factor in their computation time and utilize the root mean square error (RMSE, Equation (10)) to evaluate the overall reconstruction quality of the algorithms, as illustrated in Table 2. Benefiting from its analytical reconstruction nature, which circumvents cumbersome iterative processes, FDK enjoys a clear advantage in computation time. Conversely, SART frequently demands more time during reconstruction due to its nested iterative loop. Considering these aspects holistically, it becomes apparent that FDK ensures high-quality reconstruction outcomes and meets the demand for swift reconstruction.

$$\mathrm{RMSE} = \sqrt{\sum_{i=1}^{j} \frac{(|\nabla n|_i^{CFD} - |\nabla n|_i^{Tomo})^2}{j}} \tag{10}$$

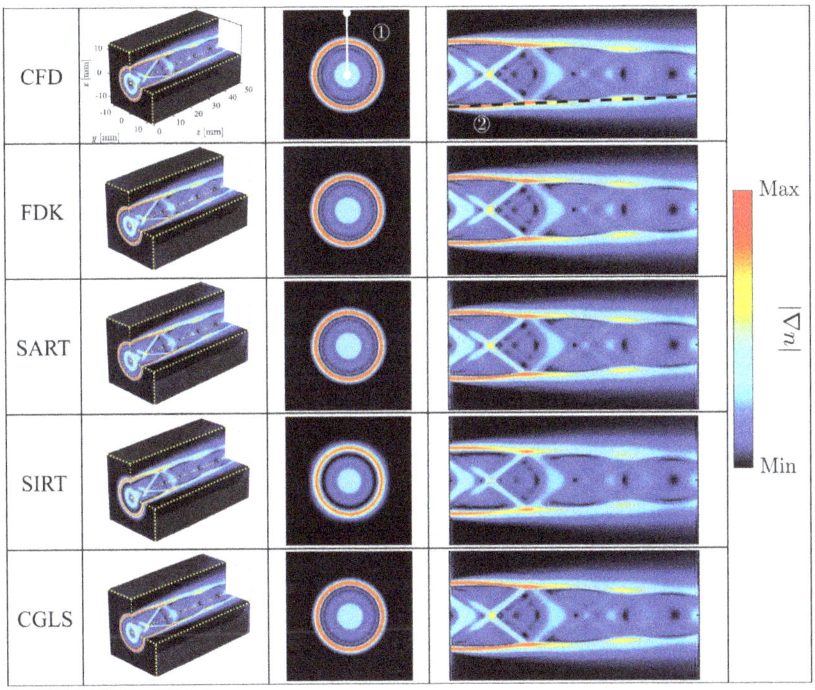

Figure 5. Comparison of $|\nabla n|$ between CFD data and reconstruction results: (**left**) the 3D spatial distribution, (**center**) $xy-$slice at $z = 2.5$ mm, and (**right**) $yz-$slice at $x = 0$ mm.

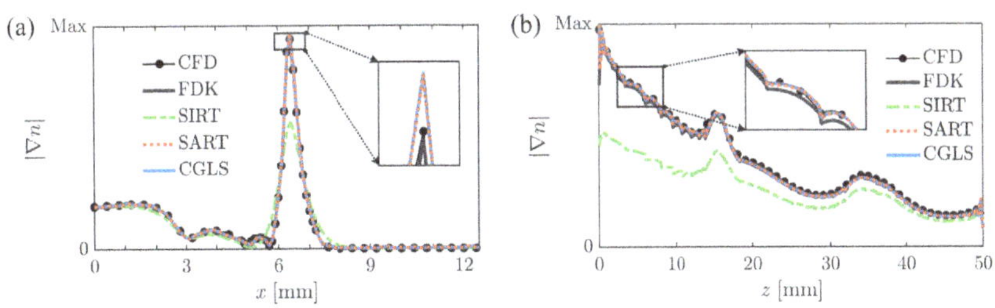

Figure 6. Comparison of $|\nabla n|$ profiles: (**a**,**b**) are sampled at ① and ② marked in Figure 5, respectively.

Table 2. Reconstruction results from different algorithms under axisymmetric and weakly non-axisymmetric conditions.

Algorithom	Axisymmetry			Non-Axisymmetry					
	Δ t(s)	Peak Error (%)	RMSE ($\times 10^{-6}$)	Δ t(s)	Peak Error (%)		RMSE ($\times 10^{-6}$)		
					No-VH	VH	No-VH	VH	VH *
FDK	13	0.3	1.07	2	7.1	7.1	10.89	3.83	16.67
SIRT	401	40	5.97	197	40	–	6.42	–	–
SART	2038	1.8	2.44	292	37	–	6.20	–	–
CGLS	77	1.9	2.02	32	36	9.3	6.24	3.67	5.06

* This case is conducted under the influence of noise pollution with an intensity ratio of 5%.

3.2. Non-Axisymmetric Jet Flow

3.2.1. Weakly Non-Axisymmetric Jet Flow

Next, we consider the reconstruction of a non-axisymmetric under-expanded jet using the $\theta = 60^\circ$ double-beveled nozzle. Due to the axisymmetry breaking in the jet flow structure, tomographic reconstruction necessitates multi-view BOS measurements, i.e., projections. As Section 2.3 outlined, we utilized 15 projections in the current scenario. Similarly, we applied the method described in the preceding subsection to analyze the reconstruction outcomes.

In this scenario, the limited projection data noticeably affect all reconstruction algorithms. As depicted in Figure 7, there is a significant decrease in the overall quality of the reconstruction results. On the one hand, noticeable streak artifacts appear in the reconstruction results, especially in the outer regions of the flow field. On the other hand, the algorithm's ability to capture the peak of $|\nabla n|$ is affected, with the iterative reconstruction algorithms demonstrating inferior performance.

Through specific numerical analysis, clear differences can be observed in the reconstruction results between analytical and iterative algorithms. Table 2 provides a comprehensive assessment of the overall reconstruction performance of these algorithms, indicating that although FDK retains a notable advantage in computation time, its overall reconstruction error is considerably higher compared with the iterative algorithms. The three iterative algorithms produce nearly identical reconstruction results, with the CGLS algorithm exhibiting significantly shorter computation time.

The disparity in $|\nabla n|$ is depicted in Figure 8a. The variation of $|\nabla n|$ from the center to the outer regions of the jet indicates a decreasing level of consistency between the reconstruction results of the four algorithms and the CFD data. Particularly noticeable are significant fluctuations near the edges of the flow field, with this phenomenon being especially prominent in the reconstruction results obtained with FDK. The red box in Figure 8a highlights the superior artifact suppression capabilities of the iterative algorithms compared with FDK, which contributes to the higher RMSE observed in the FDK results (as indicated in Table 2). Figure 8b analyzes the variation of the peak of $|\nabla n|$ along the flow direction. Despite being influenced to some extent, FDK still manages to capture the peak of $|\nabla n|$ effectively. This advantage is particularly evident in regions of high gradients upstream of the jet. The zoomed-in sections in Figure 8a,b illustrate that CGLS demonstrates the superior peak of $|\nabla n|$ capturing capabilities compared other iterative algorithms. Furthermore, Figure 8b indicates that, in the downstream region of the flow, the impact of artifacts on these four algorithms is minimal and the reconstructed results are generally consistent with CFD data, effectively capturing the flow characteristics of the jet.

In summary, while FDK demonstrates efficient computation time and effective capturing of the peak of $|\nabla n|$, it falls short in artifact suppression and overall reconstruction quality compared with iterative algorithms.

From the above reconstruction results, it is evident that the limited projection data cannot adequately facilitate the accurate reconstruction of non-axisymmetric jets, which is the main adverse effect of axisymmetry breaking on TBOS reconstruction. Currently, Equation (5) is seriously ill-posed. Without altering the experimental settings, one feasible approach to improve the reconstruction quality is constraining Equation (5) by utilizing prior information acquired through a comparative analysis between the reconstruction results and CFD data.

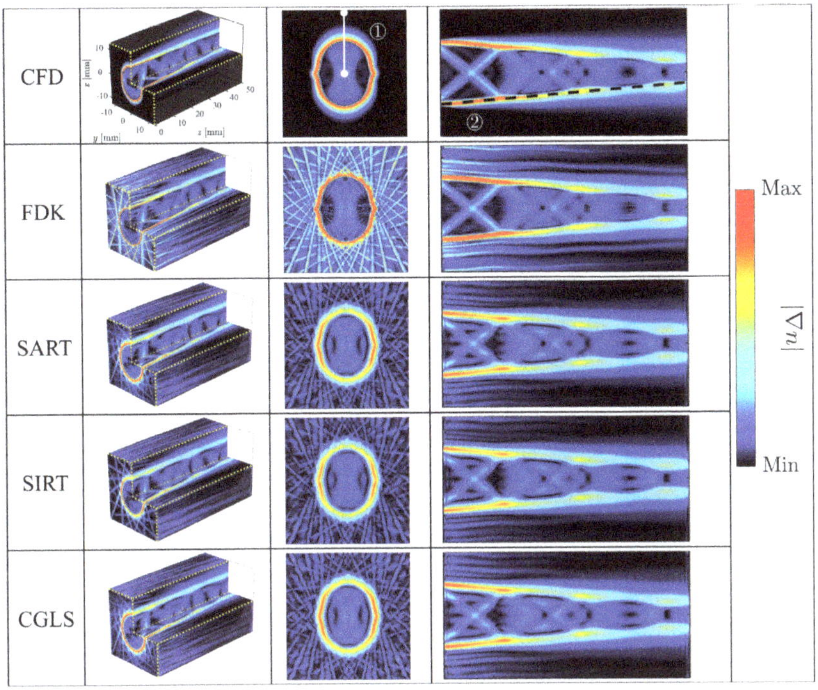

Figure 7. Comparison of $|\nabla n|$ between CFD and reconstruction results: (**left**) the 3D spatial distribution, (**center**) $xy-$slice at $z = 2.5$ mm, and (**right**) $yz-$slice at $x = 0$ mm.

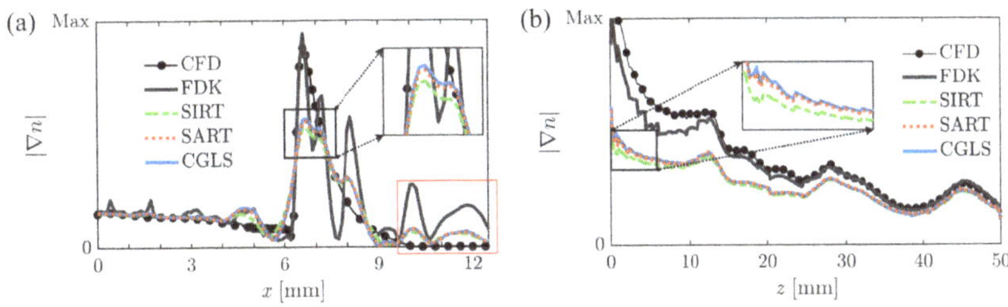

Figure 8. Comparison of $|\nabla n|$ profiles: (**a**,**b**) are sampled at ① and ② marked in Figure 7, respectively.

3.2.2. Visual Hull Technique

To ensure the completeness of the data in the test area, the typically employed reconstruction volume is a parallelepiped, covering the entire region of the test flow field. There are typically no significant changes in the refractive index in regions outside the flow field. Consequently, when the ray propagates along its whole path that does not intersect regions with $|\nabla n| \neq 0$, it will not undergo deviation, resulting in no $|\vec{\Delta}|$ in the imaging plane. However, during the tomographic reconstruction, it is inevitable that voxels without $|\nabla n|$ will receive some numerical allocation, leading to artifacts such as those observed in the reconstruction results depicted in Figure 7. Utilizing this prior information, efforts can be made to systematically exclude voxels within these regions, thereby minimizing the number of unknowns that need to be solved. This strategic approach aims to alleviate the underdetermined nature of the problem under sparse view conditions, consequently improving solution accuracy. Such a methodology is similar to the visual hull technology in computer vision and finds widespread application in tomographic reconstruction [30,44].

The main idea of constructing the VH is performed as follows: By subjecting the acquired displacement field data (Figure 9a) to thresholding and other numerical image processing techniques, the regions devoid of pixel displacement are identified (Figure 9b), thereby facilitating the construction of a 2D silhouette. As the displacement data on the background pattern reflect integrated information of light ray deflections along the path, even in cases where no displacement information is observed on the background pattern, it cannot be conclusively inferred that no deflection occurred along the propagation path of the light ray. Therefore, the 2D silhouette is selected more conservatively in digital image processing.

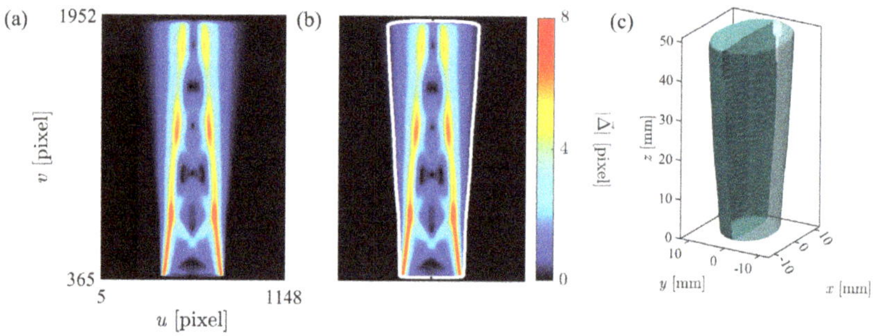

Figure 9. (**a**) Pixel displacement diagram at a certain camera viewing angle. (**b**) Pixel displacement diagram corresponding to the viewing angle under a 2D silhouette (white closed line segment). (**c**) Visual hull, the interior of which represents the flow field flow area.

Subsequently, all projection data containing the 2D silhouettes are individually back-projected onto the reconstruction volume subjected to binary encoding. The voxel values corresponding to the interior regions of the silhouette in the reconstruction volume are assigned as 1, while the voxel values corresponding to the exterior regions are assigned as 0. The summation of voxels at corresponding positions across these reconstruction volumes is conducted. Subsequently, voxels with a cumulative value of 15 after overlaying are identified as focal points for subsequent reconstruction algorithms, thereby eliminating a significant portion of voxels lying beyond the scope of the flow field region (based on prior information, $|\nabla n|$ of these voxels are assumed to be 0). Thus, we obtained a VH that closely conforms to the silhouette of the flow field, as depicted in Figure 9c.

3.2.3. Reconstruction Results with VH

In this section, we incorporate the VH into the reconstruction process to improve the reconstruction results. Based on the analysis of the results in Section 3.2.1, the well-performing FDK and CGLS will be utilized for the following research analysis. Likewise, we conduct a comparative analysis of the reconstruction results with the VH.

After employing the VH, the artifacts outside the flow field region in the reconstruction results have been effectively constrained, as depicted in Figure 10. However, residual artifacts persist within the reconstruction volume, particularly in the FDK reconstruction results. Furthermore, the VH enables the CGLS algorithm to better capture the peak of $|\nabla n|$.

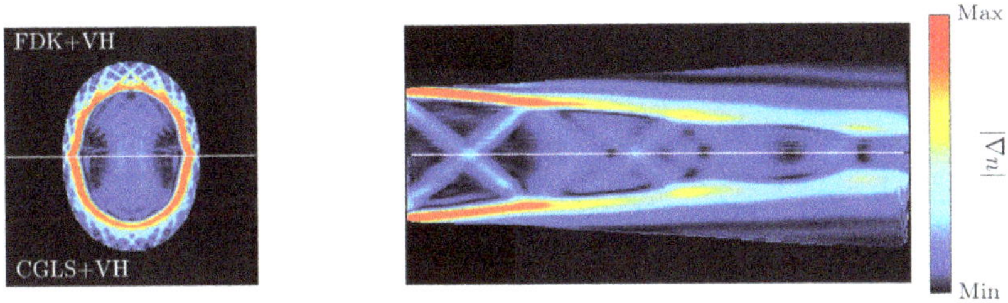

Figure 10. Comparison of $|\nabla n|$ of FDK and CGLS reconstruction results after employing VH.

Figure 11 illustrates the quantitative improvement in reconstruction accuracy achieved by employing the VH with the two algorithms. Here, δ represents the deviation between the reconstruction results and CFD data, with lower numerical values indicating superior reconstruction results. In Figure 11a, the variation trend of $|\nabla n|$ along the radial direction of the jet is depicted. It is evident that the VH significantly suppresses artifacts in the external region of the flow field for both algorithms. Particularly noteworthy is the enhanced ability of CGLS to capture internal flow field data, especially the peak of $|\nabla n|$. Incorporating the VH into CGLS brings the reconstruction results near the inner part and edge of the jet closer to the CFD data, albeit with subtle changes. Conversely, analyzing the variation of reconstruction results with the FDK algorithm, besides effectively suppressing artifacts in the external region of the flow field, there are no apparent changes in other regions. Figure 11b reflects the ability of reconstruction algorithms to capture the peak of $|\nabla n|$ along the flow direction. The improvement in this aspect is significant for the CGLS algorithm with the inclusion of the VH, whereas there is no benefit for FDK. However, the advantageous performance of the FDK algorithm in capturing the peak of $|\nabla n|$ is still evident, as reflected in the peak errors of FDK and CGLS combined with the VH, as shown in Table 2.

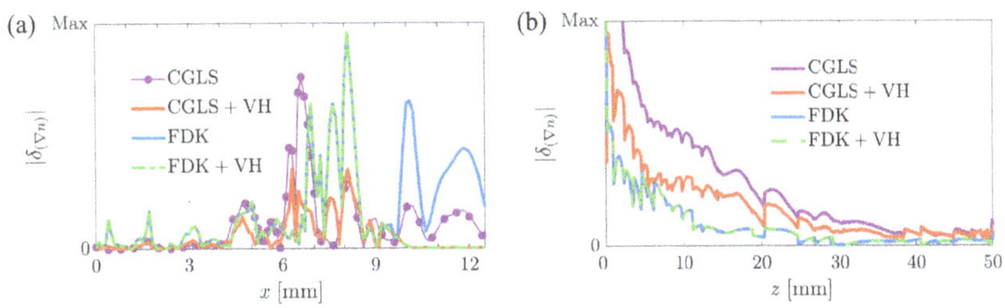

Figure 11. Comparison of $|\nabla n|$ profiles: (**a**,**b**) are sampled at ① and ② marked in Figure 7, respectively.

This observation can be attributed to the differences between the fundamental principles of FDK and the construction methodology of the VH. FDK conducts Fourier transformation on projection data and then back-projects them to reconstruct the volume. The flow field data within the 2D silhouettes mainly correspond to the flow field distribution in physical space. Therefore, the VH does not improve peak errors or reconstruction accuracy in the flow field region for FDK. Its primary function is to rigidly eliminate the influence of artifacts outside the flow field area. In contrast, for iterative algorithms, the VH can be conceptualized as forcibly reducing the weighting of voxels outside the region of interest during the reconstruction process. Consequently, voxels within the flow field region receive more numerical allocation during the iterative process, effectively enhancing the reconstruction quality of the flow field region while mitigating the impact of external artifacts.

In summary, employing the VH significantly improves the overall reconstruction accuracy of the algorithms. In such cases, it can be seen from Table 2 that the reconstruction qualities of the two algorithms are comparably close.

3.3. Noise Impact

During practical TBOS experiments, measuring noise is inevitable. For instance, when capturing background patterns with a camera, there may be a certain degree of noise, leading to discrepancies between the obtained displacement field data and the actual displacement data, ultimately impacting the results of the tomographic reconstruction. Therefore, we will consider the reconstruction results of the two algorithms when the projection data are polluted by noise. Here, we add Gaussian white noise following $N(\mu, \sigma^2)$ to the acquired displacement data, where $\mu = 0$, $\sigma = 0.333 \times R_i \times \max(|\vec{\Delta}|)$, resulting in 99.7% of the noise amplitudes being within the range of $[-R_i, R_i] \times \max(|\vec{\Delta}|)$. Three intensity ratios, $R_i = 1\%$, 2%, and 5%, are used in this work. The noisy $|\vec{\Delta}|$ is then used to generate the VH and tomographic reconstruction. It is important to note that a certain degree of filtering is necessary to reduce the impact of noise during the process of generating the VH.

The reconstructions using noisy displacement data are shown in Figure 12. It is obvious that when the noise intensity is low, both reconstruction algorithms are affected, but overall, they still achieve relatively good reconstruction results. As the noise intensity gradually increases to 5%, the reconstruction results of the FDK algorithm are severely affected, making it difficult to extract useful information. In contrast, although the CGLS algorithm is also significantly affected, it demonstrates better noise suppression than the FDK algorithm. Therefore, the CGLS algorithm exhibits better robustness, making it more conducive to reconstructing the target flow field in practical TBOS experiments. Here, we only briefly compared the robustness of the two algorithms in the presence of noise pollution in the data. Subsequently, different regularization methods will be employed to constrain the solving process for noise reduction. However, this aspect will not be extensively discussed in this paper.

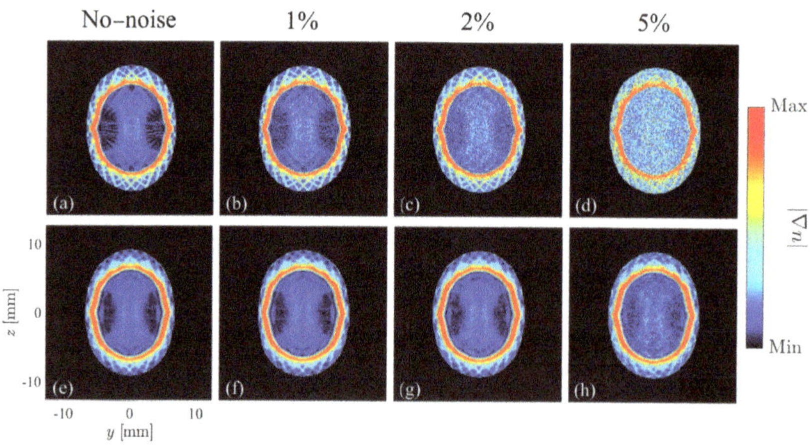

Figure 12. Upon incorporating the VH, a comparative analysis of the reconstruction results: (**a–d**) are the cases using FDK with the noise intensity ratio of 0%(no-noise), 1%, 2%, and 5% respectively, and (**e–h**) are the cases using CGLS under the same condition.

3.4. Jet Flow Structure

This subsection will briefly compare the jet flows under different nozzles based on the reconstruction results. As shown in Figure 13, we include the jet flow from the $\theta = 30°$ double-beveled nozzle in addition to the previous reconstruction results. It can be observed that the main structures of the jet flow under different nozzle conditions are well reconstructed. Therefore, we can use this to analyze the jet flow structure briefly.

It can be observed that when there are double bevels, in the case of under-expanded jet flow, the upstream portion of the jet comes into contact with the surroundings earlier. Due to the higher jet flow static pressure, the transverse pressure relief accelerates the spread of the jet, leading to a transition of the jet cross-section from circular to elongated. As the bevel angle increases, this phenomenon becomes more pronounced. Therefore, compared with a circular nozzle, the jet flow from a double-beveled nozzle may be easier to control and adjust regarding jet direction.

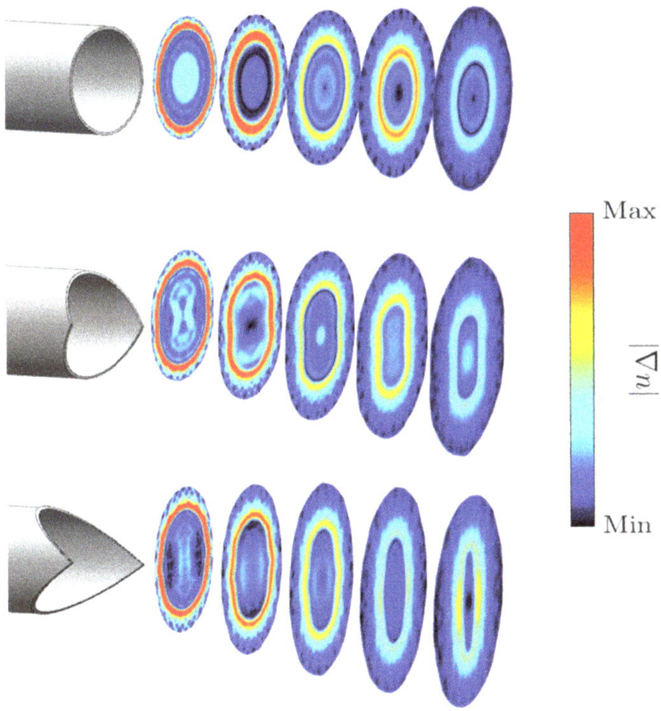

Figure 13. Cross-sectional views depicting the downstream development of jet flows from different nozzles: (**top**) circular nozzle, (**middle**) $\theta = 30^\circ$ double−beveled nozzle, and (**bottom**) $\theta = 60^\circ$ double−beveled nozzle.

At the same time, from the $yz-$slice in Figures 5 and 7, it can be observed that, within the under-expanded jet flow, two shock waves appear internally. These shock waves undergo multiple reflections along the downstream potential core of the jet, resulting in prominent diamond shocks. It is foreseeable that these shocks will dissipate downstream with convection, while the shear layers of the jet will gradually merge at the end of the potential core. With the increase in bevel angle, it can be observed that the shear layers of the jet merge more rapidly, resulting in a shorter length of the potential core. This implies better jet mixing performance. Therefore, compared with jet flows from circular nozzles, jet flows from double-beveled nozzles can achieve better jet mixing performance.

4. Conclusions

This paper uses multi-view-based TBOS to reconstruct axisymmetric and weakly non-axisymmetric under expanded supersonic jets. Several classical TBOS reconstruction algorithms are compared quantitatively, including FDK, SART, SIRT, and CGLS. For axisymmetric jet flows, all algorithms except for SIRT achieve high-quality reconstruction, while FDK outperforms other methods in terms of reconstruction speed. For weakly non-axisymmetric jet flows, view sparseness leads to the emergence of prominent streak artifacts outside the reconstructed flow field regions. A significantly improved reconstruction quality can be achieved by imposing the VH during the reconstruction. Furthermore, CGLS outperforms the other methods in suppressing reconstruction artifacts, especially when there are strong noise contaminations. Finally, by analyzing the supersonic jet emanating from the double-beveled nozzle, TBOS measurements confirm that a double-beveled nozzle can induce a wider jet spread phenomenon and enhance jet mixing.

While we have considered the practical limitation on the number of cameras in TBOS experiments, in reality, the current number is still relatively high. Under more complex observation conditions, the actual number of cameras that can be used may not exceed 10, further affecting the experimental results. However, our current research has not yet conducted an in-depth analysis of this aspect. In future studies, attempts will be made to integrate more prior information about the flow to further optimize the reconstruction process, aiming to enhance the reconstruction quality under even sparser view conditions.

Author Contributions: Conceptualization, Y.X. and J.L.; methodology, J.L.; formal analysis, T.J.; data curation, T.J. and J.W.; writing—original draft preparation, T.J.; writing—review and editing, J.L., J.W., and Y.X.; funding acquisition, Y.X.; supervision, Y.X. All authors have read and agreed to the published version of the manuscript.

Funding: This research was funded by the National Natural Science Foundation of China (NSFC Grants No. 12102028).

Data Availability Statement: The data supporting this study's findings are available from the corresponding author upon reasonable request.

Conflicts of Interest: The authors declare no conflicts of interest.

Abbreviations

The following abbreviations are used in this manuscript:

NPR	nozzle pressure ratio
IOR	index of refraction
BOS	background-oriented Schlieren
TBOS	tomographic background-oriented Schlieren
FBP	Filtered Back Projection
FDK	Feldkamp, Davis, and Kress
ART	algebraic reconstruction technique
SART	Simultaneous Algebraic Reconstruction Technique
SIRT	Simultaneous Iterative Reconstruction Technique
CGLS	Conjugate Gradient Least Square
CFD	Computational Fluid Dynamics
RANS	Reynolds-Averaged Navier–Stokes
VH	visual hull

References

1. Hadjadj, A.; Kudryavtsev, A.N.; Ivanov, M.S. Numerical Investigation of Shock-Reflection Phenomena in Overexpanded Supersonic Jets. *AIAA J.* **2004**, *42*, 570–577. [CrossRef]
2. Matsuo, S.; Setoguchi, T.; Nagao, J.; Alam, M.M.A.; Kim, H.D. Experimental study on hysteresis phenomena of shock wave structure in an over-expanded axisymmetric jet. *J. Mech. Sci. Technol.* **2011**, *25*, 2559–2565. [CrossRef]
3. Schmisseur, J.D.; Gaitonde, D.V. Numerical simulation of Mach reflection in steady flows. *Shock Waves* **2011**, *21*, 499–509. [CrossRef]
4. Shimshi, E.; Ben-Dor, G.; Levy, A. Numeric study of flow separation and shock reflection hysteresis in planar nozzles. *Int. J. Aerosp. Innov.* **2010**, *2*, 221–233. [CrossRef]
5. Tam, C.K.W. Supersonic Jet Noise. *Annu. Rev. Fluid Mech.* **1995**, *27*, 17–43. [CrossRef]

6. Tam, C.; Golebiowski, M.; Seiner, J. On the two components of turbulent mixing noise from supersonic jets. In Proceedings of the Aeroacoustics Conference, State College, PA, USA, 6–8 May 1996. [CrossRef]
7. Raman, G. Supersonic jet screech: Half-century from powell to the present. *J. Sound Vib.* **1999**, *225*, 543–571. [CrossRef]
8. Seiner, J.; Norum, T. Aerodynamic aspects of shock containing jet plumes. In Proceedings of the 6th Aeroacoustics Conference, Hartford, CT, USA, 4–6 June 1980. [CrossRef]
9. Phanindra, B.C.; Rathakrishnan, E. Corrugated Tabs for Supersonic Jet Control. *AIAA J.* **2010**, *48*, 453–465. [CrossRef]
10. New, T.H.; Tsovolos, D. Influence of nozzle sharpness on the flow fields of V-notched nozzle jets. *Phys. Fluids* **2009**, *21*, 084107. [CrossRef]
11. Shi, S.; New, T.H. Some observations in the vortex-turning behaviour of noncircular inclined jets. *Exp. Fluids* **2013**, *54*, 1614. [CrossRef]
12. Zaman, K.B.M.Q. Spreading characteristics of compressible jets from nozzles of various geometries. *J. Fluid Mech.* **1999**, *383*, 197–228. [CrossRef]
13. Gutmark, E.J.; Schadow, K.C.; Yu, K.H. Mixing Enhancement in Supersonic Free Shear Flows. *Annu. Rev. Fluid Mech.* **1995**, *27*, 375–417. [CrossRef]
14. Kim, J.H.; Samimy, M. Mixing enhancement via nozzle trailing edge modifications in a high speed rectangular jet. *Phys. Fluids* **1999**, *11*, 2731–2742. [CrossRef]
15. Wu, J.; New, T.H. An investigation on supersonic bevelled nozzle jets. *Aerosp. Sci. Technol.* **2017**, *63*, 278–293. [CrossRef]
16. Rice, E.; Raman, G. Mixing noise reduction for rectangular supersonic jets by nozzle shaping and induced screech mixing. In Proceedings of the 15th Aeroacoustics Conference, Long Beach, CA, USA, 25–27 October 1993. [CrossRef]
17. Raman, G. Screech tones from rectangular jets with spanwise oblique shock-cell structures. *J. Fluid Mech.* **1997**, *330*, 141–168. [CrossRef]
18. Richard, H.; Raffel, M. Principle and applications of the background oriented schlieren (BOS) method. *Meas. Sci. Technol.* **2001**, *12*, 1576–1585. [CrossRef]
19. Dalziel, S.B.; Hughes, G.O.; Sutherland, B.R. Whole-field density measurements by 'synthetic schlieren'. *Exp. Fluids* **2000**, *28*, 322–335. [CrossRef]
20. Tipnis, T.J.; Finnis, M.V.; Knowles, K.; Bray, D. Density measurements for rectangular free jets using background-oriented schlieren. *Aeronaut. J.* **2013**, *117*, 771–785. [CrossRef]
21. Ota, M.; Kurihara, K.; Aki, K.; Miwa, Y.; Inage, T.; Maeno, K. Quantitative density measurement of the lateral jet/cross-flow interaction field by colored-grid background oriented schlieren (CGBOS) technique. *J. Vis.* **2015**, *18*, 543–552. [CrossRef]
22. Wong, P. *Investigation of Axi-Symmetric Laminar and Turbulent Jets with Background Oriented Schlieren (BOS)*; VKI Stagiaire Report 6; von Karman Institute: Sint-Genesius-Rode, Belgium, 2001.
23. Kirmse, T. Weiterentwicklung des Messsystems BOS (Background Oriented Schlieren) zur Quantitativen Bestimmung axialsymmetrischer Dichtefelder. Ph.D. Thesis, Institute of Aerodynamics and Flow Technology, Göttingen, Germany, 2003. Available online: https://elib.dlr.de/49602/ (accessed on 7 April 2024).
24. Venkatakrishnan, L.; Meier, G.E.A. Density measurements using the Background Oriented Schlieren technique. *Exp. Fluids* **2004**, *37*, 237–247. [CrossRef]
25. Tan, D.J.; Edgington-Mitchell, D.; Honnery, D. Measurement of density in axisymmetric jets using a novel background-oriented schlieren (BOS) technique. *Exp. Fluids* **2015**, *56*, 204. [CrossRef]
26. Xiong, Y.; Kaufmann, T.; Noiray, N. Towards robust BOS measurements for axisymmetric flows. *Exp. Fluids* **2020**, *61*, 178. [CrossRef]
27. Kirby, R. Tomographic Background-Oriented Schlieren for Three-Dimensional Density Field Reconstruction in Asymmetric Shock-containing Jets. In Proceedings of the 2018 AIAA Aerospace Sciences Meeting, Kissimmee, FL, USA, 8–12 January 2018. [CrossRef]
28. Nicolas, F.; Donjat, D.; Léon, O.; Le Besnerais, G.; Champagnat, F.; Micheli, F. 3D reconstruction of a compressible flow by synchronized multi-camera BOS. *Exp. Fluids* **2017**, *58*, 46. [CrossRef]
29. Ihrke, I. Reconstruction and Rendering of Time-Varying Natural Phenomena. Ph.D. Thesis, Universitat des Saarlandes, Saarbrücken, Germany, 2007.
30. Nicolas, F.; Todoroff, V.; Plyer, A.; Le Besnerais, G.; Donjat, D.; Micheli, F.; Champagnat, F.; Cornic, P.; Le Sant, Y. A direct approach for instantaneous 3D density field reconstruction from background-oriented schlieren (BOS) measurements. *Exp. Fluids* **2016**, *57*, 13. [CrossRef]
31. Raffel, M.; Willert, C.E.; Scarano, F.; Kähler, C.J.; Wereley, S.T.; Kompenhans, J. *Particle Image Velocimetry: A Practical Guide*; Springer International Publishing: Cham, Switzerland, 2018. [CrossRef]
32. Grauer, S.J.; Mohri, K.; Yu, T.; Liu, H.; Cai, W. Volumetric emission tomography for combustion processes. *Prog. Energy Combust. Sci.* **2023**, *94*, 101024. [CrossRef]
33. Kak, A.C.; Slaney, M. *Principles of Computerized Tomographic Imaging*; Number 33 in Classics in Applied Mathematics; Society for Industrial and Applied Mathematics: Philadelphia, PA, USA, 2001.
34. Feldkamp, L.A.; Davis, L.C.; Kress, J.W. Practical cone-beam algorithm. *J. Opt. Soc. Am. A* **1984**, *1*, 612. [CrossRef]
35. Andersen, A.H.; Kak, A.C. Simultaneous Algebraic Reconstruction Technique (SART): A Superior Implementation of the Art Algorithm. *Ultrason. Imaging* **1984**, *6*, 81–94. [CrossRef] [PubMed]

36. Sabaté Landman, M.; Biguri, A.; Hatamikia, S.; Boardman, R.; Aston, J.; Schönlieb, C.B. On Krylov methods for large-scale CBCT reconstruction. *Phys. Med. Biol.* **2023**, *68*, 155008. [CrossRef]
37. Rajendran, L.K.; Bane, S.P.M.; Vlachos, P.P. PIV/BOS synthetic image generation in variable density environments for error analysis and experiment design. *Meas. Sci. Technol.* **2019**, *30*, 085302. [CrossRef]
38. CFX-Solver, A. *Theory Guide, Release ll*; ANSYS Inc.: Canonsburg, PA, USA, 2006.
39. Aswin, G.; Chakraborty, D. Numerical simulation of transverse side jet interaction with supersonic free stream. *Aerosp. Sci. Technol.* **2010**, *14*, 295–301. [CrossRef]
40. Wu, J.; Lim, H.D.; Wei, X.; New, T.H.; Cui, Y.D. Flow Characterization of Supersonic Jets Issuing From Double-Beveled Nozzles. *J. Fluids Eng.* **2019**, *141*, 011202. [CrossRef]
41. Lang, H.M.; Oberleithner, K.; Paschereit, C.O.; Sieber, M. Measurement of the fluctuating temperature field in a heated swirling jet with BOS tomography. *Exp. Fluids* **2017**, *58*, 88. [CrossRef]
42. Atcheson, B.; Ihrke, I.; Heidrich, W.; Tevs, A.; Bradley, D.; Magnor, M.; Seidel, H.P. Time-resolved 3d capture of non-stationary gas flows. *ACM Trans. Graph.* **2008**, *27*, 1–9. [CrossRef]
43. Pan, C.; Xue, D.; Xu, Y.; Wang, J.; Wei, R. Evaluating the accuracy performance of Lucas-Kanade algorithm in the circumstance of PIV application. *Sci. China Phys. Mech. Astron.* **2015**, *58*, 104704. [CrossRef]
44. Liu, H.; Shui, C.; Cai, W. Time-resolved three-dimensional imaging of flame refractive index via endoscopic background-oriented Schlieren tomography using one single camera. *Aerosp. Sci. Technol.* **2020**, *97*, 105621. [CrossRef]

symmetry

MDPI

Article

A Lagrangian Analysis of Tip Leakage Vortex in a Low-Speed Axial Compressor Rotor

Jiexuan Hou [1,2], Yangwei Liu [1,2,*] and Yumeng Tang [1,2]

[1] School of Energy and Power Engineering, Beihang University, Beijing 100191, China; houjx@buaa.edu.cn (J.H.); tangyumeng@buaa.edu.cn (Y.T.)
[2] National Key Laboratory of Science and Technology on Aero-Engine Aero-Thermodynamics, Beihang University, Beijing 100191, China
* Correspondence: liuyangwei@126.com or liuyangwei@buaa.edu.cn

Abstract: A Lagrangian method is introduced to analyze the tip leakage vortex (TLV) behavior in a low-speed axial compressor rotor. The finite-time Lyapunov exponent (FTLE) fields are calculated based on the delayed detached-eddy simulation (DDES) results and identifying the FTLE ridges as Lagrangian coherent structures (LCSs). The computational method of the FTLE field in three-dimensional unsteady flow fields is discussed and then applied to the instantaneous flow fields at both the design and near-stall conditions. Results show that the accuracy of the particle trajectory and the density of the initial grid of the particle trajectory greatly affect the results of the FTLE field and, thus, the LCSs. Compared to the Eulerian *Q* method, which is calculated based on the symmetric and anti-symmetric components of the local velocity gradient tensor, the Lagrangian method has great potential in unraveling the mechanism of complex vortex structures. The LCSs show a transport barrier between the TLV and the secondary TLV, indicating two separate vortices. The aLCSs show the bubble-like and bar-like structure in the isosurfaces corresponding to the bubble and spiral breakdown patterns.

Keywords: delayed detached-eddy simulation; Lagrangian coherence structure; tip leakage vortex; compressor rotor

1. Introduction

The tip leakage flow (TLF), which is driven by the pressure difference between the pressure side and the suction side of the blade, and which passes through the gap in turbomachinery [1–3], is characterized by large-scale vortices, such as the tip leakage vortex (TLV). The TLF has a significant impact on the compressor's pressure rise, efficiency, aerodynamic noise, and stall margin, and it is influenced by many factors [4–6]. Although researchers have made many efforts to mitigate the undesirable adverse effects of the TLF over several decades, the complexity of its mechanism still limits the understanding and makes it a major concern in the design of modern axial compressors [7–9]. Therefore, a better understanding of these complexities and the details of fluid transport should be made and will undoubtedly help to improve the design of turbomachinery in engineering practice.

The analysis of the flow structures in the blade tip region is mainly based on conventional Eulerian methods, by which the flow structures are identified from the spatial structure of quantities derived from the velocity field, pressure field, or their gradients [10]. Helmholtz proposed the vorticity or the curl of velocity, which is the anti-symmetric component of the velocity gradient, to identify the vortex structures in the flow field [11]. Indeed, most vortex identifications are based on the symmetric and anti-symmetric components of the local velocity gradient. Taking the Galilean invariance into consideration, vortex identifications such as the Q criterion, Δ criterion, λ_2 criterion, and λ_{ci} criterion are proposed [12–15]. New Eulerian vortex identification methods continue to be proposed in modern academia, and the Liutex criterion was proposed with the aim of solving the problem in threshold determination and vortex behavior near the wall [16], subsequently being used to analyze complex flows in turbomachinery [17,18]. These Eulerian methods have been widely applied in the studies of TLF.

A trajectory-based method from a Lagrangian perspective was first proposed by Haller and has then developed by many researchers in recent years [19]. Compared with the Eulerian methods, the Lagrangian methods are objective (not only Galilean-invariant but also frame-independent). The flow can be regarded as a dynamic system of fluid particles rather than a continuum, and the Lagrangian coherence structures (LCSs) are used to divide dynamic characteristic regions and understand fluid transport by extracting the stable and unstable manifold. These manifolds represent the exact vortex boundaries of the time-dependent separation and reattachment profiles. For steady flow, the streamline coincides with the pathline. For any point in space, only one streamline passes through that point; therefore, by observing the Eulerian velocity field, we can understand the transport structure. However, for the unsteady flow, the streamline changes with time, and the division of transport structure and dynamic characteristic area is no longer as intuitive as the steady flow, and so the advantages of the Lagrangian methods emerge.

The Lagrangian method has been applied to a variety of flows, such as geophysical flows [20–22], meteorological flows [23], and biological flows [24,25]. For the engineering flows, most of the analyses focus on the two-dimensional (2D) flows and quasi-three-dimensional (Q3D) flows based on experiment data or numerical simulation results. The LCSs in the boundary layer are identified to study its mechanism, such as the transportation properties and transition process of the boundary layer [26–28]. The flow over a cylinder is another typical flow field studied by the LCSs to reveal the formation and shedding mechanism of the vortex in the cylindrical wake [29–32]. In the external flow, the LCSs are analyzed to reveal the mechanism of flow separation, cavitation, and flow control in delta wings and airfoils [33–38]. LCS theory has great potential for unraveling the transport mechanism in engineering flows, and introducing it to the analysis of the TLF in axial compressors should enable a deeper understanding of the flow.

In this study, considering the great potential of LCS in unraveling the hidden transport phenomena in engineering flows, the Lagrangian method is introduced to a low-speed compressor rotor, and the application of LCSs identified by FTLE fields in three-dimensional (3D) unsteady flow fields is preliminarily explored based on delayed detached-eddy simulation (DDES) results. In Sections 2.1 and 2.2, the Lagrangian method and the Eulerian method of Q criterion, calculated based on the symmetric and anti-symmetric components of the local velocity gradient tensor, are introduced, respectively. Then, the numerical setups of the low-speed rotor are illustrated in Section 2.3. In Sections 3.1.1–3.1.3, three main factors of the computational method of the FTLE field are discussed. In Section 3.2, the evolution of the TLV is analyzed by the LCSs and compared with the vortex structures identified by the Q criterion, emphasizing the breakdown pattern of the TLV.

2. Methodology

2.1. Lagrangian Method

The local extremum of the finite-time Lyapunov exponent (FTLE) field is proposed to characterize LCSs [39]. Shadden proposed the precise definition of the FTLE ridge as the LCS and, based on some assumptions, derived the expression of the flow mass through the LCS, thus proving the nature of the LCS as a transport boundary [40]. The study of Haller et al. showed that the trough of the FTLE field at a certain moment marks the attracting LCS at that moment [41]. To a certain extent, this has stimulated the development of more reliable mathematical methods that can enable LCSs to identify the surface of a parameterized material surface accurately.

The essence of the FTLE method is to quantify the maximum possible distance between two particles at an infinitely small distance near the same space point at the initial time after moving with the flow for a period of time [42,43]. It is more convenient to use the Lagrangian description. The core of the Lagrangian description of fluid motion is the flow mapping $\boldsymbol{\Phi}_{t_0}^{t}: x_0 \to x(t_0, t: x_0)$ on the phase space P and the time interval $L = [t_0, t_1]$. This mapping maps the initial position $x_0 \in \boldsymbol{P}$ to its position at any time $t \in L$. If we hypothesize that two particles near a point in space are separated by a distance of $\varepsilon\xi(t_0)$, where $0 < \varepsilon \ll 1$, and $\xi(t_0)$ represents an arbitrary unit vector on $\boldsymbol{R}^n$, for the flow field, n is 2 or 3 according to the dimension of the flow field. At time $t_0 + T$, the relative position of these two particles is

$$\Delta\boldsymbol{\Phi}_{t_0}^{t}(t_0 + T, x_0) = \boldsymbol{\Phi}_{t_0}^{t_0+T}(x_0 + \varepsilon\xi(t_0)) - \boldsymbol{\Phi}_{t_0}^{t_0+T}(x_0) = J\boldsymbol{\Phi}_{t_0}^{t_0+T}(x_0)\varepsilon\xi(t_0) + \mathcal{O}(\varepsilon^2) \quad (1)$$

Here $J\boldsymbol{\Phi}_{t_0}^{t_0+T}(x_0)$ represents the Jacobian of $\boldsymbol{\Phi}_{t_0}^{t_0+T}(x_0)$. Landau notation is also used in the equation, and for a constant positive function, $g(x), f(x) = \mathcal{O}(g(x))$ means that for all $x \in \boldsymbol{R}$, $g(x)/f(x)$ is always bounded. The maximum distance between two particles is

$$\begin{aligned}\Delta_{t_0}^{t_0+T}(x_0) &= \lim_{\varepsilon\to 0}\left(\frac{1}{\varepsilon}\max_{|\xi(t_0)|=1}|\Delta\boldsymbol{\Phi}(t_0+T, x_0)|\right)\\ &= \max_{|\xi(t_0)|=1}\sqrt{\langle J\boldsymbol{\Phi}_{t_0}^{t_0+T}(x_0)\xi(t_0), J\boldsymbol{\Phi}_{t_0}^{t_0+T}(x_0)\xi(t_0)\rangle}\\ &= \max_{|\xi(t_0)|=1}\sqrt{\langle \xi(t_0), C_{t_0}^{t_0+T}(x_0)\xi(t_0)\rangle} = \sqrt{\lambda_{\max}(C_{t_0}^{t_0+T}(x_0))}\end{aligned} \quad (2)$$

where $C_{t_0}^{t_0+T}(x_0) = [J\boldsymbol{\Phi}_{t_0}^{t_0+T}(x_0)]^{\mathrm{T}}[J\boldsymbol{\Phi}_{t_0}^{t_0+T}(x_0)]$ means Cauchy–Green strain tensor, $\langle \bullet, \bullet \rangle$ represents the Euclidean inner product. Since this maximum distance tends to increase rapidly, it is more convenient to use its growth exponent $(\log(\Delta_{t_0}^{t_0+T}(x_0)))/|T|$, which is the finite-time Lyapunov exponent:

$$\sigma_{t_0}^{t_0+T}(x_0) = \frac{1}{2|T|}\log(\Delta_{t_0}^{t_0+T}(x_0)) \quad (3)$$

The FTLE is divided into a forward-time FTLE and a backward-time FTLE. The difference between the two is that the former uses forward integration time ($T > 0$) to produce repelling LCS (rLCS), while the latter uses reverse integration time ($T < 0$) to produce attracting LCS (aLCS). To facilitate the comparison of LCS structures with different integration times, the FTLE is nondimensionalized by the maximum value of the whole FTLE field, and this non-dimensional value is denoted as $\mathrm{FTLE}_{\mathrm{rel}}$.

$$\mathrm{FTLE}_{\mathrm{rel}} = \sigma_{t_0}^{t_0+T}(x_0)/\max[\sigma_{t_0}^{t_0+T}(x_0)] \quad (4)$$

In this study, the FTLE field is computed from discrete flow fields, not directly from the fluid equations. This is applied in most practical applications, and it is particularly useful when the velocity field is obtained from observations and simulations. In the following sections, we employ linear interpolation in time and space.

2.2. Eulerian Q Criterion Method

Among all the Eulerian vortex identification methods typically formulated regarding the invariants of the velocity gradient tensor $\boldsymbol{G} = \nabla\boldsymbol{u}$, if we hypothesize that the P, Q, R are the first, second and third invariant of the velocity gradient $\boldsymbol{G}$, then the characteristic equation for the eigenvalues of $\nabla\boldsymbol{u}$ reads:

$$\lambda^3 - P\lambda^2 + Q\lambda - R = 0 \quad (5)$$

The Q criterion, proposed by Hunt et al. [12] identifies the vortices using positive Q for incompressible fluid (especially the large-scale vortex in turbulent flow). Q can be expressed by the following equation for incompressible flow:

$$Q = \frac{1}{2}[tr(\nabla u)]^2 - tr[(\nabla u)^2] = \frac{1}{2}tr[(\nabla u)^2] = \frac{1}{2}(\|\Omega\|^2 - \|S\|^2) > 0 \tag{6}$$

where the strain rate tensor S and the rotation tensor Ω are the symmetric and anti-symmetric components of G, defined as $S = \frac{1}{2}(G + G^T)$ and $\Omega = \frac{1}{2}(G - G^T)$, so that $Q > 0$ is the region in which the vorticity magnitude prevails over the strain-rate magnitude.

2.3. Numerical Setups

2.3.1. Experimental Configurations

Du et al. [44] conducted experimental research in the low-speed, large-scale axial compressor test rig at Beihang University. This compressor features a 1.5-stage layout that includes an inlet guide vane, a rotor, and a stator. Calculations of the Reynolds number, which were based on the length of the rotor's tip chord (C), yielded a value of 750,000. The specific parameters of the compressor are detailed in Table 1. Velocity measurements in great detail were obtained at cross-sections that were almost perpendicular to the direction of the rotor's tip chordwise, utilizing intervals of either 5% or 10% of the chord length through the stereoscopic particle image velocimetry (SPIV) technique. For further information on the experiment, refer to ref. [44].

Table 1. Design parameters of the test facility.

Parameter	Parameter Value
Outer diameter	1.0 m
Hub-to-tip ratio	0.6
Design Speed	1200 r/min
Configuration	Inlet guide vane + Rotor + Stator
Number of rotor blades	17
Mid-span blade chord	152 mm
Mid-span Blade camber angle	40.8°
Mid-span Blade stagger angle	36.5°
Solidity (mid-span)	1.03
Aspect ratio (mid-span)	1.32
Rotor tip gap	3.5 mm
Rotor tip gap/blade height	1.75%

2.3.2. Computational Domain and Numerical Method

With the rapid development of computer resources, the computational fluid dynamics (CFD) technique was rapidly developed and has been widely used to study aerodynamic and aeroelastic problems [45,46]. The choice of the turbulence model plays a pivotal role in the Reynolds-averaged Navier–Stokes equations (RANS) method within engineering fields [47], and it represents a notable challenge in the aerodynamics of compressors [48], particularly concerning the analysis of tip leakage flows [49]. Hence, to explore the evolution of tip leakage flow in the rotor, the DDES method, a hybrid approach combining large eddy simulation (LES) and RANS, is utilized in this study to reduce the need for an extensive computational grid that an LES commonly requires. Compared to the RANS method, the DDES method performs significantly better in predicting the TLF [50]. This study employs the DDES method based on the shear stress transport (SST) k-ω model [51], a turbulent model that is frequently used in turbomachinery simulations [52].

For the simulation focusing on tip leakage flow, the study opted for an O4H-type mesh configuration to maintain high grid quality, as illustrated in Figure 1. The y^+ is set to around 0.8 of the solid walls. The simulation's inlet was positioned at 1.0 chord length ahead of the blade's leading edge, with the entire computational domain extending approximately 4.0 chord lengths in the axial direction downstream. Within the TLV region, at a spanwise height exceeding 80% of the blade, refined meshes were utilized. The mesh has an aspect ratio below 2 and an expansion ratio under 1.05 to enhance the resolution of the TLV structure. The grid spacings in the spanwise, pitchwise, and streamwise dimensions relative to the blade tip's chord length are capped at $\Delta x^+ \leq 110$, $\Delta y^+ \leq 1$, and $\Delta z^+ \leq 120$, respectively. This grid quality aligns with that of prior studies conducted under comparable flow conditions that also employed the DDES method [53,54]. The total number of grid points used in the simulation is around 6.2 million.

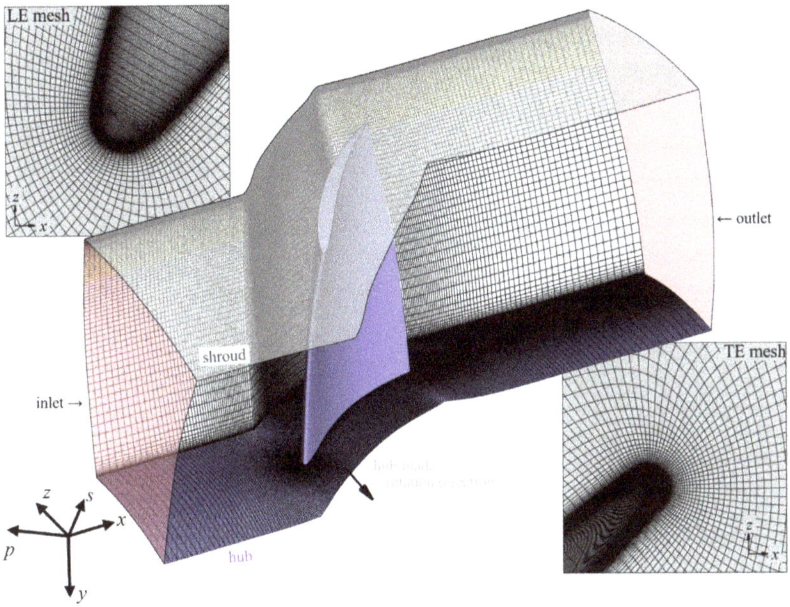

Figure 1. Computation domain and mesh of the rotor.

The simulations were performed in Fluent with a pressure-based implicit solver. The SIMPLE algorithm, an algorithm for coupling pressure and velocity, was applied in the simulations. The spatial aspect of the simulation utilized a bounded central differencing scheme for integration, while temporal discretization was achieved through second-order implicit time integration. To capture the steady total pressure distribution of the inlet flow between the inlet guide vane and the rotor, five-hole probes were strategically placed in the center of the blade passage, and the results of pressure measurements at various radii were applied to the inlet boundary condition.

In the DDES calculations, approximately 1000 time steps (Δt) cover the duration it takes for a rotor blade to pass one pitch. This time step size is selected to adequately capture the dynamics of the unsteady tip leakage flow, ensuring compliance with the Courant–Friedrichs–Lewy condition for numerical stability. The chosen time step is nondimensional, normalized by both the rotor-tip chord length and the speed of the inlet's main flow, set at 5×10^{-4}. During each time step, 30 inner iterations are executed to ensure accuracy and convergence. To comprehensively document the flow's behavior over time, 3000 instantaneous results are recorded throughout the simulation, with a saving interval of every 10 physical time steps, providing a detailed temporal resolution of the flow characteristics.

2.3.3. Validation of the Numerical Method

The simulation results of time-averaged streamwise vorticity of the representative cross-sections at design (DE) and near-stall (NS) conditions are compared with an experimental measurement in Figure 2. To better capture the TLV, the s-y-p coordinate system is set up by rotating the x-axis and z-axis about the y-axis to the position that the x-axis is parallel to the blade tip chord and perpendicular to the cross-sections in the SPIV measurements, as shown in Figure 1. The rotated x-axis direction is termed streamwise, and the rotated z-axis direction is termed pitchwise in the following sections. The flow coefficient has been checked to guarantee that the operating point is compatible with the experiment. The flow coefficients in the simulations are 0.58 with the DE condition and 0.51 with the NS condition, respectively, showing that the agreement between them is quite good. The DDES captures the location and track of TLV at different working conditions. The simulation results capture the location of TLV and show qualitatively good agreement with experimental results.

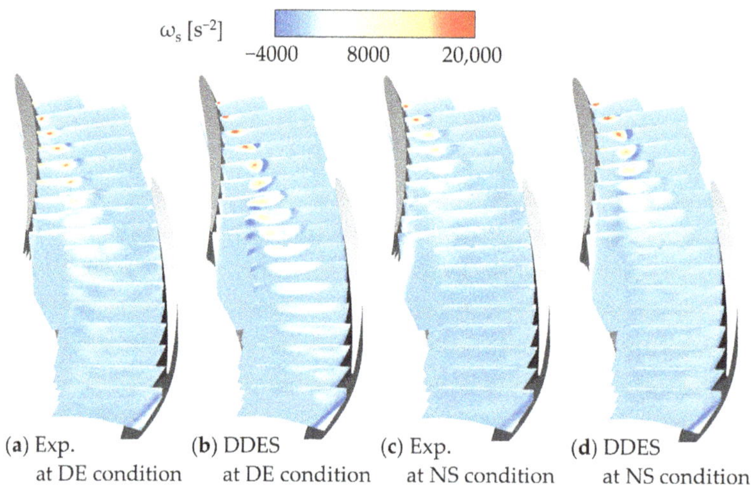

Figure 2. Time-averaged streamwise vorticity of different conditions. (**a**) Experimental result at the DE condition; (**b**) DDES result at the DE condition; (**c**) experiment result at the NS condition; (**d**) DDES result at the NS condition.

3. Results and Discussion

3.1. Effect of Parameters on the LCS Structure in the Low-Speed Rotor

As understood from the calculation process of the FTLE, there are several parameters which affect the results of the LCSs. Three of them are considered to be the main factors, namely the initial grid of the particle trajectory, the trajectory integration method, and the integration time T. In this section, these three factors are discussed separately.

3.1.1. The Initial Grid of the Particle Trajectory

Two regular grids are compared, and the calculation parameters of the FTLE field are shown in Table 2. The initial grids are located at the cross-section of $s/C = 0.5$ and cover the TLV in the time-averaged flow field at DE condition. In the streamwise cross-section plane, the grid is uniform, and two layers of the grid are set in the streamwise direction to compute the derivative along this direction. The grid cells are nearly cubic. The grid of particle trajectory is advected in time as the flow field is steady. The total number of grid points in grid-2 is 6.25 times greater than that in grid-1, attributed to a 2.5-fold increase in the grid point number of grid-1 in each spatial direction. Calculations for both negative and positive integration times are performed, which correspond to aLCSs and rLCSs, respectively. The results are shown in Figure 3.

Table 2. Calculation parameters of FTLE field with different regular grids.

Grid No.	Grid Size	Integration Time Step	Integration Time	Time Integration Method
1	$300(p) \times 180(r)$	$2.5\ \Delta t$	$\pm 2000\ \Delta t$	Fourth-order Runge-Kutta
2	$750(p) \times 450(r)$	$2.5\ \Delta t$	$\pm 2000\ \Delta t$	Fourth-order Runge-Kutta

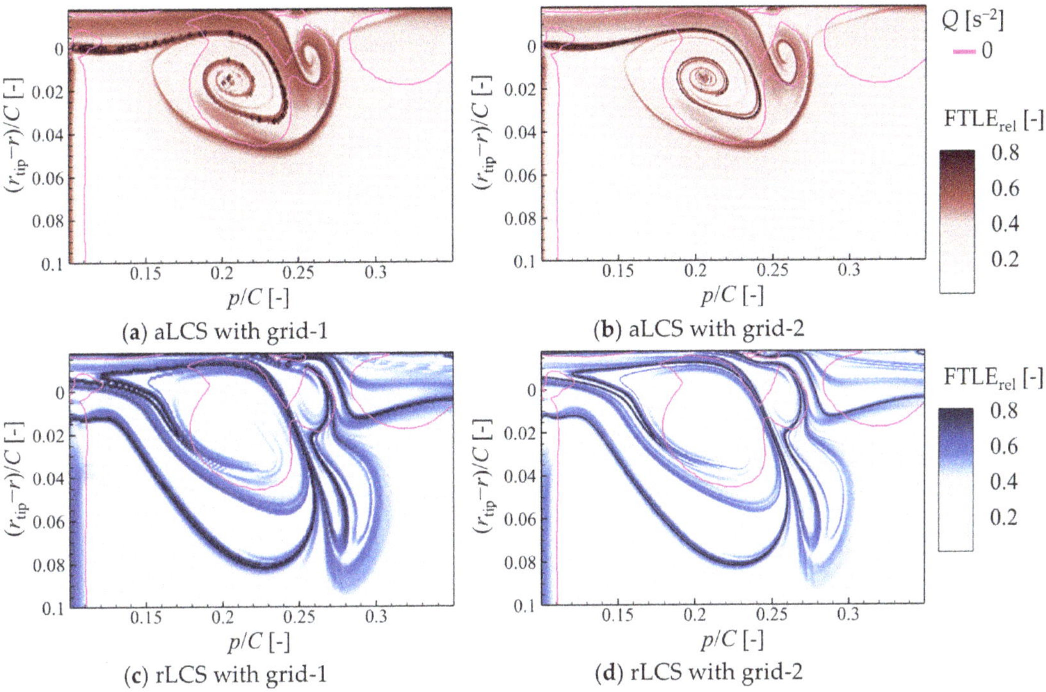

Figure 3. Sketch of initial grids and contours of an FTLE field with different regular grids. Contours of aLCS and isolines of $Q = 0\ s^{-2}$ with (**a**) grid-1 and (**b**) grid-2; contours of rLCS and isolines of $Q = 0\ s^{-2}$ with (**c**) grid-1 and (**d**) grid-2.

The general LCSs are similar in both grids, while the ridges are more evident in the higher-density grid. The LCSs are defined as the 'ridges' of the FTLE field represented by the position of local maximum values, and the exact values are not that important in illustrating the structures. For the aLCSs of the TLV core region, where the curvature of the ridge is much larger and more complicated in a small region, the differences are more significant between the two grids. For the FTLE field of forward time, the ridges are more complicated, and the lower-density initial grid mashes the ridges close to each other, forming into one.

The increasing initial grid points result in more distinct LCSs and higher computation costs simultaneously. Adaptive meshes are generated based on the FTLE fields from grid-1 to balance the calculation accuracy and costs. Regular grid points of four different densities are used to seed at different regions with various FTLE ranges and are particularly refined inside the TLV core, as shown in Table 3. Then, these seeding points are triangulated into the unstructured grid. As the grid density near the ridges stays nondecreasing, the grid numbers of the adaptive grids reduce to 27.5% and 82.7% of the grid-2 for the FTLE field of negative and positive integration time, respectively. The sketch of the adaptive grids and contours of the FTLE fields are shown in Figure 4. With the adaptive grids, the ridges are further clarified, while the general LCSs stay the same. In the core region of the TLV, where the radius of curvature of the ridges is small, the detailed aLCSs are captured with the higher-density grid. For the rLCSs, the decrement of the total grid points of the initial grid is not significant, as the structures are complicated.

Table 3. Grid densities for different FTLE ranges.

Grid Density Level	Grid Size	FTLE Range
1	$150(p) \times 90(r)$	$(-\infty, 0.27]$
2	$300(p) \times 180(r)$	$(0.27, 0.56)$
3	$750(p) \times 450(r)$	$[0.56, +\infty)$
4	$1500(p) \times 900(r)$	$[0.56, +\infty)$ (partial)

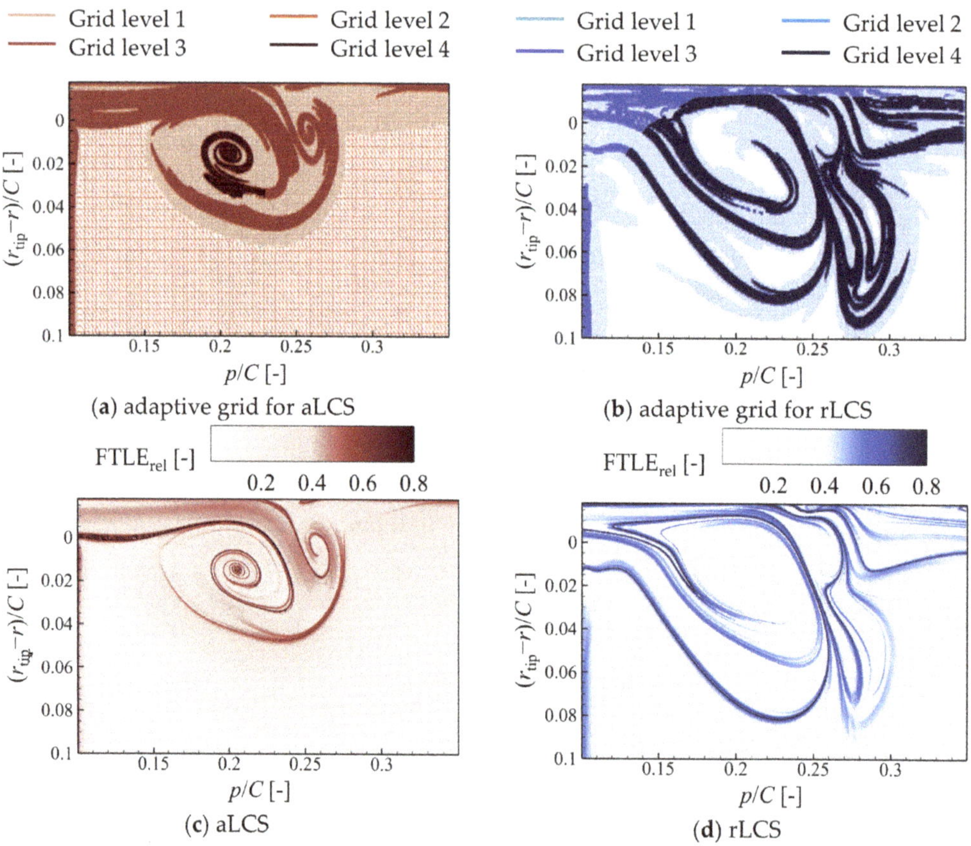

Figure 4. Sketch of adaptive grids and contours of FTLE field with different direction of integration times. (**a**) Sketch of adaptive grid for aLCS; (**b**) sketch of adaptive grid for rLCS; (**c**) contours of aLCS; (**d**) contours of rLCS.

For the calculation of the three-dimensional FTLE field, the cost increases sharply. To capture the LCS as that at the 2D plane, the grid number of the initial particle trajectory grid would be greatly larger than the mesh of the numerical simulation. To save computation resources, the initial grid of particle trajectory is limited to the region inside the blade row within a 25% span near the shroud. The sketch of the initial grid is shown in Figure 5, and the grid is referred to as grid-3D in this study. The total grid number of grid-3D is approximately 4.34 million.

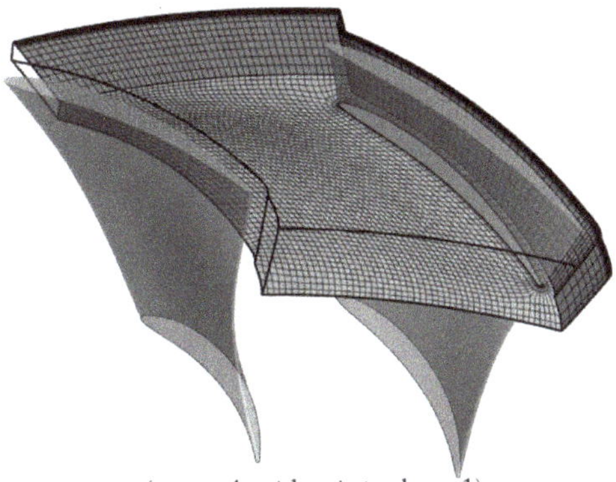

Figure 5. Sketch of the initial grid grid-3D.

Figure 6 shows the FTLE field and vorticity field of the time-averaged flow field with the DE condition. The Eulerian Q criterion is applied as well, and the isolines of $Q = 0\ s^{-2}$ are shown in Figure 6c, along with the vorticity field. Compared to the LCSs calculated from grid-1, the outlines of the general structures at the cross-section plane $s/C = 0.5$ are similar, while the ridges of grid-3d are rougher due to the lower density of the initial grid of the particle trajectory.

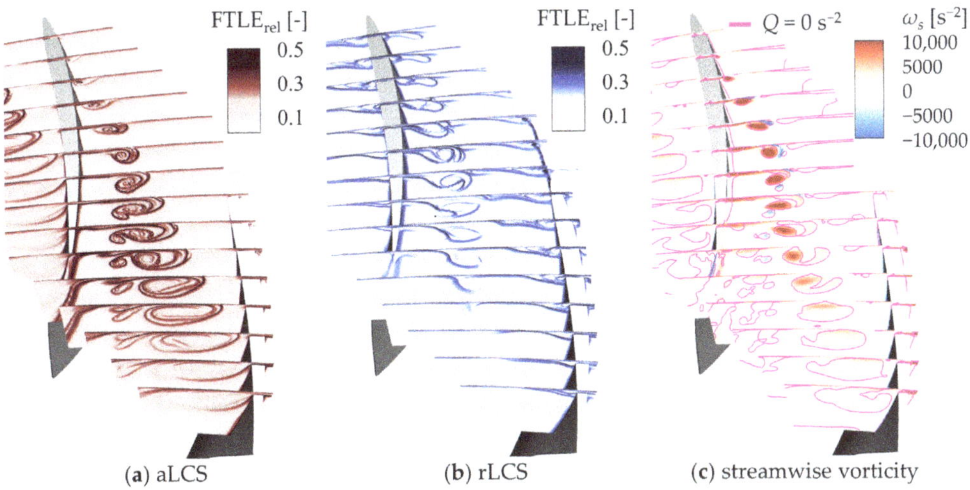

Figure 6. FTLE field and vorticity field of the time-averaged flow field at the DE condition. (**a**) Contours of aLCS; (**b**) contours of rLCS; (**c**) streamwise vorticity with $Q = 0\ s^{-2}$.

A regular grid, termed grid-3Duni, is generated and shown in Figure 7a. The total number of grid points is approximately 8.80 million, and the distributions of the grid points in each direction are uniform. The spanwise range of the grid is narrowed down to a 7.5% span to capture the TLV core and save the computational resources. The grid density at the cross-section plane $s/C = 0.5$ is nearly half of grid-1.

This grid gets clearer ridge structures of the FTLE field than the other 3D grid, especially with the high curvature. However, compared with the results from grid-1, the detailed LCSs cannot be distinguished, while the computational costs are far beyond that of grid-1. Limited by computational resources and efficiency, the identification of LCSs in

3D presents the overall LCSs in the whole flow field, and detailed LCSs are more suitable for identification in 2D.

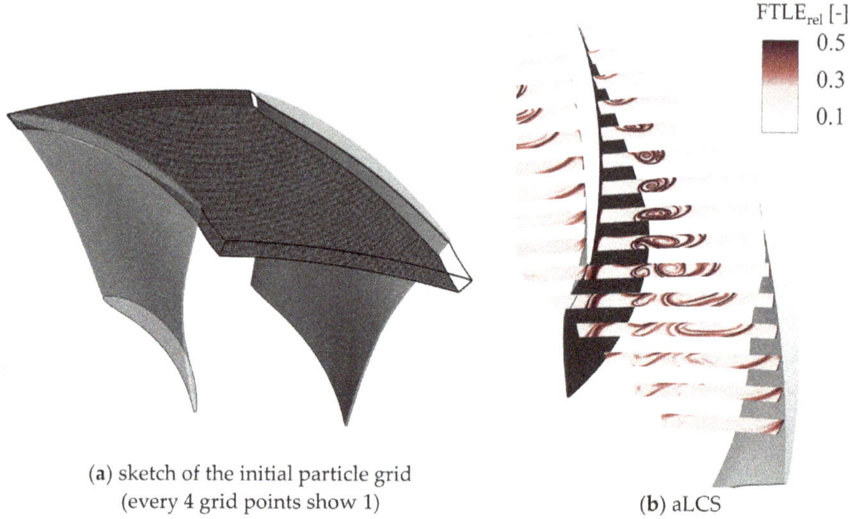

Figure 7. Sketch of grid-3Duni and FTLE field of time-averaged flow field at DE condition. (**a**) sketch of grid-3Duni; (**b**) contours of aLCS.

3.1.2. The Time Integration Method

The time integrations of all the particle trajectories results above are carried out by the fourth-order Runge–Kutta method. To test the effect of the time integration method, the time integrations are performed via the fourth-order Runge–Kutta method, with a longer integration time step, and the linear integration method with two integration time steps; the FTLE fields are shown in Figure 8. The impact of the time integration method falls primarily in the aLCSs, while the rLCSs are almost unaffected. Compared to the aLCSs in Figure 3c, the ridges of the TLV become rougher with the longer time step, while the general shape and scale of the ridges stay the same. The ridges of the induced vortex stand out when using the low-order time integration method. The ridges of both the TLV and the induced vortex shrink towards the core, and the scales of these LCSs are all reduced. The results are more sensitive to the integration time step length when using the linear method. When increasing the integration time step length from 2.5 Δt to the longer integration time step of 20 Δt, the LCSs are significantly different from the other results.

Figure 9 shows the final streamwise position of the particle with different time integration methods. When applying the time integration in a negative time direction, the ridges of the TLV and IV are related to two regions of large particle separations, respectively, with one being near the shear layer at the blade tip, while the other is near the shear layer at the shroud. For an axial rotor, the blade is relatively thin, and the clearance is small. However, the flow inside the gap is complex, featuring vortex structures and jet flows. When particles advect to the gap, a large time interval may lead them to pass the gap within a few integration steps, and thus, it is difficult to capture the flow features inside the gap. The RK4 method performs better than the linear method in tracking the particles inside the gap and thus gives a more stable structure of the LCSs.

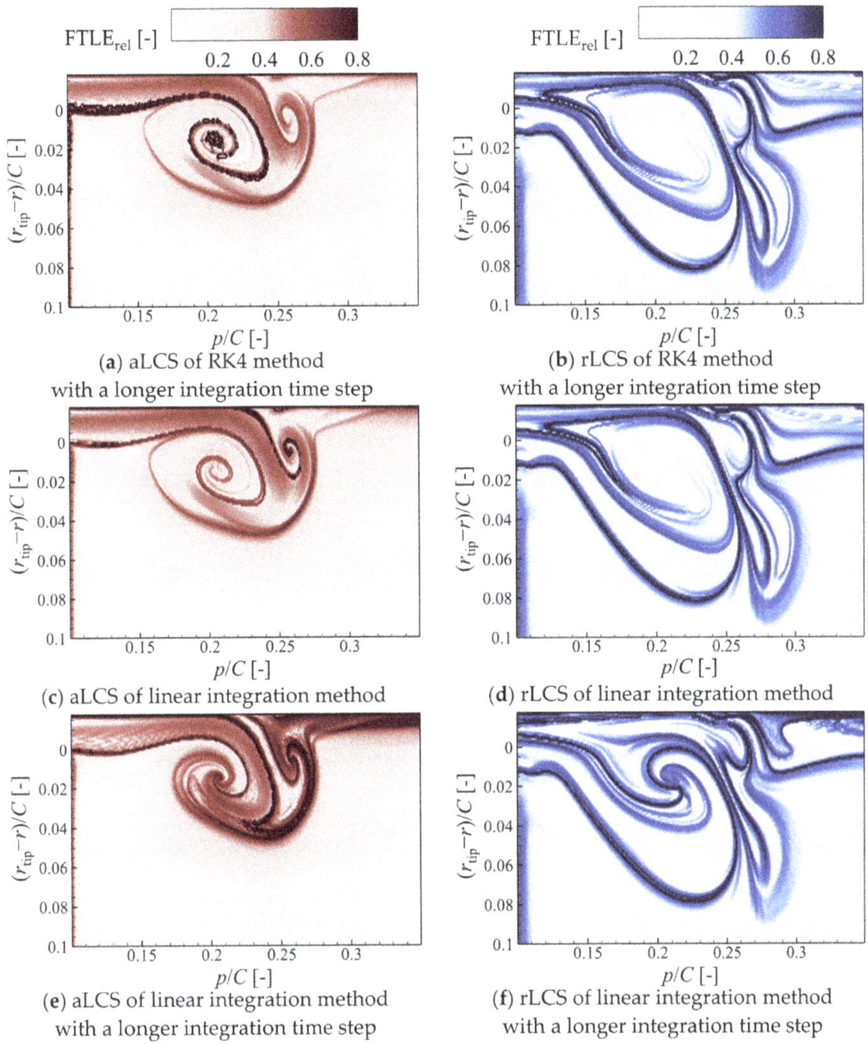

Figure 8. The FTLE field of different time integration methods. (**a**) Contours of aLCS of RK4 method with integration time step of 20 Δt; (**b**) contours of rLCS of RK4 method with an integration time step of 20 Δt; (**c**) contours of aLCS of linear integration method with an integration time step of 2.5 Δt; (**d**) contours of rLCS of linear integration method with an integration time step of 2.5 Δt; (**e**) contours of aLCS of linear integration method with an integration time step of 20 Δt; (**f**) contours of rLCS of linear integration method with an integration time step of 20 Δt.

Figure 10 shows the 3D view of the final position of the particles with different time integration methods with an integration time step of 20 Δt. The differences in the final positions of the particles between the two methods mainly exist in the particles that have passed the blade gap. For the linear method, a great part of the particles is finally located near the shroud. The particles near the shroud tend to be trapped in the shear layer at low speed or hit the shroud and stop. With a longer time step length, the particle trajectories further deviate with time-advecting, and the general LCSs are altered.

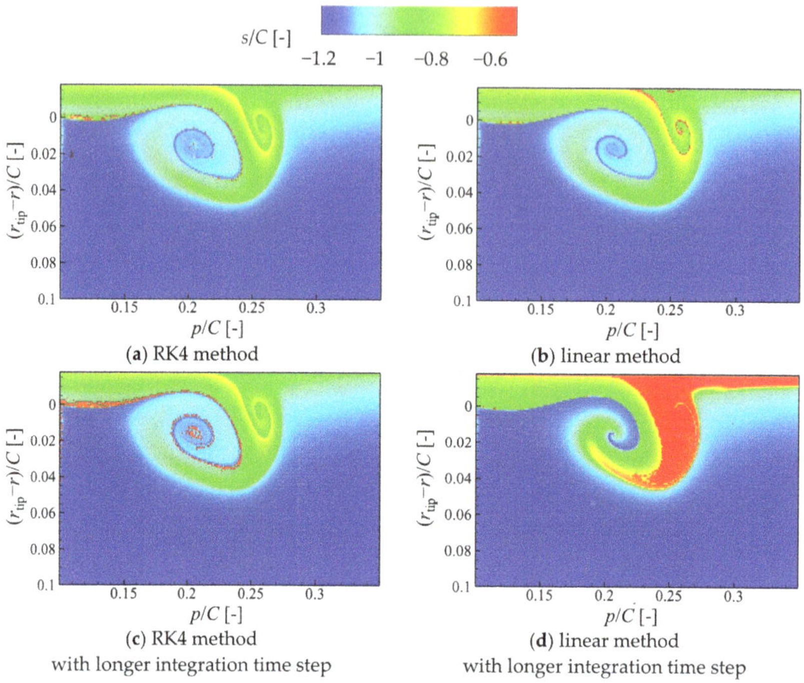

Figure 9. Final streamwise position of the particles with different time integration methods. (**a**) RK4 method with an integration time step of 2.5 Δt; (**b**) linear method with an integration time step of 2.5 Δt; (**c**) RK4 method with an integration time step of 20 Δt; (**d**) linear method with an integration time step of 20 Δt.

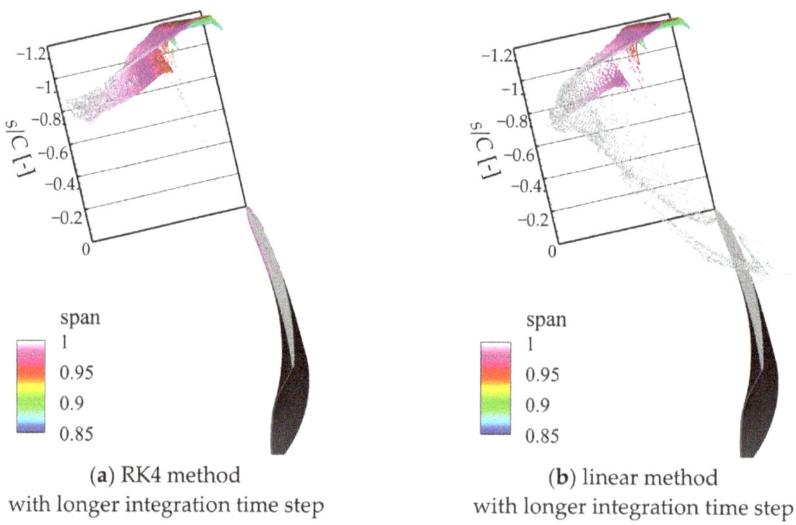

Figure 10. Final position of the particles with different time integration methods. (**a**) RK4 method with an integration time step of 20 Δt; (**b**) linear method with an integration time step of 20 Δt.

3.1.3. The Integration Time

The integration time (T) is another critical factor for the LCSs, as it determines the final position of the particle. The LCSs of a series of integration times for both negative and positive time directions are shown in Figure 11, and the corresponding final particle positions are shown in Figure 12. For the negative time direction, the particles transport ahead of the blade row where the flow is relatively undisturbed, and the general structures of the aLCSs are similar. In the positive time integration direction, the particles are transported downstream and entrained into different vortex structures, or even into the gap of the adjacent blade. Thus, more rLCSs emerge with the increase in the integration time. In this rotor, the period of a rotor blade passing through one pitch is about 1000 Δt, and the period of a particle passing through the blade row is about 1500 Δt. As the integration time length exceeds 1000 Δt, the primary LCSs tend toward stability.

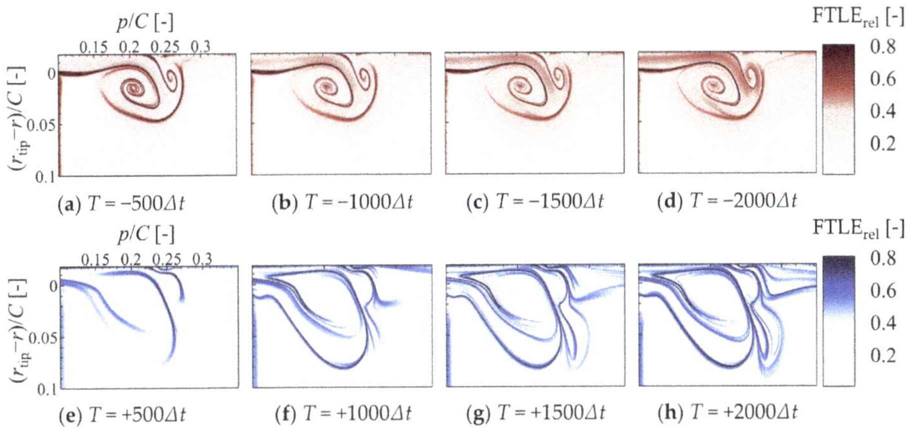

Figure 11. The LCSs of a series of integration times, both negative and positive, and the time direction of the time-averaged flow field at the DE condition. (**a**) Integration time of $T = -500\ \Delta t$; (**b**) integration time of $T = -1000\ \Delta t$; (**c**) integration time of $T = -1500\ \Delta t$; (**d**) integration time of $T = -2000\ \Delta t$; (**e**) integration time of $T = +500\ \Delta t$; (**f**) integration time of $T = +1000\ \Delta t$; (**g**) integration time of $T = +1500\ \Delta t$; (**h**) integration time of $T = +2000\ \Delta t$.

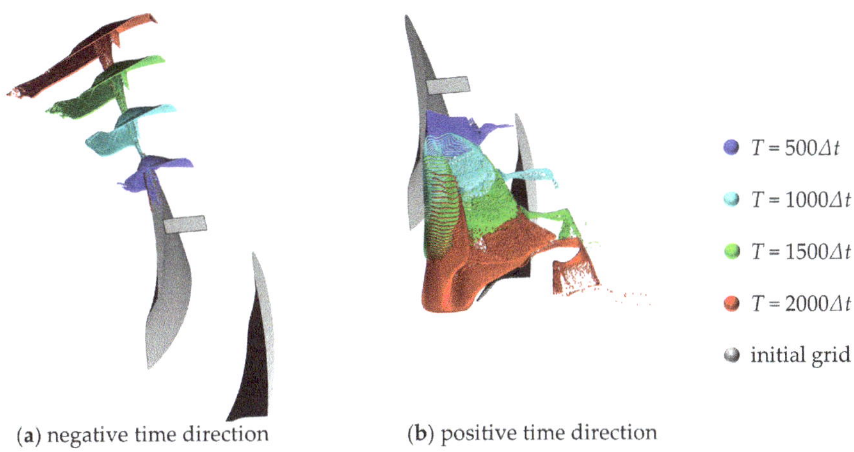

Figure 12. Final particle positions of a series of integration times. (**a**) Negative time direction for aLCS; (**b**) positive time direction for rLCS.

3.2. LCSs of TLF in the Low-Speed Rotor

The FTLE and vorticity fields of the time-averaged flow fields in the DE and NS conditions are shown in Figures 6 and 13, respectively. The Eulerian Q criterion is applied, and the isolines of $Q = 0\ s^{-2}$ are shown in Figures 6c and 13c, along with the vorticity field. For the time-averaged flow field, the aLCSs are similar to the TLV structure identified by the Q criterion, while the correlativity between the rLCS and the vortex structures seems weak. The region enclosed by the aLCS and the rLCS covers the TLV. In our previous study [55,56], the evolution of the TLV was divided into three phases, namely the generation phase, the development phase, and the dissipation phase. In the first two phases, the LCSs are clear, while, as the flow travels downstream, the TLV steps into the dissipation phase. In the first stage of the dissipation phase, the TLF jet shears across the passage flow and rolls up together as the secondary TLV (STLV). This structure can be confirmed by the aLCS, as clear ridges between the TLV and the STLV indicate two separate vortex structures. Then, the TLV breaks down. For the Eulerian Q criterion, the vortex structures it identified become unrecognizable. Meanwhile, for the LCSs, the aLCS gives the boundary of the passage flow entrainment.

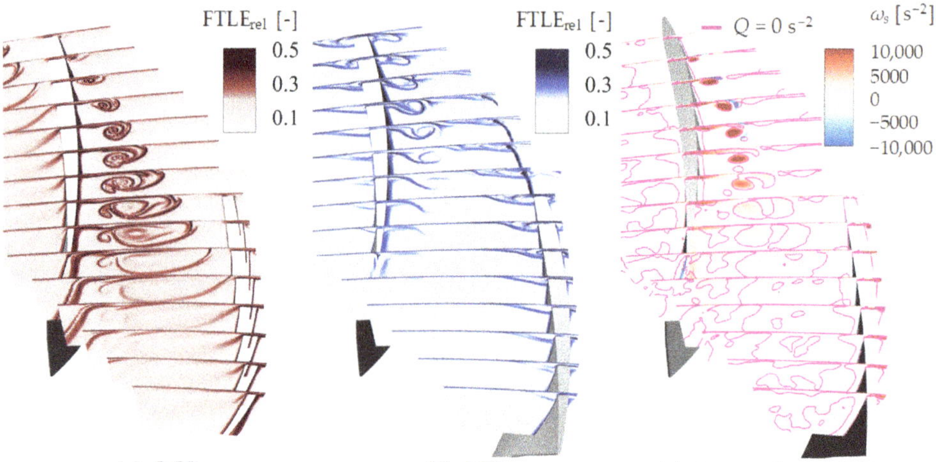

Figure 13. TLV structure in the time-averaged flow field at the NS condition. (**a**) contours of aLCS; (**b**) contours of rLCS; (**c**) contours of streamwise vorticity with isonlines of $Q = 0\ s^{-2}$.

Figure 14 shows the LCSs at the cross-section $s/C = 0.5$ of two instantaneous flow fields at the DE and NS conditions. The FTLE field is calculated using the grid-2, same as that in Figure 3b,d, employing the RK4 method for time integration with a step length of 2.5 Δt and a total integration period of 1000 Δt. The figure also depicts isolines for $Q = 0\ s^{-2}$ and $10^7 s^{-2}$. Affected by the unsteadiness, the rLCS with the DE condition in the instantaneous flow field demonstrate a higher number of ridges than that in the steady flow field, which indicates the unstableness of the flow. With the NS condition, the TLV evolves more rapidly with higher unsteadiness, and the rLCSs exhibit increased complexity. The general structures of the aLCSs maintain consistency at the DE and NS conditions. In the DE condition, the vortex region, as identified by the Q criterion, covers the helical ridges of the TLV and induced vortex, while, with the NS condition, the vortex region identified by the Q criterion only covers the helical ridges of the TLV, especially for the higher-value regions. Part of the region at the shear layer of the TLF and passage flow is recognized as the vortex. The transport of the induced vortex is delineated by the aLCS, which is not captured by the Eulerian method.

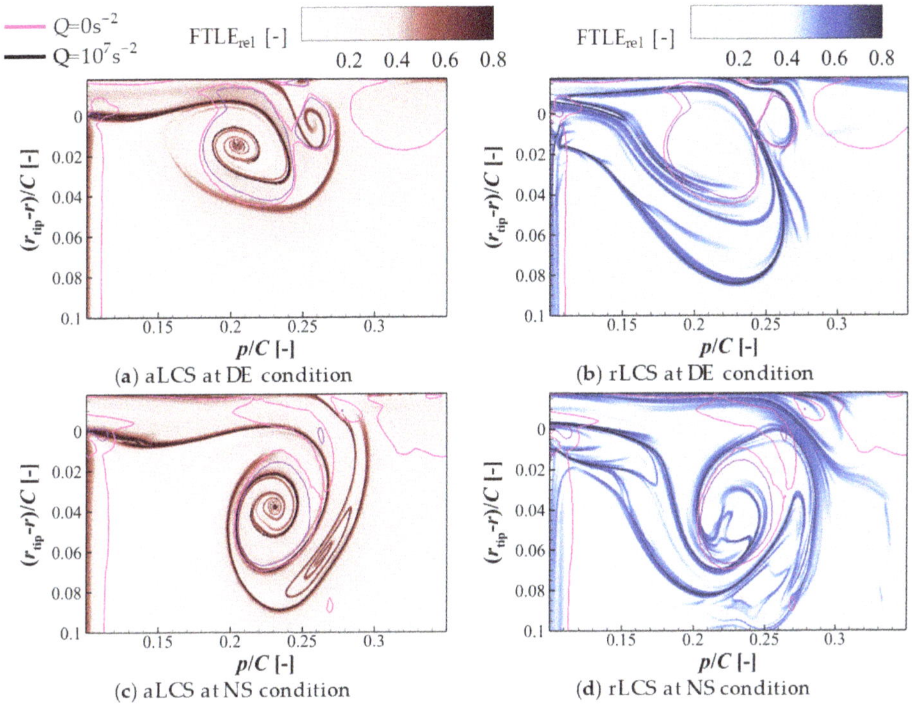

Figure 14. FTLE field with isolines of Q in the instantaneous flow field of cross-section $s/C = 0.5$ at $t = +17{,}500\ \Delta t$ at the DE and NS conditions. Contours of (**a**) aLCS and (**b**) rLCS with the DE condition; (**c**) aLCS and (**d**) rLCS with the NS condition.

The 3D LCSs of two instantaneous flow fields with the DE and NS conditions are shown in Figures 15 and 16, respectively. The aLCSs at different streamwise cross-sections are shown in (a), the aLCS isosurfaces of $FTLE_{rel}$ are shown in (b), the isosurfaces of Q are shown in (c), and the snapshots of the streaklines releasing along the TLV core are shown in (d) for both working conditions. A streakline is formed by continuously releasing fluid particles at a fixed point and observing the trajectories that the fluid particles reach over time. In this study, the streaklines are constructed by releasing particles along the TLV core at continuous time intervals of 10 Δt. The particle release begins from the initial sampled flow field, with particles advecting downstream with an integration time step of 10 Δt using the RK4 time integration method. The integration method is consistent with the LCS calculations. Compared to the aLCS in the time-averaged flow field, the differences are caused by the unsteadiness of the flow. The breakdown of the TLV features a characteristic unsteady flow structure in this rotor, and the breakdown patterns are different in the DE and NS conditions. Only the spiral breakdown pattern of the TLV exits at the DE condition, while the spiral and bubble breakdown patterns of the TLV appear alternately at the NS condition.

As shown in Figures 15c and 16c, it is hard to assess the breakdown pattern by the isosurfaces of Q. From the snapshots of the streaklines, the breakdown patterns of these two flow fields could be distinguished: a spiral breakdown pattern for the DE condition and a bubble breakdown pattern for the NS condition. This method is commonly used to determine the breakdown pattern of vortices in the flow of pipes or delta wings. In the flow field of a rotor where the secondary flow is strong, the structure of the vortex breakdown alters, and the performance of this method relies on the judgment of the researchers. An accurate depiction of the breakdown pattern is contingent upon the strategic placement of the initial seeding points. When reducing the number of seeding points along the core, or displacing the seeding points away from the TLV core, it is hard for the streaklines to construct a bubble, and the breakdown pattern becomes ambiguous. Moreover, due to the complexity of the vortex structures, streaklines are inadequate for illustrating the features of the entire flow field, resulting in particles dispersing throughout. The isosurfaces of the aLCSs show different structures in vortex breakdown patterns. For the TLV of the bubble breakdown pattern, the surface of the TLV core ends, and a bubble-like surface appears downstream. For the TLV of the spiral breakdown pattern, the surface of the TLV core ends, and a bar-like surface appears downstream. The FTLE fields of other instantaneous flow fields have been checked for both working conditions. The aLCS patterns of each breakdown pattern correspond with the above descriptions and so are not shown here.

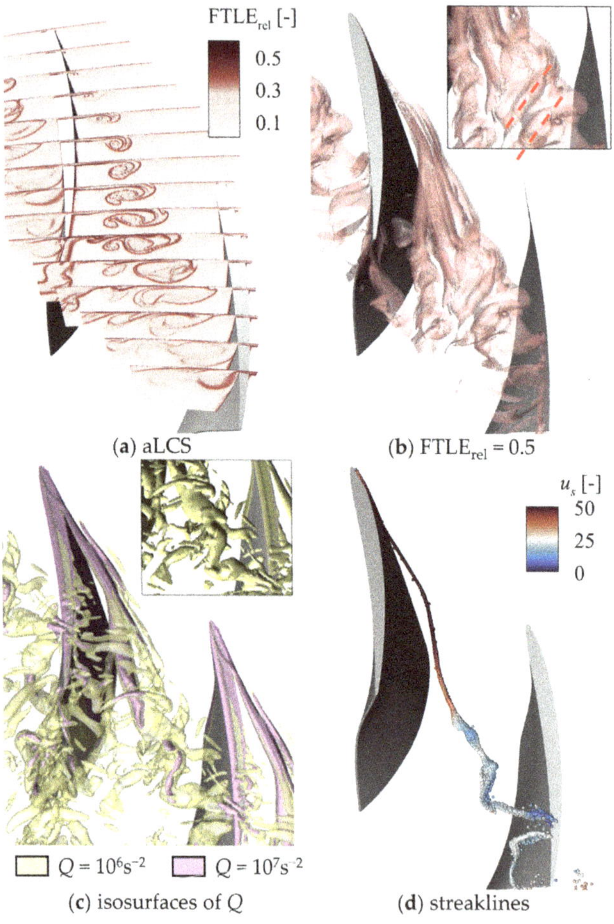

Figure 15. FTLE field, Q, and streaklines of instantaneous flow field at t = +17,500 Δt at the DE condition. (**a**) Contours of aLCS; (**b**) isosurfaces of FTLErel = 0.6; (**c**) isosurfaces of Q; (**d**) snapshot of streaklines.

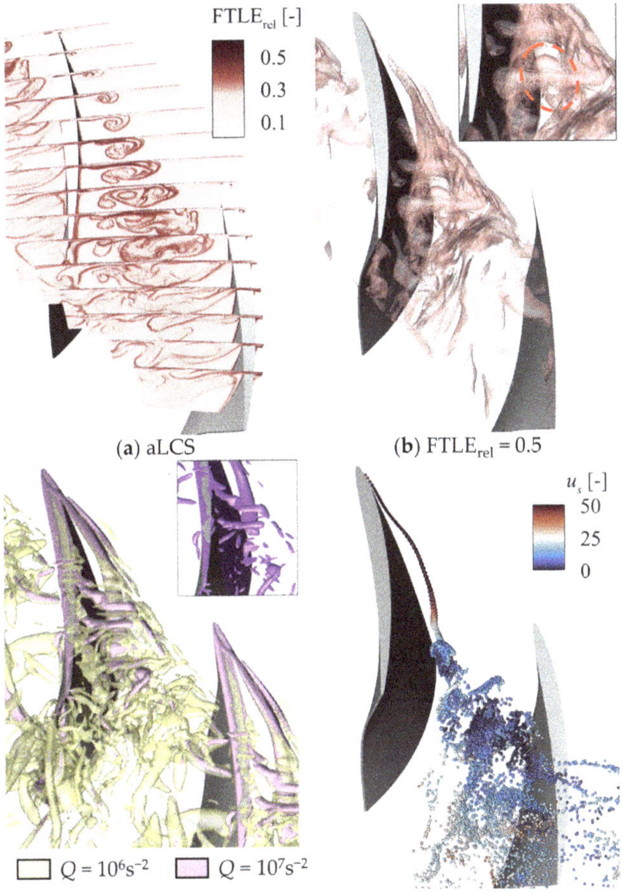

Figure 16. FTLE field, Q, and streaklines of instantaneous flow field at t = +17,500 Δt at the NS condition. (**a**) Contours of aLCS; (**b**) isosurfaces of FTLE$_{rel}$ = 0.6; (**c**) isosurfaces of Q; (**d**) snapshot of streaklines.

4. Conclusions

A Lagrangian analysis is introduced to study the LCSs of the TLF in a low-speed axial compressor rotor based on DDESs. The calculation method of the FTLE field in a 3D flow field is discussed, emphasizing three primary factors. Then, the Lagrangian method is compared with the Eulerian Q method in analyzing the vortex structures of the TLF. The main findings are summarized as follows:

(1) The accuracy of calculating the particle advecting trajectory affects the results of the FTLE field the most. A shorter integration step or higher-order integration method would improve this accuracy.
(2) The clarity of the ridges depends on the density of the initial grid near them. With the complex flow field as the low-speed axial compressor rotor, it is suggested that the general LCSs be detected on the coarse 3D initial grid while a two-layer mesh is then used to capture detailed LCSs.
(3) The LCSs have advantages in identifying the relationships and interactions between vortex structures. The LCSs show a transport barrier between the TLV and the secondary TLV, indicating two separate vortices.

(4) The breakdown patterns of the vortices in the whole flow field can be recognized clearly by the LCSs, which is not true with the Q criterion method, while the streaklines rely on subjective judgment. The aLCSs show the bubble-like and bar-like structure in the isosurfaces corresponding to the bubble and spiral breakdown patterns. The Lagrangian method has great potential in regard to unraveling the mechanism of complex vortex structures and is worth applying more in turbomachinery.

Author Contributions: Conceptualization, J.H., Y.L. and Y.T.; software, J.H.; formal analysis, J.H. and Y.T.; resources, Y.L.; data curation, J.H.; writing—original draft preparation, J.H.; writing—review and editing, Y.L. and Y.T.; visualization, J.H.; supervision, Y.L.; funding acquisition, Y.L. All authors have read and agreed to the published version of the manuscript.

Funding: This work was supported by the National Natural Science Foundation of China (Nos. 52106039, 51976006), the Industry–University–Research Cooperation Project of AECC (Grant Nos.HF- ZL2022CXY001 and HFZL2023CXY002), and the Fundamental Research Funds for the Central Universities.

Data Availability Statement: Data are contained within the article.

Conflicts of Interest: The authors declare no conflicts of interest.

Nomenclature

Δt	Physical time step of simulation
t	Time of instantaneous flow field
s,p	Streamwise and pitchwise coordinate
C	Blade tip chord length
Q	Second invariant of the velocity gradient tensor
T	Integration time
u	Velocity
ω	Vorticity
$\sigma_{t_0}^{t_0+T}(x_0)$	Finite-time Lyapunov exponent
$FTLE_{rel}$	Relative finite-time Lyapunov exponent
TLV	Tip leakage vortex
TLF	Tip leakage flow
FTLE	Finite-time Lyapunov exponent
aLCS	Attracting Lagrangian coherent structure
rLCS	Repelling Lagrangian coherent structure
DDES	Delayed detached-eddy simulation
DE	Design condition
NS	Near-stall condition

References

1. Lakshminarayana, B.; Pouagare, M.; Davino, R. Three-dimensional Flow Field in the Tip Region of a Compressor Rotor Passage-Part I: Mean Velocity Profiles and Annulus Wall Boundary Layer. *ASME J. Eng. Power* **1982**, *104*, 760–772. [CrossRef]
2. Lakshminarayana, B.; Pouagare, M.; Davino, R. Three-dimensional Flow Field in the Tip Region of a Compressor Rotor Passage-Part II: Turbulence Properties. *J. Eng. Power* **1982**, *104*, 772–781. [CrossRef]
3. Furukawa, M.; Inoue, K.; Saiki, K.; Yamada, K. The Role of Tip Leakage Vortex Breakdown in Compressor Rotor Aerodynamics. *ASME J. Turbomach.* **1999**, *121*, 469–480. [CrossRef]
4. Denton, J.D. Loss Mechanisms in Turbomachines. *ASME J. Turbomach.* **1993**, *115*, 621–656. [CrossRef]
5. Kameier, F.; Neise, W. Experimental Study of Tip Clearance Losses and Noise in Axial Turbomachines and Their Reduction. *ASME J. Turbomach.* **1997**, *119*, 460–471. [CrossRef]
6. Day, I.J. Stall, Surge, and 75 Years of Research. *ASME J. Turbomach.* **2016**, *138*, 011001. [CrossRef]
7. Cao, Z.; Zhang, X.; Liang, Y.; Liu, B. Influence of Blade Lean on Performance and Shock Wave/Tip Leakage Flow Interaction in a Transonic Compressor Rotor. *J. Appl. Fluid Mech.* **2022**, *15*, 153–167. [CrossRef]
8. Lu, H.; Yang, Y.; Guo, S.; Huang, Y.; Wang, H.; Zhong, J. Flow Control in Linear Compressor Cascades by Inclusion of Suction Side Dimples at Varying Locations. *Proc. Inst. Mech. Eng. Part A J. Power Energy* **2018**, *232*, 706–721. [CrossRef]
9. Zhong, F.; Zhou, C. Effects of Tip Gap Size on the Aerodynamic Performance of a Cavity-Winglet Tip in a Turbine Cascade. *ASME J. Turbomach.* **2017**, *139*, 101009. [CrossRef]

10. Green, M.; Rowley, C.; Haller, G. Detection of Lagrangian Coherent Structures in 3D Turbulence. *J. Fluid Mech.* **2007**, *572*, 111–120. [CrossRef]
11. Helmholtz, H. On the Integrals of the Hydrodynamic Equations Corresponding to Vortex Motions. *Lond. Edinb. Dublin Philos. Mag. J. Sci.* **1867**, *33*, 485–512. [CrossRef]
12. Hunt, J.C.; Wray, A.A.; Moin, P. Eddies, Streams, and Convergence Zones in Turbulent Flows. *Cent. Turbul. Res. Rep.* **1988**, *CTR-S88*, 193–208.
13. Duda, D.; Yanovych, V.; Uruba, V. Vortex Profiles in Grid Turbulence Observed by PIV. *AIP Conf. Proc.* **2023**, *2672*, 020002. [CrossRef]
14. Chong, M.S.; Perry, A.E.; Cantwell, B.J. A General Classification of Three-Dimensional Flow Fields. *Phys. Fluids* **1990**, *2*, 765–777. [CrossRef]
15. Jeong, J.; Hussain, F. On the Identification of a Vortex. *J. Fluid Mech.* **1995**, *285*, 69–94. [CrossRef]
16. Liu, C.; Xu, H.; Cai, X.; Gao, Y. *Liutex and Its Applications in Turbulence Research*, 1st ed.; Elsevier: Amsterdam, The Netherlands, 2020; ISBN 978-012-819-023-4.
17. Zhong, W.; Liu, Y.; Tang, Y. Unsteady Flow Structure of Corner Separation in a Highly Loaded Compressor Cascade. *ASME J. Turbomach.* **2024**, *146*, 031003. [CrossRef]
18. Hou, J.; Liu, Y. Effect of Moving End Wall on Tip Leakage Flow in a Compressor Cascade with Different Clearance Heights. *AIP Adv.* **2024**, *14*, 015327. [CrossRef]
19. Haller, G.; Yuan, G. Lagrangian Coherent Structures and Mixing in Two-Dimensional Turbulence. *Phys. D* **2000**, *147*, 352–370. [CrossRef]
20. D'Ovidio, F.; Fernández, V.; Hernández-García, E.; López, C. Mixing Structures in the Mediterranean Sea from Finite-Size Lyapunov Exponents. *Geophys. Res. Lett.* **2004**, *31*, L17203. [CrossRef]
21. Prants, S. Transport Barriers in Geophysical Flows: A Review. *Symmetry* **2023**, *15*, 1942. [CrossRef]
22. Lekien, F.; Coulliette, C.; Mariano, A.J.; Ryan, E.H.; Shay, L.K.; Haller, G.; Marsden, J. Pollution Release Tied to Invariant Manifolds: A Case Study for The Coast of Florida. *Physica D* **2005**, *210*, 1–20. [CrossRef]
23. Sapsis, T.; Haller, G. Inertial Particle Dynamics in a Hurricane. *J. Atmos. Sci.* **2009**, *66*, 2481–2492. [CrossRef]
24. Shadden, S.C.; Astorino, M.; Gerbeau, J.-F. Computational Analysis of An Aortic Valve Jet with Lagrangian Coherent Structures. *Chaos* **2010**, *20*, 017512. [CrossRef] [PubMed]
25. Shadden, S.C.; Taylor, C.A. Characterization of Coherent Structures in the Cardiovascular System. *Ann. Biomed. Eng.* **2008**, *36*, 1152–1162. [CrossRef] [PubMed]
26. He, G.; Pan, C.; Feng, L.; Gao, Q.; Wang, J. Evolution of Lagrangian Coherent Structures in a Cylinder-Wake Disturbed Flat Plate Boundary Layer. *J. Fluid Mech.* **2016**, *792*, 274–306. [CrossRef]
27. Pan, C.; Wang, J.; Zhang, C. Identification of Lagrangian Coherent Structures in the Turbulent Boundary Layer. *Sci. China Ser. G-Phys. Mech. Astron.* **2009**, *52*, 248–257. [CrossRef]
28. Li, S.; Jiang, N.; Yang, S.; Huang, Y.; Wu, Y. Coherent Structures over Riblets in Turbulent Boundary Layer Studied by Combining Time-Resolved Particle Image Velocimetry (TRPIV), Proper Orthogonal Decomposition (POD), and Finite-Time Lyapunov Exponent (FTLE). *Chin. Phys. B.* **2018**, *27*, 104701. [CrossRef]
29. Rockwood, M.P.; Taira, K.; Green, M.A. Detecting Vortex Formation and Shedding in Cylinder Wakes Using Lagrangian Coherent Structures. *AIAA J.* **2017**, *55*, 15–23. [CrossRef]
30. Olcay, A.B. Investigation of a wake formation for flow over a cylinder using Lagrangian coherent structures. *Prog. Comput. Fluid Dyn.* **2016**, *16*, 2. [CrossRef]
31. Wang, W.; Prants, S.V.; Zhang, J.; Zhang, J.; Wang, L. A Lagrangian Analysis of Vortex Formation in the Wake behind a Transversely Oscillating Cylinder. *Regul. Chaot. Dyn.* **2018**, *23*, 583–594. [CrossRef]
32. Sun, P.N.; Colagrossi, A.; Marrone, S.; Zhang, A.M. Detection of Lagrangian Coherent Structures in the SPH framework. *Comput. Methods Appl. Mech. Eng.* **2016**, *305*, 849–868. [CrossRef]
33. Wang, L.; Wang, P.; Chang, Z.; Huang, B.; Wu, D. A Lagrangian Analysis of Partial Cavitation Growth and Cavitation Control Mechanism. *Phys. Fluids* **2022**, *34*, 113329. [CrossRef]
34. Ahmad, R.; Zhang, J.; Farooqi, A.; Nauman Aslam, M. Transport Phenomena and Mixing Induced by Vortex Formation in Flow Around Airfoil Using Lagrangian Coherent Structures. *Numer. Math. Theor. Meth. Appl.* **2019**, *12*, 1231–1245. [CrossRef]
35. Tang, J.; Tseng, C.; Wang, N. Lagrangian-Based Investigation of Multiphase Flows by Finite-Time Lyapunov Exponents. *Acta. Mech. Sin.* **2012**, *28*, 612–624. [CrossRef]
36. Tu, H.; Marzanek, M.; Green, M.A.; Rival, D.E. FTLE and Surface-Pressure Signature of Dynamic Flow Reattachment During Delta-Wing Axial Acceleration. *AIAA J.* **2022**, *60*, 2178–2194. [CrossRef]
37. Li, K.; Savari, C.; Barigou, M. Computation of Lagrangian Coherent Structures from Experimental Fluid Trajectory Measurements in a Mechanically Agitated Vessel. *Chem. Eng. Sci.* **2022**, *254*, 117598. [CrossRef]
38. Tseng, C.; Hu, H. Flow Dynamics of a Pitching Foil by Eulerian and Lagrangian Viewpoints. *AIAA J.* **2016**, *54*, 712–727. [CrossRef]
39. Haller, G. Distinguished Material Surfaces and Coherent Structures in Three-Dimensional Fluid Flows. *Phys. D* **2001**, *149*, 248–277. [CrossRef]
40. Shadden, S.C. *A Dynamical Systems Approach to Unsteady Systems*; California Institute of Technology: Pasadena, CA, USA, 2006.

41. Haller, G.; Sapsis, T. Lagrangian Coherent Structures and the Smallest Finite-Time Lyapunov Exponent. *Chaos* **2011**, *21*, 023115. [CrossRef]
42. Shadden, S.C.; Lekien, F.; Marsden, J.E. Definition and Properties of Lagrangian Coherent Structures from Finite-Time Lyapunov Exponents in Two-Dimensional Aperiodic Flows. *Phys. D* **2005**, *212*, 271–304. [CrossRef]
43. Lekien, F.; Shadden, S.C.; Marsden, J.E. Lagrangian Coherent Structures in n-Dimensional Systems. *J. Math. Phys.* **2007**, *48*, 065404. [CrossRef]
44. Du, H.; Yu, X.; Zhang, Z.; Liu, B. Relationship Between the Flow Blockage of Tip Leakage Vortex and Its Evolutionary Procedures inside the Rotor Passage of a Subsonic Axial Compressor. *J. Therm. Sci.* **2013**, *22*, 522–531. [CrossRef]
45. Liu, Y.; Zhao, S.; Wang, F.; Tang, Y. A Novel Method for Predicting Fluid-Structure Interaction with Large Deformation Based on Masked Deep Neural Network. *Phys. Fluids* **2024**, *36*, 027103. [CrossRef]
46. Liu, Y.; Wang, F.; Zhao, S.; Tang, Y. A Novel Framework for Predicting Active Flow Control by Combining Deep Reinforcement Learning and Masked Deep Neural Network. *Phys. Fluids* **2024**, *36*, 037112. [CrossRef]
47. Spalart, P.R. Philosophies and Fallacies in Turbulence Modeling. *Prog. Aerosp. Sci.* **2015**, *74*, 1–15. [CrossRef]
48. Liu, Y.; Wei, X.; Tang, Y. Investigation of Unsteady Rotor–Stator Interaction and Deterministic Correlation Analysis in a Transonic Compressor Stage. *ASME J. Turbomach.* **2023**, *145*, 071004. [CrossRef]
49. Lee, G.H.; Park, J.I.; Baek, J.H. Performance Assessment of Turbulence Models for the Quantitative Prediction of Tip Leakage Flow in Turbomachines. In Proceedings of the ASME Turbo Expo 2004: Power for Land, Sea, and Air, Volume 5: Turbo Expo 2004, Parts A and B, Vienna, Austria, 14–17 June 2004; pp. 1501–1512. [CrossRef]
50. Liu, Y.; Zhong, L.; Lu, L. Comparison of DDES and URANS for Unsteady Tip Leakage Flow in an Axial Compressor Rotor. *ASME J. Fluids Eng.* **2019**, *141*, 121405. [CrossRef]
51. Gritskevich, M.S.; Garbaruk, A.V.; Schütze, J.; Menter, F.R. Development of DDES and IDDES Formulations for the k-ω Shear Stress Transport Model. *Flow Turbul. Combust.* **2012**, *88*, 431–449. [CrossRef]
52. Menter, F.R.; Kuntz, M.; Langtry, R. Ten Years of Industrial Experience with the SST Turbulence Model. In *Turbulence, Heat and Mass Transfer 4*; Hanjalic, K., Nagano, Y., Tummers, M.J., Eds.; Begell House: New York, NY, USA; Wallingford, UK, 2003; Volume 4, pp. 625–632.
53. Tucker, P. *Unsteady Computational Fluid Dynamics in Aeronautics*; Fluid Mechanics and Its Applications; Springer: Dordrecht, The Netherlands, 2013; Volume 104. [CrossRef]
54. Li, H.; Su, X.; Yuan, X. Entropy Analysis of the Flat Tip Leakage Flow with Delayed Detached Eddy Simulation. *Entropy* **2019**, *21*, 21. [CrossRef]
55. Hou, J.; Liu, Y.; Zhong, L.; Zhong, W.; Tang, Y. Effect of Vorticity Transport on Flow Structure in the Tip Region of Axial Compressors. *Phys. Fluids* **2022**, *34*, 055102. [CrossRef]
56. Hou, J.; Liu, Y. Evolution of Unsteady Vortex Structures in the Tip Region of an Axial Compressor Rotor. *Phys. Fluids* **2023**, *35*, 045107. [CrossRef]

symmetry MDPI

Article
Low-Order Moments of Velocity Gradient Tensors in Two-Dimensional Isotropic Turbulence

Chensheng Luo [1], Ping-Fan Yang [2,3] and Le Fang [1,*]

[1] Laboratory of Complex System, Ecole Centrale de Pékin/School of General Engineering, Beihang University, Beijing 100191, China; chensheng_luo@buaa.edu.cn
[2] The State Key Laboratory of Nonlinear Mechanics, Institute of Mechanics, Chinese Academy of Sciences, Beijing 100190, China; yangpingfan@imech.ac.cn
[3] School of Engineering Sciences, University of Chinese Academy of Sciences, Beijing 100049, China
* Correspondence: le.fang@buaa.edu.cn

Abstract: In isotropic turbulence, symmetry of different directions can reduce the number of independent components for velocity gradient tensors. In three-dimensional isotropic turbulence, the independent components under either incompressible or compressible conditions have already been analyzed in the literature. However, for two-dimensional isotropic turbulence, they are still unclear. We derive rigorously the independent components for velocity gradient tensors of two-dimensional isotropic turbulence and give physical explanations. These theoretical results are validated using high-resolution direct numerical simulations (DNSs) of two-dimensional compressible turbulence. Results show that the present DNS setup is still not sufficient to capture the isotropy of third-order moments, suggesting that more investigations on determining the smallest scale and improving the numerical schemes for two-dimensional compressible turbulence are required.

Keywords: two-dimensional turbulence; compressible turbulence; velocity gradient

1. Introduction

Homogeneous and isotropic turbulence (HIT) is the simplest type of turbulent flow. It provides a cornerstone for us to study more complex turbulence systems. For example, the local isotropy assumption for small-scale dynamics is one of the main hypotheses in the celebrated K41 theory [1,2], from which we could obtain the important $-5/3$ energy spectrum, the four-fifth law for third-order structure functions, etc. In the small-scale dynamics of turbulence, the velocity gradient tensor $\mathbf{m} \equiv \nabla \boldsymbol{u}$ (or $m_{ij} \equiv \frac{\partial u_i}{\partial x_j}$ in component form, where $\boldsymbol{u}$ denotes the velocity) characterize the local flow pattern. For example, the deformation and rotational motion can be described by the symmetric part $\mathbf{S}$ and the anti-symmetric part $\mathbf{w}$ of $\mathbf{m}$, where $\mathbf{S} \equiv \frac{1}{2}(\mathbf{m}+\mathbf{m}^T)$ and $\mathbf{w} \equiv \frac{1}{2}(\mathbf{m}-\mathbf{m}^T)$ [3–5]. Specifically, the second-order moments of the velocity gradient $\mathbf{m}$, defined as $M^{(2)}_{ipjq} \equiv \left\langle \frac{\partial u_i}{\partial x_p}\frac{\partial u_j}{\partial x_q} \right\rangle$, describe the strength of the local motions, like the enstrophy $\mathcal{D} \equiv \frac{1}{2}\langle \omega_i \omega_i \rangle$, which describes the strength of local rotational motion, where $\omega_i = \epsilon_{ijk} w_{jk}$ is the vorticity vector and ϵ_{ijk} is the Levi-Civita symbol. Furthermore, the viscosity dissipation in turbulence, $\nu \left\langle \frac{\partial u_i}{\partial x_j}\frac{\partial u_i}{\partial x_j} \right\rangle$, with ν, the kinematic viscosity, is obviously determined by $\mathbf{M}^{(2)}$. Then, in the dynamic equation of the second-order moment $\mathbf{M}^{(2)}$, the third-order moment of the velocity gradient $M^{(3)}_{ipjqkr} \equiv \left\langle \frac{\partial u_i}{\partial x_p}\frac{\partial u_j}{\partial x_q}\frac{\partial u_k}{\partial x_r} \right\rangle$ appears due to the nonlinearity of the Navier–Stokes equation. These third-order moments $\mathbf{M}^{(3)}$ determine the key nonlinear process such as the vortex stretching rate $\omega_i S_{ij} \omega_j$ [3,5,6], which is crucial for the generation of small-scale motions in turbulence. As a consequence, a complete understanding of the properties of the moments of the velocity gradient is important in the studies of small-scale dynamics of turbulence.

In this work, we will focus on the kinematics properties of $\mathbf{M}^{(2)}$ and $\mathbf{M}^{(3)}$, especially their invariants and independent components.

Betchov's relation is one of the important theoretical results in the studies of kinematic properties of velocity gradients. In 1956, Betchov derived two constraints on the second- and third-order invariants of $\mathbf{m}$ for homogeneous and incompressible flows [7]. In three-dimensional (3D) isotropic incompressible turbulence, with the help of Betchov's relations, we can determine all the components of $\mathbf{M}^{(2)}$ and $\mathbf{M}^{(3)}$ with, respectively, only one scalar quantity [8,9]. Recently, Yang et al. [10] generalized the Betchov's relation to compressible turbulence and showed that one needs two and four scalars to fully determine $\mathbf{M}^{(2)}$ and $\mathbf{M}^{(3)}$, respectively, in isotropic compressible turbulence. Furthermore, we note that Carbone and Wilczek [11] have proved that Betchov's relations are the only constraints on the invariants of the velocity gradient under the homogeneity condition. In addition to the second- and third-order moments, the structure of higher-order moments, such as fourth-order, has also been studied previously for 3D homogeneous incompressible and compressible turbulence [12–14].

Up to now, most studies on the velocity gradient focus on 3D turbulence. At the same time, a reduction in dimensionality leads to the emergence of new phenomena. For example, in two-dimensional (2D) incompressible HIT, all components of the third-order moments are equal to zero because of the absence of vortex stretching [15–17]. This results in the conservation of enstrophy and a net energy transfer from small to large scales, which is known as the inverse energy cascade [18,19]. Furthermore, 2D isotropic compressible turbulence exhibits distinct phenomena, such as the energy flux loop between incompressible and compressible components [20]. However, to the best of our knowledge, the structure of the velocity gradient in 2D turbulence, especially in the compressible case, is yet to be explored. The current studies on 2D compressible turbulence [20–23] are rather phenomenological and mainly focus on the dual cascade of energy and enstrophy, as well as the interchange of energy between the dilation and divergence-free parts in the spectral space. There is still no strict theory explaining the energy transfer and energy flux loop. We recall that in 3D isotropic incompressible turbulence, the energy transfer can be represented by using the Kármán–Howarth equation, in which the analysis of independent scalars plays an essential role. A study of the velocity gradient tensor in this case would be beneficial for understanding the local flow motion and energy transfer in 2D turbulence.

In this work, we investigate the structure of the velocity gradient tensor in 2D turbulence, including the components of its low-order moments, i.e., $M^{(2)}_{ipjq}$ and $M^{(3)}_{ipjqkr}$, and the corresponding invariants, e.g., $\langle tr(\mathbf{m}^2)\rangle$, $\langle tr(\mathbf{m}^3)\rangle$. The rest of the paper is structured as follows. In Section 2, we present the theoretical derivations, focusing first on the decomposition of velocity gradient tensor $\mathbf{m}$, and then using the isotropy and homogeneity constraints to determine the independent components of low-order moments. Furthermore, we express the second- and third-order invariants in terms of these independent components and explain the underlying physical meaning of them. The above theoretical derivation is verified in Section 3 by direct numerical simulations (DNSs) of 2D homogeneous and isotropic compressible turbulence. Finally, in Section 4 we give further discussions on the analytical relations and the numerical results.

2. Analytical Studies on the Components of Velocity Gradient Tensors in 2D Turbulence

2.1. Decomposition of the Velocity Gradient Tensor

In general, the velocity gradient tensor $\mathbf{m}$ can be decomposed into dilatational, symmetric-deviatoric, and antisymmetric parts, as follows:

$$\mathbf{m} = \frac{1}{n}\theta\mathbf{I} + \mathbf{s} + \mathbf{w}, \tag{1}$$

where $\frac{1}{n}\theta\mathbf{I}$ is the dilatational part with $\theta = \nabla \cdot \boldsymbol{u} = \overline{\mathbf{m}}$ and n is the dimension of the flow field. Here, for simplicity, we used the overbar symbol to denote the trace operation: $\overline{\mathbf{X}} = \mathrm{tr}(\mathbf{X})$. The tensor $\mathbf{s}$ is the deviatoric part of the rate-of-strain tensor:

$$\mathbf{s} = (\mathbf{m} + \mathbf{m}^T)/2 - \frac{1}{n}\theta\mathbf{I}, \tag{2}$$

where $\mathbf{X}^T$ is the transpose of the tensor. The tensor $\mathbf{w}$ is the antisymmetric rate-of-rotation tensor defined as in the introduction, which can be equivalently expressed by the vorticity vector $\boldsymbol{\omega}$ with $\omega_i = \varepsilon_{ijk}\omega_{jk}$.

In a 3D space generated by an orthogonal basis $(\boldsymbol{e}_1, \boldsymbol{e}_2, \boldsymbol{e}_3)$, if the flow is confined in the plane $(\boldsymbol{e}_1, \boldsymbol{e}_2)$, then the velocity gradient tensor can be decomposed as:

$$\mathbf{m} = \frac{1}{2}\begin{bmatrix} \varphi & \psi \\ \psi & -\varphi \end{bmatrix}_{(\boldsymbol{e}_1,\boldsymbol{e}_2)} + \frac{1}{2}\omega\begin{bmatrix} 0 & -1 \\ 1 & 0 \end{bmatrix}_{(\boldsymbol{e}_1,\boldsymbol{e}_2)} + \frac{1}{2}\theta\begin{bmatrix} 1 & 0 \\ 0 & 1 \end{bmatrix}_{(\boldsymbol{e}_1,\boldsymbol{e}_2)}, \tag{3}$$

where $\varphi, \psi, \omega, \theta$ are four independent components of the local velocity gradient tensor, and the subscripts of each matrix means that the matrix is expressed in the $(\boldsymbol{e}_1, \boldsymbol{e}_2)$ basis. More precisely, φ and ψ are defined as $\varphi \equiv \frac{\partial u_1}{\partial x_1} - \frac{\partial u_2}{\partial x_2}$ and $\psi \equiv \frac{\partial u_1}{\partial x_2} + \frac{\partial u_2}{\partial x_1}$, respectively, which represent the local shear motion of the flow. ω denotes the vorticity and because of the constraints from 2D space, it only has the $\boldsymbol{e}_3$ components, i.e., $\boldsymbol{\omega} = \omega\boldsymbol{e}_3$, where $\omega = \frac{\partial u_2}{\partial x_1} - \frac{\partial u_1}{\partial x_2}$. Finally, θ denotes the dilatation as mentioned above.

For 2D homogeneous turbulence, the four independent components $\theta(\boldsymbol{x}), \omega(\boldsymbol{x}), \psi(\boldsymbol{x}), \varphi(\boldsymbol{x})$ are locally independent, but they are globally related in spectral space. This relation is shown in Appendix A.

2.2. Isotropic Expressions for $\mathbf{M}^{(2)}$ and $\mathbf{M}^{(3)}$

Under the isotropic condition, second- and third-order moments of the velocity gradient, $\mathbf{M}^{(2)}$ and $\mathbf{M}^{(3)}$, can be expressed, respectively, as the following [9]:

$$M^{(2)}_{ipjq} = a_1\delta_{ip}\delta_{jq} + a_2\delta_{ij}\delta_{pq} + a_3\delta_{iq}\delta_{pj}, \tag{4}$$

$$\begin{aligned} M^{(3)}_{ipjqkr} =& b_1\delta_{ip}\delta_{jq}\delta_{kr} + b_2(\delta_{ip}\delta_{jk}\delta_{qr} + \delta_{jq}\delta_{ik}\delta_{pr} + \delta_{kr}\delta_{ij}\delta_{pq}) \\ &+ b_3(\delta_{ip}\delta_{jr}\delta_{qk} + \delta_{jq}\delta_{ir}\delta_{pk} + \delta_{kr}\delta_{iq}\delta_{pj}) + b_4(\delta_{iq}\delta_{pk}\delta_{jr} + \delta_{ir}\delta_{pj}\delta_{qk}) \\ &+ b_5(\delta_{ij}\delta_{pk}\delta_{qr} + \delta_{ij}\delta_{qk}\delta_{pr} + \delta_{ik}\delta_{pj}\delta_{qr} + \delta_{ik}\delta_{rj}\delta_{pq} + \delta_{jk}\delta_{qi}\delta_{pr} + \delta_{jk}\delta_{ri}\delta_{pq}), \end{aligned} \tag{5}$$

where δ is the Kronecker tensor ($\delta_{ij} = 1$ if $i = j$, and 0 otherwise), and a_1, a_2, a_3 and b_1, b_2, b_3, b_4, b_5 are scalar quantities.

These two formulas generally hold in 2D and 3D (or even in higher-dimension flows, which do not exist physically). However, if the flow is confined to 2D, the subscripts in Equations (4) and (5), (i, j, k, p, q, r), can choose only from two different values, 1 and 2. As the second-order moment $M^{(2)}_{ipjq}$ has the specific form shown in Equation (4), its components would be non-zero only if there are even numbers of 1 and even numbers of 2 among the four subscripts (i, p, j, q). As a consequence, there are only four types of non-zero components of $M^{(2)}_{ipjq}$, as shown in Table 1. For example, if a second-order moment has the form $M^{(2)}_{\alpha\alpha\alpha\alpha}$, such as $M^{(2)}_{1111}$ with $\alpha = 1$ and $M^{(2)}_{2222}$ with $\alpha = 2$, its value will be $a_1 + a_2 + a_3$. This can be shown by setting all subscripts (i, p, j, q) in Equation (4) to 1 or 2.

Table 1. All non-zero types of $M^{(2)}_{ipjq}$. The subscripts α, β in the type name only show different numbers without applying the Einstein summation convention.

Type	Examples	Expression with Independent Scalar
$M^{(2)}_{\alpha\alpha\alpha\alpha}$	$M^{(2)}_{1111}, M^{(2)}_{2222}$	$a_1 + a_2 + a_3$
$M^{(2)}_{\alpha\alpha\beta\beta}$	$M^{(2)}_{1122}, M^{(2)}_{2211}$	a_1
$M^{(2)}_{\alpha\beta\alpha\beta}$	$M^{(2)}_{1212}, M^{(2)}_{2121}$	a_2
$M^{(2)}_{\alpha\beta\beta\alpha}$	$M^{(2)}_{1221}, M^{(2)}_{2112}$	a_3

Similarly, $M^{(3)}_{ipjqkr}$ is non-zero only if there are even numbers of 1 and even numbers of 2 among the six subscripts (i, p, j, q, k, r). We should also note that the indices pairs $(i, p), (j, q), (k, r)$ are interchangeable due to the symmetry of $\mathbf{M}^{(3)}$. $M^{(3)}_{ipjqkr}$ thus has four non-zero types of components, as shown in Table 2. For example, $M^{(3)}_{111122}$, $M^{(3)}_{222211}$ and $M^{(3)}_{112211}$ all belong to the same type $M^{(3)}_{\alpha\alpha\alpha\alpha\beta\beta}$, as the first two are just the cases $\alpha = 1, \beta = 2$ and $\alpha = 2, \beta = 1$, and the third $M^{(3)}_{112211} = M^{(3)}_{111122}$ because the indices pairs (j, q) and (k, r) are interchangeable. Thus, from Equation (5), these three components all have the same value $b_1 + b_2 + b_3$.

Table 2. All non-zero types of $M^{(3)}_{ipjqkr}$. The subscripts α, β in the type name only show different numbers without applying the Einstein summation convention. The order of three pairs, i.e., $(i, p), (j, q), (k, r)$, could be interchanged symmetrically.

Type	Examples	Expression with Independent Scalar
$M^{(3)}_{\alpha\alpha\alpha\alpha\alpha\alpha}$	$M^{(3)}_{111111}, M^{(3)}_{222222}$	$b_1 + 3b_2 + 3b_3 + 2b_4 + 6b_5$
$M^{(3)}_{\alpha\alpha\alpha\alpha\beta\beta}$	$M^{(3)}_{111122}, M^{(3)}_{222211}, M^{(3)}_{112211}$, *etc.*	$b_1 + b_2 + b_3$
$M^{(3)}_{\alpha\alpha\beta\alpha\beta}$ and $M^{(3)}_{\alpha\alpha\beta\alpha\beta\alpha}$	$M^{(3)}_{111212}, M^{(3)}_{222121}, M^{(3)}_{112121}, M^{(3)}_{221212}$, *etc.*	$b_2 + 2b_5$
$M^{(3)}_{\alpha\alpha\alpha\beta\beta\alpha}$	$M^{(3)}_{111221}, M^{(3)}_{221221}, M^{(3)}_{121121}$, *etc.*	$b_3 + b_4 + b_5$

It can be shown from Table 2 that

$$\begin{aligned} M^{(3)}_{\alpha\alpha\alpha\alpha\alpha\alpha} =& 2(b_2 + 2b_5) + 2(b_3 + b_4 + b_5) + (b_1 + b_2 + b_3) \\ =& 2M^{(3)}_{\alpha\alpha\alpha\beta\alpha\beta} + 2M^{(3)}_{\alpha\alpha\alpha\beta\beta\alpha} + M^{(3)}_{\alpha\alpha\alpha\alpha\beta\beta}. \end{aligned} \quad (6)$$

Thus, for 2D compressible isotropic turbulence, $M^{(3)}_{ipjqkr}$ could be fully determined by three scalars, i.e., $B_1 = b_1 + b_2 + b_3$, $B_2 = b_2 + 2b_5$, and $B_3 = b_3 + b_4 + b_5$. It should be remarked that this is only applicable in 2D and b_1, b_2, b_3, b_4, b_5 are all indeed independent in 3D or in higher dimensions. This derivation is a novel process in 2D that has not been and can not be applied in 3D cases to reduce the independent components.

Next, in Tables 3 and 4, we express the second- and third-order invariants in terms of the scalars in Equations (4) and (5). In these two tables, the first column gives the tensor notation of the invariants and the second column gives the corresponding subscript representation in terms of the components of $\mathbf{M}^{(2)}$ and $\mathbf{M}^{(3)}$. By substituting Equation (3) into the definitions of the invariants, i.e., the expressions in the first columns of those tables, we could obtain the expression of these invariants in terms of $\theta, \omega, \psi, \varphi$, the results of which are presented in the third columns of Tables 3 and 4. Furthermore, with the help of Equations (4) and (5), we can express these invariants in terms of the scalars, a_1, a_2, a_3 and B_1, B_2, B_3, which is shown in the fourth columns of those tables.

Table 3. Second-order invariants of the velocity gradient under the isotropic constraints.

Tensor Notation	Subscript Notation	Expression of θ, ω, ψ and φ	Independent Scalar Expression
$\left\langle \overline{\mathbf{m}^2} \right\rangle$	$M^{(2)}_{iijj}$	$\langle \theta^2 \rangle$	$4a_1 + 2a_2 + 2a_3$
$\left\langle \mathbf{m}^2 \right\rangle$	$M^{(2)}_{ijji}$	$\frac{1}{2}(\langle \psi^2 \rangle + \langle \varphi^2 \rangle) - \frac{1}{2}\langle \omega^2 \rangle + \frac{1}{2}\langle \theta^2 \rangle$	$2a_1 + 2a_2 + 4a_3$
$\left\langle \mathbf{m}\mathbf{m}^T \right\rangle$	$M^{(2)}_{ijij}$	$\frac{1}{2}(\langle \psi^2 \rangle + \langle \varphi^2 \rangle) + \frac{1}{2}\langle \omega^2 \rangle + \frac{1}{2}\langle \theta^2 \rangle$	$2a_1 + 4a_2 + 2a_3$

Table 4. Third-order invariants of the velocity gradient under the isotropic constraints.

Tensor Notation	Subscript Notation	Expression of θ, ω, ψ and φ	Independent Scalar Expression
$\left\langle \overline{\mathbf{m}^3} \right\rangle$	$M^{(3)}_{iijjkk}$	$\langle \theta^3 \rangle$	$8B_1 + 4B_2 + 4B_3$
$\left\langle \mathbf{m}^3 \right\rangle$	$M^{(3)}_{ijjkki}$	$\frac{3}{4}(\langle \psi^2\theta \rangle + \langle \varphi^2\theta \rangle) - \frac{3}{4}\langle \omega^2\theta \rangle + \frac{1}{4}\langle \theta^3 \rangle$	$2B_1 + 4B_2 + 10B_3$
$\left\langle \mathbf{m}\mathbf{m}^2 \right\rangle$	$M^{(3)}_{iijkkj}$	$\frac{1}{2}(\langle \psi^2\theta \rangle + \langle \varphi^2\theta \rangle) - \frac{1}{2}\langle \omega^2\theta \rangle + \frac{1}{2}\langle \theta^3 \rangle$	$4B_1 + 4B_2 + 8B_3$
$\left\langle \mathbf{m}\mathbf{m}\mathbf{m}^T \right\rangle$	$M^{(3)}_{iijkjk}$	$\frac{1}{2}(\langle \psi^2\theta \rangle + \langle \varphi^2\theta \rangle) + \frac{1}{2}\langle \omega^2\theta \rangle + \frac{1}{2}\langle \theta^3 \rangle$	$4B_1 + 8B_2 + 4B_3$
$\left\langle \mathbf{m}^2\mathbf{m}^T \right\rangle$	$M^{(3)}_{ijjkik}$	$\frac{3}{4}(\langle \psi^2\theta \rangle + \langle \varphi^2\theta \rangle) + \frac{1}{4}\langle \omega^2\theta \rangle + \frac{1}{4}\langle \theta^3 \rangle$	$2B_1 + 8B_2 + 6B_3$

The identifications in Tables 3 and 4 also allow us to make a connection between the independent scalars and $\theta, \omega, \psi, \varphi$. For the second-order values,

$$\left\langle \theta^2 \right\rangle = 4a_1 + 2a_2 + 2a_3, \tag{7a}$$

$$\left\langle \psi^2 \right\rangle + \left\langle \phi^2 \right\rangle = 4a_2 + 4a_3, \tag{7b}$$

$$\left\langle \omega^2 \right\rangle = 2a_2 - 2a_3. \tag{7c}$$

For the third-order values,

$$\left\langle \psi^2\theta \right\rangle + \left\langle \varphi^2\theta \right\rangle = 8B_2 + 8B_3, \tag{8a}$$

$$\left\langle \omega^2\theta \right\rangle = 4B_2 - 4B_3, \tag{8b}$$

$$\left\langle \theta^3 \right\rangle = 8B_1 + 4B_2 + 4B_3. \tag{8c}$$

The identifications in Table 4 and Equations (8a)–(8c) are also consistent with the limiting case when the turbulence is incompressible, i.e., $\theta = 0$. The left-hand side of Equations (8a)–(8c) will all be zero and will lead to:

$$B_1 = B_2 = B_3 = 0. \tag{9}$$

As a result, velocity gradient skewness $\mathcal{S} = \left\langle (\partial u_x/\partial x)^3 \right\rangle / \left\langle (\partial u_x/\partial x)^2 \right\rangle^{3/2}$ is zero for 2D incompressible isotropic flows.

2.3. Homogeneity Constraints to Second- and Third-Order Invariants

According to Yang et al. [10], under the homogeneity condition, i.e., $(\partial/\partial x_i)\langle \bullet \rangle = 0$, one can obtain two invariants relation for second- and third-order moments of velocity gradient, i.e.,

$$M^{(2)}_{iijj} = M^{(2)}_{ijji}, \tag{10}$$

$$M^{(3)}_{ijjkki} = \frac{3}{2}M^{(3)}_{iijkkj} - \frac{1}{2}M^{(3)}_{iijjkk}. \tag{11}$$

For the second-order invariant, by substituting the results of Table 3 into Equation (10), we have:

$$a_1 = a_3 = M^{(2)}_{\alpha\alpha\beta\beta} = M^{(2)}_{\alpha\beta\beta\alpha}. \tag{12}$$

As a result, the second-order velocity gradient moment only has two independent components. From Equations (7a)–(7c), we can show that:

$$\left\langle \omega^2 \right\rangle + \left\langle \theta^2 \right\rangle = \left\langle \varphi^2 \right\rangle + \left\langle \psi^2 \right\rangle, \tag{13}$$

which could also be proved from the spectral approach—see Equation (A13) in the Appendix A. This shows the consistency between independent scalar analysis and spectral space analysis.

However, if we combine Table 4 and Equation (11) for the third-order invariants, we find that Equation (11) is automatically satisfied without giving any new constraint to B_1, B_2, B_3. This is different from the 3D case analysed by Yang et al. [10], where Equation (11) gives a new constraint $b_1 = 3b_3 - 2b_4$ to the five independent components. This shows that the constraints on independent components for the third-order moment are from different sources. In the 2D case, the constraints are from isotropy, while in the 3D case, the constraints are from the homogeneity. Consequently, copying simply the same method of 3D tensor derivation in reference [10] will lead to incorrect independent scalars and will result in a contraction that the third-order moment has more independent components in 2D than in 3D.

We also note that $\langle \theta^2 \rangle$ is positive definite ($\langle \theta^2 \rangle \geq 0$), thus, from Equations (7a) and (12), we can readily show that $3a_1 + a_2 \geq 0$, and the equal sign holds when $\langle \theta^2 \rangle = 0$, that is, the incompressible case. Similarly, as $\langle \omega^2 \rangle \geq 0$, from Equations (7c) and (12), we have $a_2 \geq a_1$, and the equal sign holds when the flow is irrotational. These two inequalities provide restrictions on the ratio between the components of $\mathbf{M}^{(2)}$:

$$\frac{1}{3} = \frac{a_2}{2a_2 + a_2} \leq \frac{M^{(2)}_{\alpha\beta\alpha\beta}}{M^{(2)}_{\alpha\alpha\alpha\alpha}} = \frac{a_2}{2a_1 + a_2} = 3\frac{a_2}{2(3a_1 + a_2) + a_2} \leq 3. \tag{14}$$

The minimum and maximum of the ratio are achieved, respectively, at irrotational flow and incompressible flow.

3. Numerical Validation of Independent Components

To verify the theoretical results presented above, we performed a 2D compressible isotropic DNS. We numerically solve the following two-dimensional compressible Navier–Stokes equations:

$$\frac{\partial \rho}{\partial t} + \frac{\partial \rho u_i}{\partial x_i} = 0, \tag{15a}$$

$$\frac{\partial \rho u_i}{\partial t} + \frac{\partial}{\partial x_j}(\rho u_i u_j + p\delta_{ij}) - \frac{\partial}{\partial x_j}\sigma_{ij} = 0, \tag{15b}$$

$$\frac{\partial E}{\partial t} + \frac{\partial}{\partial x_j}[(E + p)u_j] - \frac{\partial}{\partial x_j}(\sigma_{ij}u_i - Q_j) = 0, \tag{15c}$$

in which ρ is the density, p is the pressure, $E = \frac{1}{2}\rho u_i u_i + p/(\gamma - 1)$ is the total energy with $\gamma = 1.4$, the ratio of the specific heats, $\sigma_{ij} = \mu(\frac{\partial u_i}{\partial x_j} + \frac{\partial u_j}{\partial x_i} - \frac{2}{3}\frac{\partial u_k}{\partial x_k}\delta_{ij})$ is the viscous stress tensor with the effect of bulk viscosity neglected [24], and $Q_j = -\kappa(\frac{\partial T}{\partial x_j})$ is the heat flux. The ideal gas law $p = \rho RT$ is applied in calculation to connect the temperature T, pressure p, and density ρ through the ideal gas law, and R is the ideal gas constant. The viscosity μ is connected with the local temperature through Sutherland's law $\mu = \mu_{ref}(T/T_{ref})^{3/2}((T_{ref} + T_s)/(T + T_s))$ with μ_{ref}, T_{ref}, and T_s constants. The thermal conductivity κ is determined from the viscosity by a constant Prandtl number $Pr = \mu C_p/\kappa = 0.72$, with $C_p = \frac{\gamma}{\gamma-1}R$ being the specific heat at constant pressure.

In terms of the numerical methods, the set of Equations (15a)–(15c) are solved with a high-order finite difference method. Specifically, the convective terms are calculated by a seventh-order low-dissipative monotonicity-preserving scheme [25] such that shock waves in a compressible flow can be captured and the capabilities of resolving small-scale turbulent structures are preseved. The diffusion terms are discretized by a sixth-order compact central scheme [26] with a domain decoupling scheme for parallel computation [27]. The time integration is computed by a three-step third-order total variation diminishing Runge–Kutta method [28]. For the flow solution of the present study, we use an open-source solver called ASTR (The code is accessible in https://github.com/astr-code/astr (accessed on 30 December 2023)). This solver has been widely validated in DNSs of various compressible turbulent flows with and without shock waves [10,14,25,29–31].

The flow we study is a decaying flow. The computational domain is a $(2\pi)^2$ square domain discretized with a 4096^2 uniform grid. Periodic boundary conditions are applied in both directions. The initial velocity field is divergence-free with a spectrum $E(k) = Ak^4e^{-2k^2/k_0^2}$. The maximum wavenumber is set at $k_0 = 25$ to include both directions of the 2D dual-cascade effect in the simulation. The initial kinetic energy is determined by a constant A. The density, pressure, and temperature are all initialized to constant values. The divergence-free nature of the initial velocity field leads to the emergence of strong compression during the initial stage before transitioning to a stable decaying state. To measure the compressibility, we use the turbulent Mach number, which is defined basing on the root-mean-square velocity $u' = \sqrt{u_iu_i}$ and the speed of sound $c = \sqrt{\gamma RT}$, i.e., $\mathrm{Ma}_t = u'/c$. Its initial value is set as $\mathrm{Ma}_t = 1.10$. Figure 1a shows the evolution of Ma_t. Except for a small fluctuation during the initial stage, Ma_t decreases continuously to about 0.56 at the end of the simulation. In terms of the time normalization, we adopt the initial large-eddy-turnover time $\tau_0 = (\int_0^\infty E(k)/k\mathrm{d}k)/u'^3$. Throughout the simulation, the Kolmogorov scale first decreases and then increases continuously once the turbulent regime is well-established. The minimum Kolmogorov scale verifies $\eta/\Delta x = 1.8$ with Δx being the grid size; therefore, the flow is well-resolved down to the dissipation scale.

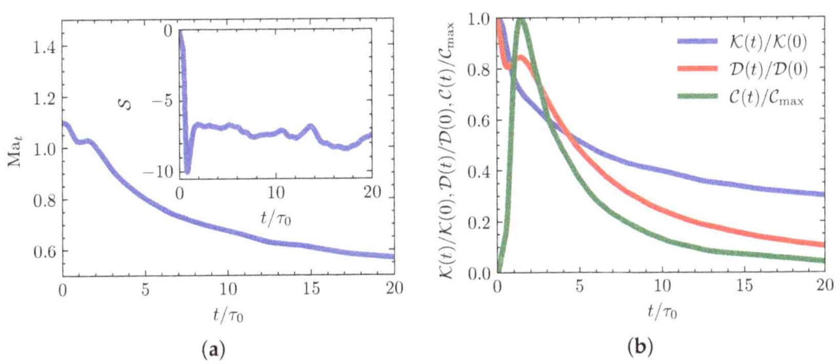

Figure 1. (**a**) Time evolution of the turbulent Mach number Ma_t and the skewness of the longitudinal velocity derivative $\mathcal{S}$ (shown in inset). (**b**) Time evolution of the kinetic energy normalized by its initial value $\mathcal{K}(t)/\mathcal{K}(0)$, the enstrophy normalized by its initial value $\mathcal{D}(t)/\mathcal{D}(0)$ and the average squared dilatation normalized by its maximum value $\mathcal{C}(t)/\mathcal{C}_{\max}$.

The inset of Figure 1a represents the evolution of the skewness of the longitudinal velocity gradient, calculated as $\mathcal{S} = \langle(\partial u_1/\partial x_1)^3 + (\partial u_2/\partial x_2)^3\rangle/[\langle(\partial u_1/\partial x_1)^2\rangle^{3/2} + \langle(\partial u_2/\partial x_2)^2\rangle^{3/2}]$. Figure 1b shows the evolution of the kinetic energy $\mathcal{K} = \frac{1}{2}\langle\rho u_iu_i\rangle$ normalized by its initial value $\mathcal{K}(0)$, the enstrophy $\mathcal{D} = \frac{1}{2}\langle\omega^2\rangle$ normalized by its initial value $\mathcal{D}(0)$, and the average squared dilatation $\mathcal{C} = \frac{1}{2}\langle\theta^2\rangle$ normalized by its maximum value $\mathcal{C}_{\max}$. The latter is defined similarly to enstrophy to show the compressibility effect. These figures illustrate the flow field evolution. In the initial stage, a strong compression emerges due to the divergence-free uniform-density initial condition, leading to a rapid increase in the average squared dilatation $\mathcal{C}$, a sharp decrease in the enstrophy $\mathcal{D}$, and a distinct negative peak in the skewness $\mathcal{S}$. Both $\mathcal{C}$ and $\mathcal{S}$ reach their extrema at $t/\tau_0 \sim 1$. The flow field attains a relatively stable state after $t/\tau_0 \sim 2$, from which point, the skewness becomes stable with small oscillations between -8 and -6. The kinetic energy $\mathcal{K}$, the enstrophy $\mathcal{D}$, and the average squared dilatation $\mathcal{C}$ all show a smooth decaying curve caused by the viscous dissipation. Notably, $\mathcal{D}$ and $\mathcal{C}$ decay more rapidly than the kinetic energy $\mathcal{K}$.

In the following paragraphs, we will calculate the quantities relating to velocity gradients to validate the analytical results in Section 2.

Figure 2 shows the time evolution of quantities relating to the independent components of the second-order moments. Figure 2a shows the equality of a_1 and a_3 normalized by the average squared Frobenius norm of the velocity gradient tensor $\langle ||\mathbf{m}||_F^2 \rangle = \left\langle \overline{\mathbf{m}\mathbf{m}^T} \right\rangle$. The initial value of a_1 (and a_3) reaches the incompressible limit of $-\frac{1}{8}$, which can be proved by using Table 3 and Equation (7a)

$$\frac{a_1}{\left\langle \overline{\mathbf{m}\mathbf{m}^T} \right\rangle} = \frac{a_1}{4a_1 + 4a_2} = \frac{a_1}{4a_1 - 12a_1} = -\frac{1}{8}. \tag{16}$$

Figure 2b shows the evolution of the ratio $M^{(2)}_{\alpha\beta\alpha\beta}/M^{(2)}_{\alpha\alpha\alpha\alpha}$, which should lie between 1/3 and 3 according to Equation (14). The initial value reaches the incompressible limit value 3.

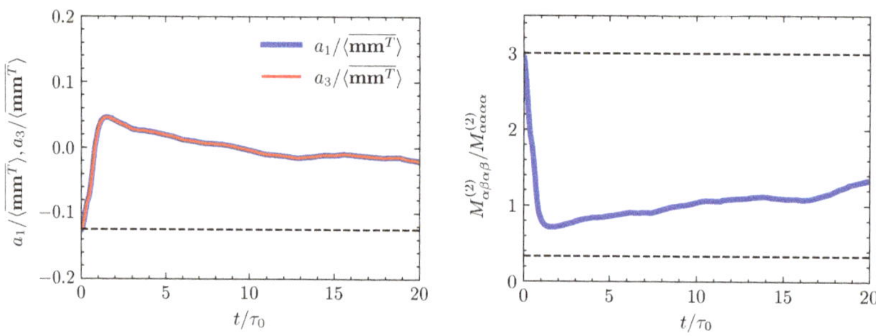

Figure 2. (**a**) Time evolution of second-order independent components $a_1 = M^{(2)}_{\alpha\alpha\beta\beta}$ and $a_3 = M^{(2)}_{\alpha\beta\beta\alpha}$ normalized by $\left\langle \overline{\mathbf{m}\mathbf{m}^T} \right\rangle$, which should be equal in 2D isotropic compressible turbulence. The dashed line in black is $-1/8$. (**b**) Time evolution of the ratio $M^{(2)}_{\alpha\beta\alpha\beta}/M^{(2)}_{\alpha\alpha\alpha\alpha}$, which should lie between 1/3 and 3 in 2D isotropic turbulence. The dashed lines in black are, respectively, 1/3 and 3.

Figure 3 shows the time evolution of quantities relating to the second-order invariants or moments of the velocity gradient tensor. The evolution of the third-order invariants $M^{(3)}_{ijjkki}$ and $\frac{3}{2}M^{(3)}_{iijkkj} - \frac{1}{2}M^{(3)}_{iijjkk}$ is presented in Figure 3a, normalized by $\left\langle \overline{\mathbf{m}\mathbf{m}^T} \right\rangle^{3/2}$. The equality in Equation (11), obtained by homogenous constraints, is validated perfectly. For Equation (6), the time evolution of the left-hand side, i.e., $M^{(3)}_{\alpha\alpha\alpha\alpha\alpha\alpha}$, and of the right-hand side, i.e., $2M^{(3)}_{\alpha\alpha\alpha\beta\alpha\beta} + 2M^{(3)}_{\alpha\alpha\alpha\beta\beta\alpha} + M^{(3)}_{\alpha\alpha\alpha\alpha\beta\beta}$, is presented in Figure 3b. Their values are both normalized by $\left\langle \overline{\mathbf{m}\mathbf{m}^T} \right\rangle^{3/2}$. To mitigate directional bias, each type of velocity gradient moment is computed by considering all possible expressions. For instance, $M^{(3)}_{\alpha\alpha\alpha\alpha\alpha\alpha}$ is calculated using $\frac{1}{2}(M^{(3)}_{111111} + M^{(3)}_{222222})$. In Figure 3b, both values exhibit similar trends and fall to the same level, but slight differences occur occasionally. These small disagreements are due to the strict isotropy condition required for Equation (6), which is difficult to achieve numerically. In compressible flow, strong local shocks are likely to occur and cause local anisotropy, as a single shock can only be oriented in a specific direction; while global isotropy is theoretically attainable in a sufficiently large flow field, practical limitations emerge due to bounded domains and finite grid sizes, resulting in unavoidable random anisotropy in numerical results. This random anisotropy leads to disagreements in results requiring isotropic conditions. In particular, third-order moments of the velocity gradient are more sensitive than second-order moments, since they are more influenced by rare but extreme values of the local velocity gradient. These extreme values are exactly caused by local strong shocks. As our DNS is performed by the exact use of classical numerical criteria, i.e., the grid resolution set to $\eta/\Delta x > 0.5$ with η the Kolmogorov length, in this sense, the present analytical study on the moments of the velocity gradient tensor reveals the limitation of current numerical methods in capturing isotropy of third-order moments in 2D compressible turbulence. This calls for more investigations on the smallest scale of two-dimensional compressible turbulence and the numerical schemes of shock capturing in the future.

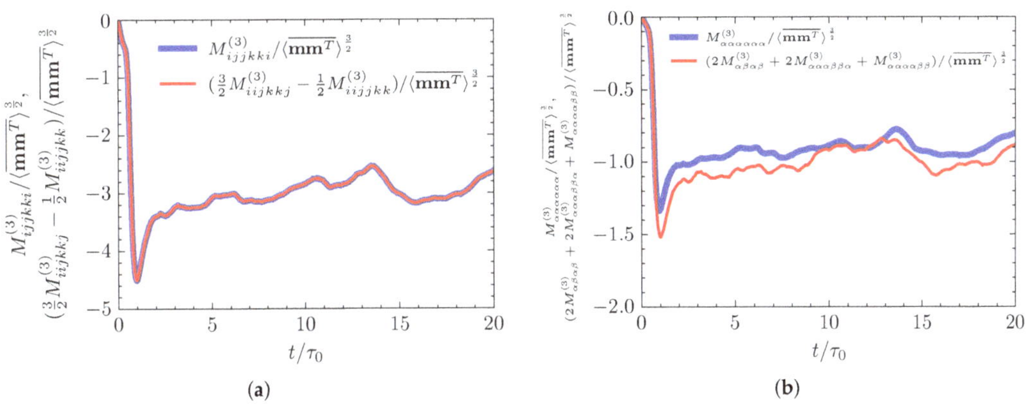

Figure 3. (**a**) Time evolution of third-order invariants $M^{(3)}_{ijjkki}$ and $\frac{3}{2}M^{(3)}_{iijkkj} - \frac{1}{2}M^{(3)}_{iijjkk}$, normalized by $\left\langle\overline{\mathbf{mm}^T}\right\rangle^{3/2}$, which should be equal in 2D isotropic compressible turbulence. (**b**) Time evolution of third-order independent components $M^{(3)}_{\alpha\alpha\alpha\alpha\alpha\alpha}$ and $2M^{(3)}_{\alpha\beta\alpha\beta} + 2M^{(3)}_{\alpha\alpha\alpha\beta\beta\alpha} + M^{(3)}_{\alpha\alpha\alpha\alpha\beta\beta}$, normalized by $\left\langle\overline{\mathbf{mm}^T}\right\rangle^{3/2}$, which should be equal in strictly isotropic 2D compressible turbulence.

4. Conclusions

In this paper, we analyze the independent components of the second- and third-order moments of the velocity gradient tensor in 2D isotropic turbulence. The second-order moments have two independent components in the 2D compressible isotropic turbulence, which is the same as in the 3D case. However, as an example of the effect of dimension reduction, the third-order moments could be fully determined by three independent components in 2D isotropic compressible turbulence, while in 3D we need four. This fact leads to an important result that the third-order moments disappear when the flow is incompressible in 2D. From the perspective of small-scale generation, which is closely related to the third-order moments of the velocity gradient, the difference between the number of independent components for 2D and 3D is a consequence of the absence of vortex stretching in two-dimensional flows. We expect that the present study on the independent scalars in 2D isotropic compressible turbulence will represent a step toward rigorously explaining the energy transfer in the future.

We further demonstrate those analytical discussions by numerical simulations. We find that all theoretical results are satisfied by the DNS data except Equation (6), where we see small discrepancies between the l.h.s. and r.h.s. of the equation. This observation is novel and shows that the isotropy of third-order moment of the velocity gradient is not satisfied, such that the derived exact relation, i.e., Equation (6), cannot be obtained by a classical DNS numerical setup. The underlying reason for this might due to the inappropriate prediction on the smallest scale of two-dimensional compressible turbulence, or the inappropriate numerical schemes of shock capturing, which requires future improvement in numerical simulations of HIT. Our result is thus expected to be a criterion for examining the appropriate resolution of small-scale structures and the isotropy of the turbulence field.

Author Contributions: Conceptualization, L.F.; methodology, P.-F.Y. and L.F.; software, C.L.; validation, C.L.; formal analysis, C.L.; investigation, C.L. and P.-F.Y.; resources, L.F.; writing—original draft preparation, C.L.; writing—review and editing, P.-F.Y. and L.F.; visualization, C.L.; supervision, P.-F.Y. and L.F.; project administration, L.F.; funding acquisition, P.-F.Y. and L.F. All authors have read and agreed to the published version of the manuscript.

Funding: This work was supported by the National Natural Science Foundation of China (Project Approval Nos. 12372214 and 12202452) and the Science Center for Gas Turbine Project (Grant No. P2022-C-III-001-001).

Data Availability Statement: For more specific information about the simulated data, please email Le Fang at le.fang@buaa.edu.cn.

Conflicts of Interest: The authors declare no conflicts of interest. The funders had no role in the design of the study; in the collection, analyses, or interpretation of data; in the writing of the manuscript; or in the decision to publish the results.

Appendix A. The Relation of $\theta, \omega, \psi, \varphi$ in Fourier Space

In this appendix, we will show the global relation of the locally independent velocity gradient components, i.e., $\theta(\boldsymbol{x}), \omega(\boldsymbol{x}), \psi(\boldsymbol{x}), \varphi(\boldsymbol{x})$, in the Fourier space.

In a 3D space generated by an orthogonal basis $(\boldsymbol{e}_1, \boldsymbol{e}_2, \boldsymbol{e}_3)$, where the flow is confined in the plane $(\boldsymbol{e}_1, \boldsymbol{e}_2)$, we first apply the Helmholtz decomposition to the velocity field $\boldsymbol{u}(\boldsymbol{x})$, as

$$\hat{\boldsymbol{u}}(\boldsymbol{k}) = \hat{\boldsymbol{u}}^s(\boldsymbol{k}) + \hat{\boldsymbol{u}}^d(\boldsymbol{k}) = \hat{u}^s(\boldsymbol{k})\tilde{\boldsymbol{k}}^\perp + \hat{u}^d(\boldsymbol{k})\tilde{\boldsymbol{k}}, \boldsymbol{k} \neq \boldsymbol{0}, \tag{A1}$$

where $\hat{\boldsymbol{u}}(\boldsymbol{k})$ denotes the Fourier transform of the velocity field, $\tilde{\boldsymbol{k}} = \frac{\boldsymbol{k}}{|\boldsymbol{k}|} = \frac{\boldsymbol{k}}{k}$ denotes the unit vector in the direction of wave vector $\boldsymbol{k}$, and $\tilde{\boldsymbol{k}}^\perp = \boldsymbol{e}_3 \times \tilde{\boldsymbol{k}}$ denotes the unit vector with which an orthogonal right-hand coordinate system $(\tilde{\boldsymbol{k}}, \tilde{\boldsymbol{k}}^\perp, \boldsymbol{e}_3)$ can be formed. In this equation, $\hat{u}^s(\boldsymbol{k})$ and $\hat{u}^d(\boldsymbol{k})$ represent, respectively, the solenoidal and the dilatational component of the velocity fluctuation field. The $\boldsymbol{k} = \boldsymbol{0}$ component of $\hat{\boldsymbol{u}}$ is exactly the ensemble average of velocity field

$$\hat{\boldsymbol{u}}(\boldsymbol{k} = \boldsymbol{0}) = \langle \boldsymbol{u} \rangle, \tag{A2}$$

which will be dropped out in the following text as it will not contribute to the velocity gradient. It is also worth noting that the solenoidal component can only follow one direction, i.e., $\tilde{\boldsymbol{k}}^\perp$, in 2D, while in 3D, it can be in the Craya plane and gives two independent components.

The counterpart of the velocity gradient tensor in Fourier space can be expressed as

$$\hat{\mathbf{m}}(\boldsymbol{k}) = (i\boldsymbol{k} \otimes \hat{\boldsymbol{u}}(\boldsymbol{k}))^T, \boldsymbol{k} \neq \boldsymbol{0} \tag{A3}$$

Furthermore, we set $\hat{\mathbf{m}}(\boldsymbol{k} = \boldsymbol{0}) = \boldsymbol{0}$. Specifically, the vorticity and the dilatation in Fourier space can be uniquely expressed as

$$\hat{\boldsymbol{\omega}}(\boldsymbol{k}) = \hat{\omega}(\boldsymbol{k})\boldsymbol{e}_3 = i\boldsymbol{k} \times \hat{\boldsymbol{u}}^s(\boldsymbol{k}) = i\boldsymbol{k} \times \tilde{\boldsymbol{k}}^\perp \hat{u}^s(\boldsymbol{k}) = ik\hat{u}^s(\boldsymbol{k})\boldsymbol{e}_3, \tag{A4}$$

$$\hat{\theta}(\boldsymbol{k}) = i\boldsymbol{k} \cdot \hat{\boldsymbol{u}}^d(\boldsymbol{k}) = i\boldsymbol{k} \cdot \tilde{\boldsymbol{k}}\hat{u}^d(\boldsymbol{k}) = ik\hat{u}^d(\boldsymbol{k}). \tag{A5}$$

The velocity field can also be expressed as

$$\hat{u}(k) = \hat{u}^s(k)\tilde{k}^{\perp} + \hat{u}^d(k)\tilde{k} = \frac{\hat{\omega}(k)}{ik}\tilde{k}^{\perp} + \frac{\hat{\theta}(k)}{ik}\tilde{k}, k \neq 0. \tag{A6}$$

As a result, the velocity gradient tensor in Fourier space can be expressed with only $\hat{\theta}(k)$ and $\hat{\omega}(k)$

$$\hat{\mathbf{m}}(k) = \left[ik \otimes \left(\frac{\hat{\omega}(k)}{ik}\tilde{k}^{\perp} + \frac{\hat{\theta}(k)}{ik}\tilde{k}\right)\right]^T = \hat{\omega}(k)\tilde{k}^{\perp} \otimes \tilde{k} + \hat{\theta}(k)\tilde{k} \otimes \tilde{k}, k \neq 0. \tag{A7}$$

The Fourier space velocity gradient tensor can also be decomposed into dilatational, symmetric-deviatoric, and antisymmetric parts. When it is expressed in the global basis (e_1, e_2), it can result in a similar expression to Equation (3) with $\hat{\varphi}(k), \hat{\psi}(k), \hat{\omega}(k), \hat{\theta}(k)$. However, when we express it in the local wave-number basis $(\tilde{k}, \tilde{k}^{\perp})$, it gives

$$\hat{\mathbf{m}}(k) = \frac{1}{2}\begin{bmatrix} \hat{\theta}(k) & \hat{\omega}(k) \\ \hat{\omega}(k) & -\hat{\theta}(k) \end{bmatrix}_{(\tilde{k},\tilde{k}^{\perp})} + \frac{1}{2}\hat{\omega}(k)\begin{bmatrix} 0 & -1 \\ 1 & 0 \end{bmatrix}_{(\tilde{k},\tilde{k}^{\perp})} + \frac{1}{2}\hat{\theta}(k)\begin{bmatrix} 1 & 0 \\ 0 & 1 \end{bmatrix}_{(\tilde{k},\tilde{k}^{\perp})}, k \neq 0. \tag{A8}$$

It should be noted that $(\tilde{k}, \tilde{k}^{\perp})$ is a local basis. For different wave numbers k_a, k_b , unless they are in the same direction, $(\tilde{k}_a, \tilde{k}_a^{\perp})$ and $(\tilde{k}_b, \tilde{k}_b^{\perp})$ give two different bases. If we carry out a change of basis for Equation (A8) to (e_1, e_2), we can derive for $k \neq 0$

$$\hat{\theta}(k) = 2\hat{\psi}(k)\tilde{k}_1\tilde{k}_2 + \hat{\varphi}(k)(\tilde{k}_1^2 - \tilde{k}_2^2), \tag{A9}$$

$$\hat{\omega}(k) = \hat{\psi}(k)(\tilde{k}_1^2 - \tilde{k}_2^2) - 2\hat{\varphi}(k)\tilde{k}_1\tilde{k}_2, \tag{A10}$$

with $\tilde{k}_i = \frac{k_i}{k}$ being the coordinate of $\tilde{k}$ in $e_i (i = 1, 2)$ direction. These two equations link $\hat{\omega}(k), \hat{\theta}(k)$ and $\hat{\varphi}(k), \hat{\psi}(k)$, showing the consistency between different decompositions.

It can also be concluded from Equations (A9) and (A10) that

$$\hat{\omega}^2(k) + \hat{\theta}^2(k) = \hat{\varphi}^2(k) + \hat{\psi}^2(k), k \neq 0 \tag{A11}$$

and $k = 0$ also verifies this equation. With an integral in the whole wave number space and Parseval's identity, this leads to

$$\int_{x \in \mathbb{R}^2} \omega^2(x)dx + \int_{x \in \mathbb{R}^2} \theta^2(x)dx = \int_{x \in \mathbb{R}^2} \varphi^2(x)dx + \int_{x \in \mathbb{R}^2} \psi^2(x)dx. \tag{A12}$$

In the case of homogeneous turbulence, we then obtain

$$\left\langle \omega^2 \right\rangle + \left\langle \theta^2 \right\rangle = \left\langle \varphi^2 \right\rangle + \left\langle \psi^2 \right\rangle, \tag{A13}$$

which is also proved in Equation (13) using the independent components of the second-order moment.

References

1. Kolmogorov, A. Dissipation of energy in locally isotropic turbulence. *Dokl. Akad. Nauk SSSR* **1941**, *32*, 16–18.
2. Kolmogorov, A. The local structure of turbulence in incompressible viscous fluid for very large Reynolds numbers. *Dokl. Akad. Nauk SSSR* **1941**, *30*, 301–305.
3. Tsinober, A. *An Informal Conceptual Introduction to Turbulence*; Springer: Berlin/Heidelberg, Germany, 2009. [CrossRef]
4. Meneveau, C. Lagrangian Dynamics and Models of the Velocity Gradient Tensor in Turbulent Flows. *Annu. Rev. Fluid Mech.* **2011**, *43*, 219–245. [CrossRef]
5. Davidson, P.A. *Turbulence: An Introduction for Scientists and Engineers*; Oxford University Press: Oxford, UK, 2015. [CrossRef]
6. Taylor, G.I. The spectrum of turbulence. *Proc. R. Soc. Lond. Ser. A-Math. Phys. Sci.* **1938**, *164*, 476–490. [CrossRef]
7. Betchov, R. An inequality concerning the production of vorticity in isotropic turbulence. *J. Fluid Mech.* **1956**, *1*, 497–504. [CrossRef]
8. Champagne, F.H. The fine-scale structure of the turbulent velocity field. *J. Fluid Mech.* **1978**, *86*, 67–108. [CrossRef]

9. Pope, S.B. *Turbulent Flows*; Cambridge University Press: Cambridge, UK, 2000.
10. Yang, P.F.; Fang, J.; Fang, L.; Pumir, A.; Xu, H. Low-order moments of the velocity gradient in homogeneous compressible turbulence. *J. Fluid Mech.* **2022**, *947*, R1. [CrossRef]
11. Carbone, M.; Wilczek, M. Only two Betchov homogeneity constraints exist for isotropic turbulence. *J. Fluid Mech.* **2022**, *948*, R2. [CrossRef]
12. Siggia, E.D. Invariants for the one-point vorticity and strain rate correlation functions. *Phys. Fluids* **1981**, *24*, 1934–1936. [CrossRef]
13. Hierro, J.; Dopazo, C. Fourth-order statistical moments of the velocity gradient tensor in homogeneous, isotropic turbulence. *Phys. Fluids* **2003**, *15*, 3434–3442. [CrossRef]
14. Fang, L.; Zhang, Y.; Fang, J.; Zhu, Y. Relation of the fourth-order statistical invariants of velocity gradient tensor in isotropic turbulence. *Phys. Rev. E* **2016**, *94*, 023114. [CrossRef] [PubMed]
15. Tabeling, P. Two-dimensional turbulence: A physicist approach. *Phys. Rep.* **2002**, *362*, 1–62. [CrossRef]
16. Clercx, H.; van Heijst, G. Two-Dimensional Navier–Stokes Turbulence in Bounded Domains. *Appl. Mech. Rev.* **2009**, *62*, 020802. [CrossRef]
17. Boffetta, G.; Ecke, R.E. Two-dimensional turbulence. *Annu. Rev. Fluid Mech.* **2012**, *44*, 427–451. [CrossRef]
18. Kraichnan, R.H. Inertial ranges in two-dimensional turbulence. *Phys. Fluids* **1967**, *10*, 1417–1423. [CrossRef]
19. Kraichnan, R.H. Inertial-range transfer in two- and three-dimensional turbulence. *J. Fluid Mech.* **1971**, *47*, 525–535. [CrossRef]
20. Falkovich, G.; Kritsuk, A.G. How vortices and shocks provide for a flux loop in two-dimensional compressible turbulence. *Phys. Rev. Fluids* **2017**, *2*, 092603. [CrossRef]
21. Dahlburg, J.; Dahlburg, R.; Gardner, J.; Picone, J. Inverse cascades in two-dimensional compressible turbulence. I. Incompressible forcing at low Mach number. *Phys. Fluids A Fluid Dyn.* **1990**, *2*, 1481–1486. [CrossRef]
22. Alexakis, A.; Biferale, L. Cascades and transitions in turbulent flows. *Phys. Rep.* **2018**, *767–769*, 1–101. [CrossRef]
23. Fouxon, I.; Kritsuk, A.G.; Mond, M. Compressible two-dimensional turbulence: Cascade reversal and sensitivity to imposed magnetic field. *New J. Phys.* **2023**, *25*, 113005. [CrossRef]
24. Pan, S.; Johnsen, E. The role of bulk viscosity on the decay of compressible, homogeneous, isotropic turbulence. *J. Fluid Mech.* **2017**, *833*, 717–744. [CrossRef]
25. Fang, J.; Li, Z.; Lu, L. An optimized low-dissipation monotonicity-preserving scheme for numerical simulations of high-speed turbulent flows. *J. Sci. Comput.* **2013**, *56*, 67–95. [CrossRef]
26. Lele, S.K. Compact finite difference schemes with spectral-like resolution. *J. Comput. Phys.* **1992**, *103*, 16–42. [CrossRef]
27. Fang, J.; Gao, F.; Moulinec, C.; Emerson, D. An improved parallel compact scheme for domain-decoupled simulation of turbulence. *Int. J. Numer. Methods Fluids* **2019**, *90*, 479–500. [CrossRef]
28. Gottlieb, S.; Shu, C.W. Total variation diminishing Runge–Kutta schemes. *Math. Comput.* **1998**, *67*, 73–85. [CrossRef]
29. Fang, J.; Yao, Y.; Li, Z.; Lu, L. Investigation of low-dissipation monotonicity-preserving scheme for direct numerical simulation of compressible turbulent flows. *Comput. Fluids* **2014**, *104*, 55–72. [CrossRef]
30. Fang, J.; Yao, Y.; Zheltovodov, A.A.; Li, Z.; Lu, L. Direct numerical simulation of supersonic turbulent flows around a tandem expansion-compression corner. *Phys. Fluids* **2015**, *27*, 125104 . [CrossRef]
31. Fang, J.; Zheltovodov, A.A.; Yao, Y.; Moulinec, C.; Emerson, D.R. On the turbulence amplification in shock-wave/turbulent boundary layer interaction. *J. Fluid Mech.* **2020**, *897*, A32. [CrossRef]

symmetry

MDPI

Article

Symmetry Analysis of Mean Velocity Distribution in Stratified Atmospheric Surface Layers

Yong Ji [1] and Xi Chen [2,*]

1 School of Mathematics and Statistics, Ningxia University, Yinchuan 750021, China; jiyong@pku.edu.cn
2 Key Laboratory of Fluid Mechanics of Ministry of Education, Beihang University (Beijing University of Aeronautics and Astronautics), Beijing 100191, China
* Correspondence: chenxi97@outlook.com

Abstract: The mean velocity distributions of unstably and stably stratified atmospheric surface layers (ASLs) are investigated here using the symmetry approach. Symmetry groups for the mean momentum and the Reynolds stress equations of ASL are searched under random dilation transformations, which, with different leading order balances in different flow regions, lead to a set of specific scalings for the characteristic length ℓ_{13} (defined by Reynolds shear stress and mean shear). In particular, symmetry analysis shows that in the shear-dominated region, ℓ_{13} scales linearly with the surface height z, which corresponds to the classical log law of mean velocity. In the buoyancy-dominated region, $\ell_{13}/L \sim (z/L)^{4/3}$ for unstably stratified ASL and $\ell_{13}/L \sim const$ for stably stratified ASL, where L is the Obukhov length. The specific formula of the celebrated Monin–Obukhov similarity function is obtained, and hence an algebraic model of mean velocity profiles in ASL is derived, showing good agreement with the datum from the QingTu Lake observation array (QLOA) in China.

Keywords: atmospheric surface layer; mean velocity profile; symmetry analysis

1. Introduction

The atmospheric surface layer (ASL) is a specific turbulent boundary layer (TBL) in which atmosphere exchanges momentum and energy with the earth's surface through nearly constant momentum and heat flux [1]. Due to its large dimension in the horizontal and surface-normal directions, ASL typically reaches a (friction) Reynolds number of several millions [2], making it a candidate for the study of high Reynolds number canonical wall turbulence. Also, as a result of the diurnal release and absorption of the heat of the earth's surface, the turbulent flow states of ASL in the daytime and at night are significantly different [3]. Notable observation arrays, such as QLOA [4] and SLTEST (Surface Layer Turbulence and Environmental Science) [5], have been established to experimentally study the high Reynolds number properties of turbulent boundary layers [2,5–13]. Due to the limitation of grid spatial resolution, the numerical weather research and forecasting model (WRF) cannot solve the flow field in the ASL [14]. Therefore, studies on the unified description of mean velocity distribution (MVD) under various momentum and heat flux conditions are important for applications in the weather forecasting, pollutant prediction, and wind energy industry [1,3,15–17].

Since the boundary layer concept of Prandlt [18], the Monin–Obukhov (MO) similarity theory [19] has been a milestone for understanding the ASL and is regarded as the starting point of modern micrometeorology [16]. The MO theory relies on statistical stationarity and horizontal homogeneity, which results in the constant momentum and heat flux in the surface-height (or wall-normal) direction. In particular, four physical parameters are proposed in MO to depict the mean flow of the ASL: heat flux $H_w = \overline{w'T'}$, friction velocity $u_\tau = \sqrt{-\overline{u'w'}}$, buoyancy force factor $g/\overline{T}$, and the height z. Dimensional analysis leads to the Obukhov length scale $L = -u_\tau^3/\left[\kappa\overline{w'T'}g/\overline{T}\right]$, where $\overline{T}$ is the mean temperature, g is the acceleration of gravity, κ is the von Kármán constant, and w' and T' are the fluctuating wall-normal velocity and temperature, respectively. In practical terms, L can be explained as a critical height, below which turbulence is dominated by the wall-induced shear effect, and above which it is dominated by buoyancy. Consequently, a non-dimensional similarity variable $\zeta = z/L$ is defined by MO to quantify the ratio between the shear production and buoyancy effect in the ASL. Since then, statistical quantities have all been expressed as certain functions of ζ, validated by many atmospheric experimental studies [20,21].

Particularly for the mean velocity distribution, MO proposed investigating the following dimensionless similarity function:

$$\varphi_m(\zeta) = \frac{\kappa z}{u_\tau}\frac{dU}{dz}. \tag{1}$$

To determine φ_m, an implicit function equation $\varphi_m^4 - \gamma\varphi_m^3\zeta = 1$ has been derived under an approximation of the kinetic energy balance equation, known as the O'KEYPS equation (Obukhov, Kazansky, Ellison, Yamamoto, Panofsky, and Sellers) [22]. However, the constant eddy diffusivity assumed in the O'KEYPS equation remains a subject of doubt, and the empirical parameter γ, which varies from 5 to 18 for different data sets, indicates that the form of φ_m in the O'KEYPS equation may not be universal [23].

On the other hand, based on the Kansas experimental measurements, $\varphi_m = (1-16\zeta)^{-1/4}$ for $\zeta < 0$ and $\varphi_m = 1 + 4.7\zeta$ for $\zeta > 0$ are well-known as the Businger–Dyer (BD) function [24,25]. However, the asymptotic scaling of BD function at the free convection limit, i.e., $\varphi_m \sim (-\zeta)^{-1/4}$ for $-\zeta \gg 1$, differs from the prediction by teh O'KEYPS equation. To address this, Carl et al. [26] proposed a modified $\varphi_m = (1-15\zeta)^{-1/3}$, to conform with the O'KEYPS equation. Moreover, Kader and Yaglom [27] argued that in the O'KEYPS equation, $\gamma \gg 1$, as the buoyancy acts only on the direction normal to the earth's surface and hence contributes mostly to momentum transfer. They also proposed the same scaling $\varphi_m \sim (-\zeta)^{1/3}$, validated by their measured data in the top convective sublayer of the ASL [27].

More recently, Katul et al. developed a heuristic model of MVD in ASL under the attached eddy hypothesis. The characteristic velocity u_τ is derived from the assumed inertial-range spectrum, which further leads to a relation between buoyancy and momentum flux [23,28]. This model yields a similar prediction to that of the O'KEYPS equation, but it incorporates more physical considerations. For example, the large value of γ is attributed to the effects of turbulent transport, pressure redistribution, and the anisotropy of turbulent eddies. However, it should also be noted that the functional form of the similarity function becomes more complicated, and its prediction for the stable ASL deviates from observations. To account for this, Li et al. revisited Katul's model by introducing the Ozmidov scale [29] and obtained a better description of the data. Liu et al. developed an analytical model describing the vertical structure of conventionally neutral atmospheric boundary layers, providing predictions of wind and turbulent shear stress profiles [17,30]. In addition to the works mentioned, there are also efforts to extend the MO theory by further consideration of large-scale coherent structures [31], turbulence anisotropy [32], and non-zero vertical turbulent transport effects [33].

This paper aims to understand the MVD from a symmetry perspective. It is worth noting that symmetry analysis is a well-developed method for finding the similarities or invariant solutions of differential equations. A typical example is the Blasius similarity solution of a laminar boundary layer flow, the symmetry analysis of which involves three steps. First, search the dilation symmetry transformation that keeps the governing equation unchanged; second, obtain the dilation invariants for the independent and dependent variables; finally, use these invariants as variables to transfer the partial differential equation to an ordinary differential equation. When dealing with turbulence, there is a challenge in these steps because of the unknown Reynolds stresses. However, by searching the symmetry of characteristic length scales, one can construct candidate invariant solutions by using group invariants as similarity variables. This approach has been utilized to describe turbulent mean flows in wall flows, with details provided in [34–36].

Specifically, for canonical turbulent boundary layers, a multi-layer description of the mean velocity profiles has been obtained in [34–36] through a random dilation analysis of Reynolds-averaged Navier–Stokes equations. In comparison to previous works, the novelty in [34–36] is that different leading order balances are considered in the symmetry analysis, and a general ansatz is proposed to connect the local dilation invariants. In the past few years, a large set of experimental and direct numerical simulation (DNS) data of canonical wall flows (channel, pipe, and turbulent boundary layer—TBL) have verified the multi-layer description, with successful extensions to complex boundary conditions, including pressure gradient effects, heat flux effects, and surface roughness [37,38]. Note that, as ASL is also a wall-induced shear flow, the dilation symmetry proposed in [34–36] along the wall-normal direction (z) may also exist. Thus, we plan to develop a similar modeling of the ASL in this paper based on the symmetry analysis approach.

Before proceeding further, it is important to note that a theoretical description of thermally stratified ASL flow is crucial for climate modeling of the near-surface wind fields. For instance, in the Weather Research and Forecasting (WRF) model, due to computational power cost, the WRF cannot resolve the near wall flow details. Therefore the Monin–Obukhov similarity theory (MOST) is used to estimate the exchanges of heat, momentum and humidity between the earth's surface and the ASL. Specially for the wind speed at approximately 10 m above the earth's surface, the accuracy of MOST is crucial for WRF predictions. Various corrections of MOST have been proposed in the WRF [14]. In this context, we present a theoretical framework for describing thermally stratified ASLs based on the symmetry approach, which offers a more comprehensive understanding compared to the models and corrections based on MOST. The results can be integrated into the WRF to enhance the prediction of near-surface wind fields.

The rest of this paper is organized as follows. The balance equations and the dilation symmetry analysis are introduced in Section 2. Section 3 introduces the experimental data, the comparison of which with our theory is provided in Section 4. Final conclusions are presented in Section 5.

2. Methods

This study is inspired by the symmetry approach for canonical TBL developed by She et al. and Chen et al. [35,36]. Note that to obtain the mean velocity distributions in ASL, one needs to address the closure problem of the unknown Reynolds shear stress. In the literature, this is usually resolved by the hypothesis of eddy viscosity or mixing length (the so-called stress length here), which builds a relation between the mean shear and Reynolds stress. However, these hypotheses have no link with the balance equations. The current paper aims to develop a procedure to determine the mixing length (or stress length) based on the balance equations. As we demonstrate below, the balance equations allow for a set of random dilations, which define the dilation invariant of the stress length. By assuming a constant dilation invariant, we can thus obtain the power-law scaling exponent for the stress length function, and hence the mean velocity distributions in ASL. Therefore, the key contribution of our paper is to present the rationale behind the scaling exponent of stress length from the symmetry consideration of the balance equations. This is reminiscent of Monin and Obukhov, who also derived the logarithmic law of wind distribution based on the concept of dilation symmetry [19]. However, our approach contains more mathematical details and extends to different flow regions.

The ensemble averaged momentum and Reynolds stresses equations are [27,36]

$$\frac{\partial \overline{-u'w'}}{\partial z} + \nu \frac{\partial^2 U}{\partial z^2} = 0, \quad (2)$$

$$\underbrace{\overline{-u'w'}\frac{\partial U}{\partial z}}_{SP} + \underbrace{\overline{p'\frac{\partial u'}{\partial x}}}_{R_u} - \underbrace{\frac{1}{2}\frac{\partial \overline{u'^2 w'}}{\partial z}}_{T_u} - \underbrace{\nu\overline{|\nabla u'|^2}}_{\epsilon_u} + \underbrace{\nu \frac{\partial^2}{\partial z^2}\left(\frac{1}{2}\overline{u'^2}\right)}_{D_u} = 0, \quad (3)$$

$$\underbrace{\overline{p'\frac{\partial v'}{\partial y}}}_{R_v} - \underbrace{\frac{1}{2}\frac{\partial \overline{v'^2 w'}}{\partial z}}_{T_v} - \underbrace{\nu\overline{|\nabla v'|^2}}_{\epsilon_v} + \underbrace{\nu \frac{\partial^2}{\partial z^2}\left(\frac{1}{2}\overline{v'^2}\right)}_{D_v} = 0, \quad (4)$$

$$\underbrace{\overline{w'T'}\frac{g}{T}}_{B} + \underbrace{\overline{p'\frac{\partial w'}{\partial z}}}_{R_v} - \underbrace{\frac{1}{2}\frac{\partial \overline{w'^3 + 2p'w'}}{\partial z}}_{T_w} - \underbrace{\nu\overline{|\nabla w'|^2}}_{\epsilon_w} + \underbrace{\nu \frac{\partial^2}{\partial z^2}\left(\frac{1}{2}\overline{w'^2}\right)}_{D_w} = 0, \quad (5)$$

$$-\frac{1}{2}\overline{w'w'}\frac{\partial U}{\partial z} - \frac{1}{2}\overline{u'\frac{\partial p'}{\partial z}} - \frac{1}{2}\overline{w'\frac{\partial p'}{\partial x}} - \frac{1}{2}\frac{\partial \overline{u'w'w'}}{\partial z} - \nu\overline{|\nabla u' \bullet \nabla w'|} + \nu \frac{\partial^2}{\partial z^2}\left(\frac{1}{2}\overline{u'w'}\right) = 0, \quad (6)$$

where overline denotes the time-spatial ensemble average; p' is the pressure fluctuation and u', w', v' are the streamwise, vertical (or surface-height), and spanwise velocity fluctuations with U the mean streamwise velocity; SP is the shear production; $R_{u,w,v}$ are pressure redistribution terms; $T_{u,w,v}$ are vertical spatial turbulent transports; $\epsilon_{u,w,v}$ are turbulent dissipation rates; $D_{u,w,v}$ are diffusion terms; subscript u, w, v denote three velocity fluctuation components.

The above equations for ASL are similar to the canonical boundary layer flows except for the additional buoyancy term (i.e., the heat flux) $B = \overline{w'T'}g/\overline{T}$ in Equation (5). The buoyancy term could be an energy source for upward heat flux ($B > 0$), or a sink for downward heat flux ($B < 0$). Since the shear production (SP) decreases as z^{-1} and the buoyancy (B) is invariant as the height Z increases, the shear production dominates the balance equations for small z while the buoyancy dominates for large z. As shown below, the alteration of dominant balance would result in different dilation transformations, leading to different scalings in different flow regions.

To proceed, normalizing the above Equations (2)–(6) by the friction velocity u_τ, the wall heat flux H_w, and the Obukhov length L yields

$$\frac{\partial \overline{-u'^+w'^+}}{\partial \zeta_z} + \frac{1}{L^+}\frac{\partial^2 U^+}{\partial \zeta_z^2} = 0, \quad (7)$$

$$\overline{-u'^+w'^+}\frac{\partial U^+}{\partial \zeta_z} + \overline{p'^+\frac{\partial u'^+}{\partial \zeta_x}} - \frac{1}{2}\frac{\partial \overline{u'^{+2}w'^+}}{\partial \zeta_z} - \frac{1}{L^+}\overline{|\nabla_L u'^+|^2} + \frac{1}{L^+}\frac{\partial^2}{\partial \zeta_z^2}\left(\frac{1}{2}\overline{u'^{+2}}\right) = 0, \quad (8)$$

$$\overline{p'^+\frac{\partial v'^+}{\partial \zeta_y}} - \frac{1}{2}\frac{\partial \overline{v'^{+2}w'^+}}{\partial \zeta_z} - \frac{1}{L^+}\overline{|\nabla_L v'^+|^2} + \frac{1}{L^+}\frac{\partial^2}{\partial \zeta_z^2}\left(\frac{1}{2}\overline{v'^{+2}}\right) = 0, \quad (9)$$

$$\overline{w'^+T'^+} + \overline{p'^+\frac{\partial w'^+}{\partial \zeta_z}} - \frac{1}{2}\frac{\partial \overline{w'^{+3} + 2p'^+w'^+}}{\partial \zeta_z} - \frac{1}{L^+}\overline{|\nabla_L w'^+|^2} + \frac{1}{L^+}\frac{\partial^2}{\partial \zeta_z^2}\left(\frac{1}{2}\overline{w'^{+2}}\right) = 0, \quad (10)$$

$$-\frac{1}{2}\overline{w'^+w'^+}\frac{\partial U^+}{\partial \zeta_z} - \frac{1}{2}\overline{u'^+\frac{\partial p'^+}{\partial \zeta_z}} - \frac{1}{2}\overline{w'^+\frac{\partial p'^+}{\partial \zeta_x}} - \frac{1}{2}\frac{\partial \overline{u'^+w'^+w'^+}}{\partial \zeta_z} -$$
$$\frac{1}{L^+}\overline{|\nabla_L u'^+ \bullet \nabla_L w'^+|} + \frac{1}{L^+}\frac{\partial^2}{\partial \zeta_z^2}\left(\frac{1}{2}\overline{u'^+w'^+}\right) = 0, \quad (11)$$

where superscript plus + means being normalized by wall variables, i.e.,

$$[u'^+, v'^+, w'^+, U^+, p'^+, L^+, T'^+] = \left[\frac{u'}{u_\tau}, \frac{v'}{u_\tau}, \frac{w'}{u_\tau}, \frac{U}{u_\tau}, \frac{p'}{u_\tau^2}, \frac{L}{\nu/u_\tau}, \frac{T'}{\kappa H_w/u_\tau}\right],$$

and $\nabla_L = \partial/\partial\zeta_x + \partial/\partial\zeta_y + \partial/\partial\zeta_z$ is gradient operator, with $\zeta_x = x/L$, $\zeta_y = y/L$, and $\zeta_z \equiv \zeta = z/L$.

Following the step in [35,36], we define the Reynolds shear stress length function $\ell_{13}^{\wedge} = \frac{\sqrt{-\overline{u'^+w'^+}}}{\partial U^+/\partial\zeta}$, which characterizes the size of eddies responsible for vertical momentum transport. This enables us to solve the mean velocity profile from Equation (7):

$$U^+(h/L) = \int_{h_0/L}^{h/L} \frac{\sqrt{-\overline{u'^+w'^+}}}{\ell_{13}^{\wedge}} d\zeta. \tag{12}$$

Here, h is the height of interest, h_0 is the typical roughness height, and $\ell_{13}^{\wedge} = \ell_{13}/L$ with $\ell_{13} = \frac{\sqrt{-\overline{u'w'}}}{\partial U/\partial z}$. Note that ASL is a constant momentum flux layer, which means $-\overline{u'^+w'^+} = 1$; hence, Equation (12) is written

$$U(h) = u_\tau \int_{h_0}^{h} \frac{dz}{\ell_{13}}. \tag{13}$$

Once ℓ_{13} is known, the mean velocity $U(h)$ can also be determined.

To obtain the formula of ℓ_{13}, a random dilation transformation [36] is introduced as follows:

$$\zeta_i^* = q_i\zeta_i, \quad U^* = \lambda_U U \quad L^* = \lambda_L L \quad u_i'^* = \lambda_i u_i', \quad p'^* = \lambda_p p', \quad T'^* = \lambda_T T', \tag{14}$$

where q_i and λ_U, λ_L are regular dilation factors, while λ_i , λ_p, and λ_T are random factors with zero means; and $i = (1,2,3)$ denotes (x,y,z). Note that superscript '+' is neglected here. Substituting (14) into the balance Equations (7)–(11), the symmetry requires that equations under dilation remain invariant, which leads to the following relationships among dilation parameters:

$$\frac{\overline{\lambda_1\lambda_3}}{q_3} = \frac{1}{\lambda_L}\frac{\lambda_U}{q_3^2}, \tag{15}$$

$$\overline{\lambda_1\lambda_3}\frac{\lambda_U}{q_3} = \frac{\overline{\lambda_p\lambda_1}}{q_1} = \frac{\overline{\lambda_1\lambda_1\lambda_3}}{q_3} = \frac{1}{\lambda_L}\frac{\overline{\lambda_1\lambda_1}}{q_1^2} = \frac{1}{\lambda_L}\frac{\overline{\lambda_1\lambda_1}}{q_2^2} = \frac{1}{\lambda_L}\frac{\overline{\lambda_1\lambda_1}}{q_3^2}, \tag{16}$$

$$\frac{\overline{\lambda_p\lambda_2}}{q_2} = \frac{\overline{\lambda_2\lambda_2\lambda_3}}{q_3} = \frac{1}{\lambda_L}\frac{\overline{\lambda_2\lambda_2}}{q_1^2} = \frac{1}{\lambda_L}\frac{\overline{\lambda_2\lambda_2}}{q_2^2} = \frac{1}{\lambda_L}\frac{\overline{\lambda_2\lambda_2}}{q_3^2}, \tag{17}$$

$$\overline{\lambda_3\lambda_T} = \frac{\overline{\lambda_p\lambda_3}}{q_3} = \frac{\overline{\lambda_3\lambda_3\lambda_3}}{q_3} = \frac{1}{\lambda_L}\frac{\overline{\lambda_3\lambda_3}}{q_1^2} = \frac{1}{\lambda_L}\frac{\overline{\lambda_3\lambda_3}}{q_2^2} = \frac{1}{\lambda_L}\frac{\overline{\lambda_3\lambda_3}}{q_3^2}, \tag{18}$$

$$\overline{\lambda_3\lambda_3}\frac{\lambda_U}{q_3} = \frac{\overline{\lambda_p\lambda_1}}{q_3} = \frac{\overline{\lambda_p\lambda_3}}{q_1} = \frac{\overline{\lambda_1\lambda_3\lambda_3}}{q_3} = \frac{1}{\lambda_L}\frac{\overline{\lambda_1\lambda_3}}{q_1^2} = \frac{1}{\lambda_L}\frac{\overline{\lambda_1\lambda_3}}{q_2^2} = \frac{1}{\lambda_L}\frac{\overline{\lambda_1\lambda_3}}{q_3^2}. \tag{19}$$

An important fact is that in different flow regions, there are different leading order balances in (7)–(11) so that we can define a locally valid dilation group by neglecting unimportant terms. This provides extra freedoms for the group parameters in Equations (15)–(19), as practised below in different flow layers.

2.1. *Homogeneous Dilations in the Shear Dominated Layer*

In the shear dominated layer, the shear production (SP) is balanced by dissipation from all three spatial directions. Hence, all the spatial derivatives in (3)–(6) are important, leading to a homogeneous dilation in three directions, i.e., $q_1 = q_2 = q_3 = q$ in (15)–(19). Furthermore, from $\overline{\lambda_1\lambda_3}\frac{\lambda_U}{q} = \frac{\overline{\lambda_p\lambda_1}}{q} = \frac{1}{\lambda_L}\frac{\overline{\lambda_1\lambda_3}}{q^2}$, we obtain $\lambda_U\lambda_L = q^{-1}$; considering $\overline{\lambda_3\lambda_T} = 1$ for constant heat flux, from Equations (15), (18), and (19), we obtain $\lambda_L^2/\lambda_U = q^{-3}$, and hence

$$\lambda_L = q^{-4/3}, \tag{20}$$

$$\lambda_U = q^{1/3}, \tag{21}$$

$$\overline{\lambda_1\lambda_1} = \overline{\lambda_1\lambda_3} = \overline{\lambda_3\lambda_3} = q^{2/3}, \tag{22}$$

$$\overline{\lambda_p\lambda_1} = \overline{\lambda_p\lambda_3} = \overline{\lambda_1\lambda_1\lambda_3} = \overline{\lambda_2\lambda_3\lambda_3} = \overline{\lambda_3\lambda_3\lambda_3} = q. \tag{23}$$

Therefore, the dilation factor for length function $\ell^{\wedge}_{13}$ is

$$\lambda_{13} = \frac{\overline{\lambda_1\lambda_3}^{1/2}}{\lambda_U/q} = \frac{q^{1/3}}{q^{1/3}/q} = q, \tag{24}$$

whose corresponding dilation invariant is

$$\mathrm{I}_{13} = \frac{\ell^{\wedge}_{13}}{\zeta}. \tag{25}$$

It is a normal result that length function takes its ordinary dimension if no direction is preferred, the same as the classical dimensional analysis result. In [36], the dilation invariant for stress length is found to be $\mathrm{I}_{13} = 0.45$, and hence $\ell^{\wedge}_{13} = 0.45\zeta$ for canonical boundary layer flows. In ASL, the value of I_{13} may slightly change, but the functional form is the same.

2.2. Inhomogeneous Dilation in the Convective Layer ($B > 0$)

Let us consider the region where height $z > L$ and $B > 0$, and hence the buoyancy term overtakes the shear production to balance dissipation in Equations (16)–(18). Subsequently, the shear production and turbulent transport terms could be neglected in Equations (16)–(18). On the other hand, in the Reynolds shear stress Equation (11) or (19), the dominant balance is still between the first term of production and the fifth term of dissipation. Based on these considerations, Equations (15), (18) and (19) could be, respectively, simplified to

$$\frac{\overline{\lambda_1\lambda_3}}{q} = \frac{1}{\lambda_L}\frac{\lambda_U}{q^2}, \tag{26}$$

$$\overline{\lambda_3\lambda_T} = = \frac{1}{\lambda_L}\frac{\overline{\lambda_3\lambda_3}}{q^2}, \tag{27}$$

$$\overline{\lambda_3\lambda_3}\frac{\lambda_U}{q} = \frac{1}{\lambda_L}\frac{\overline{\lambda_1\lambda_3}}{q^2}. \tag{28}$$

Furthermore, considering that momentum and heat flux are constant in the buoyancy dominant layer, we obtain $\overline{\lambda_3\lambda_T} = 1$ and $\overline{\lambda_1\lambda_3} = 1$, and hence $\lambda_U = q^{-1/3}$, $\lambda_T = q^{-4/3}$ and $\overline{\lambda_3\lambda_3} = q^{2/3}$. Thus, the dilation factor of $\ell^{\wedge}_{13}$ is

$$\lambda_{13} = \frac{1}{\lambda_U/q} = q^{4/3}. \tag{29}$$

Therefore, in the convective layer, the stress length satisfies $\ell^{\wedge}_{13} \propto \zeta^{4/3}$.

2.3. Inhomogeneous Dilation in the Stably Stratified Layer ($B < 0$)

When $B < 0$, buoyancy, acting as a sink, would absorb kinetic energy. Hence, the dominant balance for the Reynolds normal stresses are between SP and B (instead of dissipation). In other words, when summing Equations (8)–(10) all together, the balance is between SP and B. Under this condition, the dilation parameters are simplified to be

$$\overline{\lambda_3\lambda_T} = \overline{\lambda_1\lambda_3}\frac{\lambda_U}{q}. \tag{30}$$

Again, as $\overline{\lambda_3\lambda_T} = 1$ and $\overline{\lambda_1\lambda_3} = 1$, we obtain $\lambda_U = q$. Therefore, the dilation parameter of ℓ_{13} is

$$\lambda_{13} = \frac{1}{\lambda_U/q} = 1, \tag{31}$$

which means that $\ell_{13}^{\wedge}$ is a constant.

2.4. Composite Formula of $\ell_{13}^{\wedge}$ in ASL

According to the dilation analysis above, the scaling of $\ell_{13}^{\wedge}$ in different flow layers is obtained. Following the same matching procedure introduced in [35], a two-layer formula of stress length connecting the adjacent power-law scalings can be obtained:

$$\ell_{13}^{\wedge} \propto \begin{cases} \zeta\left(1-\frac{\zeta}{\zeta_{UC}}\right)^{1/3}, & \zeta < 0 \\ \zeta\left(1+\frac{\zeta}{\zeta_{SC}}\right)^{-1}, & \zeta > 0 \end{cases}. \tag{32}$$

Here, ζ_{UC} and ζ_{SC} are empirical parameters to be determined from data. In [33], they are given as $\zeta_{UC} = 1/6.3$ and $\zeta_{SC}1/2$, respectively. For ζ below the critical height ζ_{UC} and ζ_{SC}, the linear scaling of $\ell_{13}^{\wedge}$ is consistent with wall-attached eddy size in the log layer [39]. For ζ above ζ_{UC}, the scaling $\zeta^{4/3}$ for unstably stratified ASL indicates that the momentum transport eddies are stretched in the vertical direction by the buoyancy force. On the contrary, for stably stratified ASL, the buoyancy force depresses eddy size in a vertical direction, resulting in a finite value of $\ell_{13}^{\wedge}$.

2.5. Composite Formula of ϕ_m in ASL

According to the definitions of ϕ_m and ℓ_{13}, we obtain the relation between ϕ_m and ℓ_{13} as:

$$\phi_m(\zeta) = \frac{\sqrt{-\overline{u'w'}}}{\ell_{13}} \times \frac{\kappa z}{u_\tau}. \tag{33}$$

Considering that $\sqrt{-\overline{u'w'}}/u_\tau = 1$, we have

$$\phi_m \propto \begin{cases} (1-\zeta/\zeta_{UC})^{-1/3}, & \zeta < 0 \\ 1+\zeta/\zeta_{SC}, & \zeta > 0 \end{cases}. \tag{34}$$

3. Data

Data collected for the verification of our derivations include the Kansas measurements, the AHATS (advection horizontal array turbulence study) measurements, and the QLOA measurements. AHATS investigated surface-layer turbulence in the San Joaquin Valley, California, during the summer of 2008, while details on Kansas and AHATS are referred to references [24,40].

For QLOA, it is conducted on the dry lake bed located in Minqin County, Gansu Province, in the northwestern region of China. QLOA has one main tower that is 32 m high, surrounded by twenty lower towers that are 5 m high and shaped like the character 'T'. Data measured from the main tower are examined here, which are acquired from eleven sonic anemometers. Wind speed vectors and virtual temperature are sampled at a frequency of 50 Hz. QLOA has collected day and night data lasting about ten years, covering various weather conditions, and has been used to study the large-scale motion, energy spectrum, amplitude modulation between multi-scale turbulent motions, and two-phase flows during sand storms [4,41,42]. The observations of QLOA are divided into time-ensemble blocks of 1 h. This study uses 12 sets of unstably stratified data and 11 sets of stably stratified data, with details shown in Table 1. The pretreatment of the data is conducted to transform wind signals into a streamwise direction [4]. The friction velocity u_τ is estimated using Reynolds shear stress, $u_\tau = \sqrt{\sum_{i=1}^{11}\langle -u'w'\rangle_i/11}$. Here i denotes sonic anemometer at various heights and $\langle * \rangle_i$ means averaging one hour at ith height. Similarly, the wall heat flux is estimated by $H_w = \sum_{i=1}^{11} \langle w'T' \rangle_i / 11$. To obtain shear production, dU/dz is calculated from mean velocity data using the log-polynomial fitting [43], specifically, $U = c_0 + c_1 \log z + c_2 (\log z)^2$. At each height z, five adjacent points are used to determine the coefficients c_0, c_1, and c_2, so that $dU/dz = c_1/z + 2c_2 \log z/z$. The turbulent dissipation rate is estimated using $\epsilon \approx \frac{1}{K_1 - K_0} \int_{K_0}^{K_1} (E_{11}(k) k^{5/3} / C_{11})^{3/2} dk$, where E_{11} is the spectrum of streamwise velocity fluctuation, $C_{11} = 0.5$ is the longitudinal Kolmogorov constant, while K_0 and K_1 indicate the start and the end of the Kolmogorov initial range, and wavenumber k is calculated from frequency and local mean velocity by using the Taylor freezing hypothesis, $k = 2\pi f / U$ [44].

Table 1. The information of QLOA data used in this paper.

No.	Time and Date	u_τ (m/s)	H_w (K·m/s)	L (m)
1	2014-5-23 7:00–8:00	0.28	0.016	−96.4
2	2014-5-23 8:00–9:00	0.31	0.087	−22.8
3	2014-5-23 9:00–10:00	0.32	0.155	−13.5
4	2014-5-23 10:00–11:00	0.33	0.203	−11.4
5	2014-5-23 11:00–12:00	0.34	0.222	−11.4
6	2014-5-23 12:00–13:00	0.29	0.209	−7.7
7	2014-5-23 13:00–14:00	0.33	0.230	−10.4
8	2014-5-23 14:00–15:00	0.29	0.279	−6.1
9	2014-5-23 15:00–16:00	0.30	0.180	−9.8
10	2014-5-23 16:00–17:00	0.34	0.161	−16.9
11	2014-5-23 17:00–18:00	0.39	0.147	−27.2
12	2014-5-23 18:00–19:00	0.35	0.080	−36.8
13	2014-3-27 0:00–1:00	0.26	−0.036	31.0
14	2014-3-27 1:00–2:00	0.59	−0.057	237.6
15	2014-3-27 2:00–3:00	0.67	−0.055	351.0
16	2014-3-27 3:00–4:00	0.59	−0.039	350.5
17	2014-3-27 4:00–5:00	0.48	−0.026	263.9
18	2014-3-27 5:00–6:00	0.41	−0.018	259.4
19	2014-3-27 6:00–7:00	0.31	−0.014	134.9
20	2014-3-27 7:00–8:00	0.25	−0.007	143.8
21	2014-5-23 2:00–3:00	0.22	−0.025	27.5
22	2014-5-23 6:00–7:00	0.21	−0.013	47.2
23	2014-5-23 20:00–21:00	0.23	−0.019	40.6

4. Results

In Section 2, we demonstrate that different leading order balances of the budget equations lead to different dilation symmetries. However, due to a scarcity of data, it is impossible to check every equation for Reynolds stresses. To address this, we sum together Equations (3)–(5) and obtain the turbulent kinetic energy equation (TKE), which could also be used to verify the leading order balances in different flow conditions. Accordingly, Figure 1 shows the wall-normal (or surface-normal) variation of shear production (SP), buoyancy effect (B), dissipation rate (ϵ), and all other terms (including pressure and spatial transport effects) as the residue (i.e., $SP + B - \epsilon$). In particular, Figure 1a presents the neutrally stratified ASL where heat flux is small and the Obukhov length L is about −96.4 m. It is clear that for lower heights (e.g., $z < 10$ m), the dominant balance is between the shear production and dissipation, which is consistent with the analysis in Section 2.1. Conversely, for $z > 10$ m, the residue is comparable with the dissipation, but the buoyancy term is always smaller than others, hence indicating the neutral stratification condition.

In contrast, Figure 1b shows the unstably stratified case in which $L = -6.1$ m. While the dominant balance is between the shear production and the dissipation for $z < 1$ m, the heat flux is much larger than the neutral case. For $z > 5$ m, the heat flux is comparable to the dissipation and the shear production, indicating a strong buoyancy effect that plays a role as an energy source—consistent with the analysis in Section 2.2. Meanwhile, the pressure and transport effect (indicated by the residue) is negative, drawing out the local kinetic energy to other flow regions.

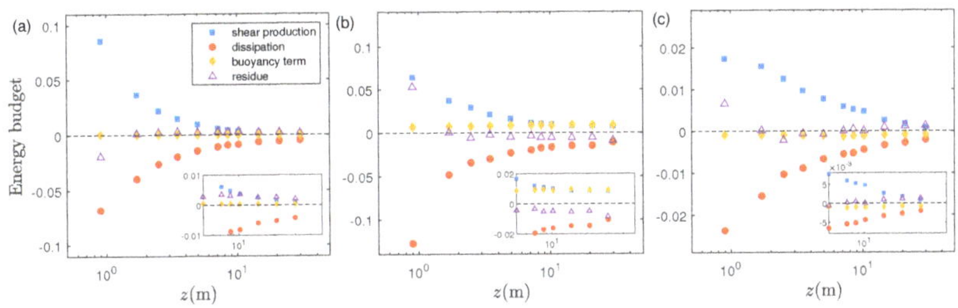

Figure 1. The measured TKE budget in QLOA for neutrally (**a**), unstably (**b**) and stably (**c**) stratified ASL. The inset shows TKE budget for $z > 5$ m.

Moreover, the TKE budget for the stably stratified ASL is shown in Figure 1c for which $L = 27.5$ m. It is obvious that when $z < 20$ m, the shear production and dissipation are the two dominant terms. However, for $z > 20$ m, the shear production, buoyancy effect, dissipation, and the transport effect are all comparable with each other. The notable point is that the buoyancy term B is negative (consistent with the analysis in Section 2.3), indicating that it drains out flow energy as the role of dissipation ϵ, a distinct feature for the stably stratified condition.

Therefore, different leading order balances specified in the above Section 2 have all been verified by ASL data. Now, let us validate the scaling of stress length derived in Section 2. Figure 2 shows the comparison between Equation (32) and the data, with panel (a) for the stable stratification and (b) for the unstable stratification of ASL. Note that ϕ_m are extracted from Kansas and AHATS data and then translated to ℓ_{13} using $\ell_{13} = \kappa z/\phi_m$. Dashed lines indicate the linear scaling, while solid lines are Equation (32) derived from our symmetry analysis. For $|\zeta| = |z/L| < 0.1$, the stress length $\ell_{13}^{\wedge}$ displays a linear variation (dashed line), indicating the dominance of the shear production and hence the log-law of mean velocity. However, for large $|\zeta|$, $\ell_{13}^{\wedge}$ deviates from linear behavior. On the one hand, for stable stratification in Figure 2a, $\ell_{13}^{\wedge}$ tends to be a constant. A close examination of the data shows the best fit of $\ell_{13}^{\wedge} = 0.35\zeta(1+2.0\zeta)^{-1}$ for QLOA measurements, while $\ell_{13}^{\wedge} = 0.35\zeta(1+4.0\zeta)^{-1}$ for the Kansas and AHATS measurements. The different value of $\zeta_{SC} = 0.5$ for QLOA compared to $\zeta_{SC} = 0.25$ for Kansas and AHATS is understandable because a non-zero pressure and spatial transport effect may bring in different heat flux, subsequently altering ζ_{SC}. Such a point has been observed and explained in [33]. On the other hand, as shown in Figure 2b for the unstable stratification case, all the data from QLOA, Kansas, and AHATS align nicely. They follow the trend of Equation (32) with $\ell_{13}^{\wedge} = 0.40\zeta(1-6.3\zeta)^{1/3}$, which is a general expression for the unstable stratification of ASL.

Finally, the mean velocity profile is obtained using Equation (13) with the stress length provided in Equation (32). While MVD data of Kansas and AHATS measurements are not available, the comparison is presented here only for the QLOA measurements, as shown in Figure 3. In total, there are twelve mean velocity profiles presented here; they are measured in different time. At $h = 30$ m (the highest data point for experimental observation), the friction Reynolds number $Re_{30\mathrm{m}} = \frac{30\,\mathrm{m} * u_\tau}{\nu}$ is calculated, along with the value of the Obukhov length L, both marked on top of the labels for each of the subplots. Notably, subplots (a), (g), and (h) pertain to the neutrally stratified ASL indicated by the very large value of $|L|$, approximately 100 m. The latter condition means that most of the data points are measured in the flow region $|z/L| < 0.1$, in which buoyancy is insignificant. Moreover, subplots (b), (c), (d), (e), and (f) are for unstably stratified ASL, where the heat flux is upward and L is negative. The rest of the subplots are for stably stratified ASL where L is positive. The dashed lines in Figure 3 indicate the log-law, which agrees with most flow data for neutrally stratified cases (subplots (a), (g), and (h)), but only depicts data at small z for other flow cases. The red lines in Figure 3 indicate the predictions using $\ell_{13}^{\wedge} = 0.35\zeta(1+2.0\zeta)^{-1}$ for unstable cases, while blue lines indicate the predictions using $\ell_{13}^{\wedge} = 0.40\zeta(1-6.3\zeta)^{1/3}$ for stable cases. It is evident that data deviation from the log-law is well captured by the formula of ℓ_{13} derived from our symmetry analysis.

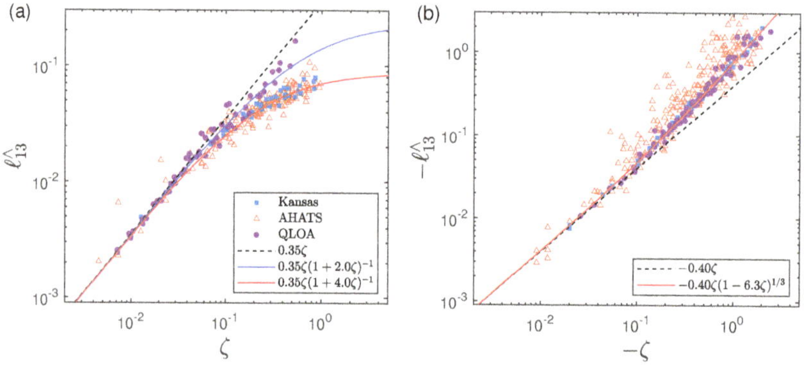

Figure 2. The normalized stress length $\ell_{13}^{\wedge} = \ell_{13}/L$ for stably (**a**) and unstably (**b**) stratified ASL. Circles are QLOA measurements; squares are Kansas experimental data [24], and triangles are AHATS experimental data [40].

It should be mentioned that, to obtain the mean velocity profile, an integration parameter h_0 is needed in Equation (13), which indicates the surface roughness height in QLOA. According to our study, we find that h_0 varies slightly for the above twelve mean velocity profiles. That is, for subplots (a)–(f), $h_0 = (0.08, 0.23, 0.57, 0.42, 0.32, 0.20)$ mm; for subplots (g)–(l), $h_0 = (0.60, 0.20, 0.28, 1.30, 1.60, 1.90)$ mm. The mechanism for the slight variation of these h_0 heights deserves future studies.

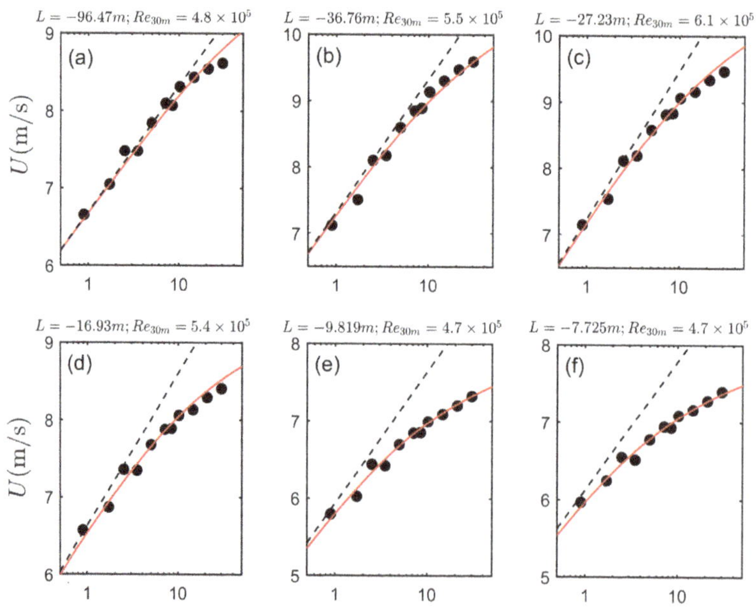

Figure 3. *Cont.*

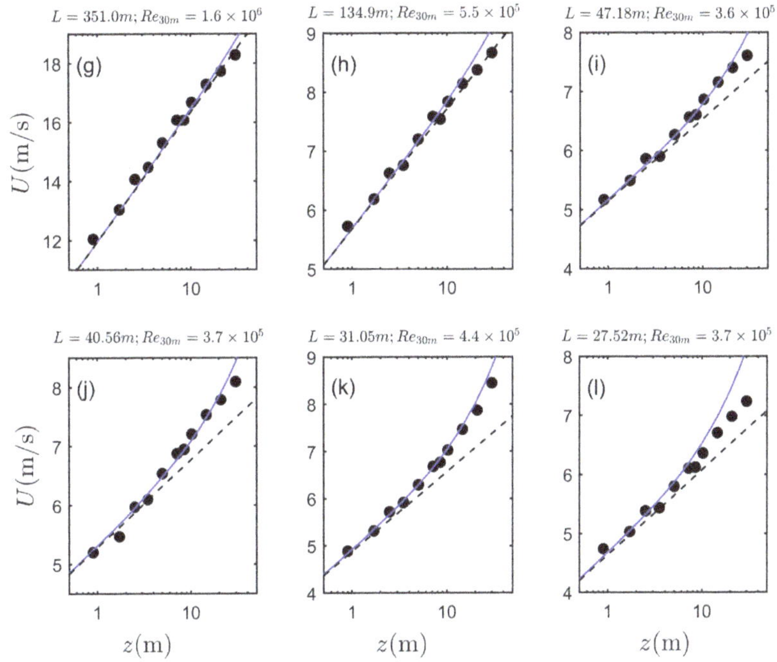

Figure 3. Mean velocity distributions measured in QLOA for neutrally (**a**,**g**,**h**), unstably (**b**–**f**), and stably (**i**–**l**) stratified ASL. Dashed lines indicate the log-law; solid lines are given by Equation (13) provided with Equation (32).

5. Conclusions

The symmetry approach developed for canonical wall turbulence [35,36] has now been extended to describe the mean velocity distribution in the stratified atmospheric surface layer. By performing the random dilation on the governing equations and further considerations on different leading order balances, the scaling of Reynolds stress length is specified in different flow layers. That is, in the shear-dominated log layer, $\ell_{13} \propto \zeta$; in the buoyancy dominated layer, $\ell_{13} \propto \zeta^{4/3}$ for unstably stratified ASL while a constant ℓ_{13} for stably stratified ASL. Using the matching procedure in [35,36], a composite formula of ℓ_{13} is obtained, i.e., $0.40\zeta(1 - 6.3\zeta)^{1/3}$ for unstable stratification and $0.35\zeta(1 + 2.0\zeta)^{-1}$ for stable stratification, which leads to a close representation for the mean velocity distributions measured in QLOA.

It should be noted that the linear coefficients are 0.40 for unstable stratification and 0.35 for stable stratification. Additionally, the value of $\zeta_{UC,SC}$ cannot be explained by the symmetry view. More effort should be devoted to determining these values from a theoretical perspective. Furthermore, the symmetry approach can also be applied to describe the intensity profiles in the streamwise, wall-normal, and spanwise directions, which will be investigated in the future.

Author Contributions: Conceptualization, Y.J. and X.C.; methodology, Y.J.; software, Y.J.; validation, Y.J.; formal analysis, Y.J.; investigation, Y.J.; resources, Y.J.; data curation, Y.J.; writing—original draft preparation, Y.J. and X.C.; writing—review and editing, Y.J. and X.C.; visualization, Y.J.; supervision, X.C.; project administration, X.C.; funding acquisition, X.C. All authors have read and agreed to the published version of the manuscript.

Funding: X. Chen acknowledges the support by the National Natural Science Foundation of China, No. 12072012, 92252201 and 91952302, and "the Fundamental Research Funds for the Central Universities".

Data Availability Statement: There is no new data used in this study.

Acknowledgments: The authors thank Zhen-Su She for helpful comments and suggestions at the early stage of this study. The authors also thank Xiaojing Zheng and Guohua Wang for sharing the QLOA data with us and for many helpful discussions.

Conflicts of Interest: The authors declare no conflict of interest.

References

1. Panofsky, H.A. The Atmospheric Boundary Layer Below 150 Meters. *Annu. Rev. Fluid Mech.* **1974**, *6*, 147–177. [CrossRef]
2. Marusic, I.; Mathis, R.; Hutchins, N. Predictive model for wall-bounded turbulent flow. *Science* **2010**, *329*, 193–196. [CrossRef] [PubMed]
3. Monin, A.S. The Atmospheric Boundary Layer. *Annu. Rev. Fluid Mech.* **1970**, *2*, 225–250. [CrossRef]
4. Wang, G.; Zheng, X. Very large scale motions in the atmospheric surface layer: A field investigation. *J. Fluid Mech.* **2016**, *802*, 464–489. [CrossRef]
5. Klewicki, J.C.; Foss, J.F.; Wallace, J.M. High Reynolds Number $[R_\theta = O(10^6)]$ Boundary Layer Turbulence in the Atmospheric Surface Layer Above Western Utah' s Salt Flats. In *Flow at Ultra-High Reynolds and Rayleigh Numbers* ; Springer: Berlin/Heidelberg, Germany, 1998; pp. 450–466.
6. Metzger, M.M.; Klewicki, J.C. A comparative study of near-wall turbulence in high and low Reynolds number boundary layers. *Phys. Fluids* **2001**, *13*, 692–701. [CrossRef]
7. Andreas, E.L.; Claffey, K.J.; Jordan, R.E.; Fairall, C.W.; Guest, P.S.; Persson, P.O.G.; Grachev, A.A. Evaluations of the von Kármán constant in the atmospheric surface layer. *J. Fluid Mech.* **2006**, *559*, 117–149. [CrossRef]
8. Marusic, I.; Kunkel, G.J. Streamwise turbulence intensity formulation for flat-plate boundary layers. *Phys. Fluids* **2003**, *15*, 2461–2464. [CrossRef]
9. Marusic, I.; Monty, J.P.;Hultmark, M.; Smits, A.J. On the logarithmic region in wall turbulence. *J. Fluid Mech.* **2013**, *716*, 1–11. [CrossRef]
10. Kunkel, G.J.; Marusic, I. Study of the near-wall-turbulent region of the high-Reynolds-number boundary layer using an atmospheric flow. *J. Fluid Mech.* **2006**, *548*, 375–402. [CrossRef]
11. Klewicki, J.C. Reynolds Number Dependence, Scaling, and Dynamics of Turbulent Boundary Layers. *J. Fluids Eng.* **2010**, *132*, 094001. [CrossRef]
12. Chen, X.; Sreenivasan, K.R. Reynolds number scaling of the peak turbulence intensity in wall flows. *J. Fluid Mech.* **2021**, *908*, R3. [CrossRef]
13. Chen, X.; Sreenivasan, K.R. Law of bounded dissipation and its consequences in turbulent wall flows. *J. Fluid Mech.* **2022**, *933*, A20. [CrossRef]
14. Lee, J.; Lee, H.J.; Kim, K.B.; Shin, H.H.; Lim, J.M.; Hong, J.; Lim, K.S.S. Height correction method based on the Monin–Obukhov similarity theory for better prediction of near-surface wind fields. *Atmos. Res.* **2023**, *292*, 106882. . [CrossRef]
15. Wyngaard, J.C. Atmospheric Turbulence. *Annu. Rev. Fluid Mech.* **1992**, *24*, 205–234. [CrossRef]
16. Foken, T. 50 Years of the Monin-Obukhov Similarity Theory. *Bound.-Layer Meteorol.* **2006**, *119*, 431–447. [CrossRef]
17. Liu, L.; Gadde, S.N.; Stevens, R.J.A.M. Universal Wind Profile for Conventionally Neutral Atmospheric Boundary Layers. *Phys. Rev. Lett.* **2021**, *126*, 104502. [CrossRef] [PubMed]
18. Prandlt, L. Meteorogische An wendung der Stromungslehre. *Beitr. Phys. At.* **1932**, *19*, 188–202
19. Monin, A.S.; Obukhov, A.M. Basic laws of turbulent mixing in the ground of the atmosphere. *Doki. Akad. Nauk SSSR* **1954**, *151*, 1963–1987.
20. Högström, U. Analysis of Turbulence Structure in the Surface Layer with a Modified Similarity Formulation for Near Neutral Conditions. *J. Atmos. Sci.* **1990**, *47*, 1949–1972. [CrossRef]
21. Högström, U. Review of some basic characteristics of the atmospheric surface layer. *Bound.-Layer Meteorol.* **1996**, *78*, 215–246. [CrossRef]
22. Priestley, C.H.B.; Panofsky, H.A. An alternative derivation of the diabatic wind profile. *Q. J. R. Meteorol. Soc.* **2010**, *87*, 437–438. [CrossRef]
23. Katul, G.G.; Konings, A.G.; Porporato, A. Mean velocity profile in a sheared and thermally stratified atmospheric boundary layer. *Phys. Rev. Lett.* **2011**, *107*, 268502. [CrossRef] [PubMed]
24. Businger, J.A. Flux profile relationships in the atmospheric surface layer. *J. Atmos. Sci.* **1971**, *28*, 181–189. [CrossRef]
25. Dyer, A.J. A review of flux-profile relationships. *Bound.-Layer Meteorol.* **1974**, *7*, 363–372. [CrossRef]
26. Carl, D.M.; Tarbell, T.C.; Panofsky, H.A. Profiles of Wind and Temperature from Towers over Homogeneous Terrain. *J. Atmos. Sci.* **1973**, *30*, 788–794. [CrossRef]

27. Kader, B.A.; Yaglom, A.M. Mean fields and fluctuation moments in unstably stratified turbulent boundary layers. *J. Fluid Mech.* **1990**, *212*, 637–662. [CrossRef]
28. Gioia, G.; Guttenberg, N.; Goldenfeld, N.; Chakraborty, P. Spectral theory of the turbulent mean-velocity profile. *Phys. Rev. Lett.* **2010**, *105*, 184501. [CrossRef]
29. Li, D.; Salesky, S.T.; Banerjee, T. Connections between the Ozmidov scale and mean velocity profile in stably stratified atmospheric surface layers. *J. Fluid Mech.* **2016**, *797*, R3. [CrossRef]
30. Liu, L.; Stevens, R.J.A.M. Vertical structure of conventionally neutral atmospheric boundary layers. *Proc. Natl. Acad. Sci. USA* **2022**, *119*, e2119369119. [CrossRef]
31. Salesky, S.T.; Anderson, W. Coherent Structures Modulate Atmospheric Surface Layer Flux-Gradient Relationships. *Phys. Rev. Lett.* **2020**, *125*, 124501. [CrossRef]
32. Stiperski, I.; Calaf, M. Generalizing Monin-Obukhov Similarity Theory (1954) for Complex Atmospheric Turbulence. *Phys. Rev. Lett.* **2023**, *130*, 124001. [CrossRef] [PubMed]
33. Ji, Y.; She, Z.S. Analytic derivation of Monin-Obukhov similarity function for open atmospheric surface layer. *Sci. China Phys. Mech. Astron.* **2021**, *64*, 34711. [CrossRef]
34. Chen, X.; Hussain, F. Similarity transformation for equilibrium boundary layers, including effects of blowing and suction. *Phys. Rev. Fluids* **2017**, *2*, 034605. [CrossRef]
35. She, Z.S.; Chen, X.; Hussain, F. Quantifying wall turbulence via a symmetry approach: A Lie group theory. *J. Fluid Mech.* **2017**, *827*, 322–356. [CrossRef]
36. Chen, X.; Hussain, F.; She, Z.S. Quantifying wall turbulence via a symmetry approach. Part 2. Reynolds stresses. *J. Fluid Mech.* **2018**, *850*, 401–438. [CrossRef]
37. She, Z.S.; Wu, Y.; Chen, X.; Hussain, F. A multi-state description of roughness effects in turbulent pipe flow. *New J. Phys.* **2012**, *14*, 093054. [CrossRef]
38. Wu, B.; Bi, W.; Hussain, F.; She, Z.S. On the invariant mean velocity profile for compressible turbulent boundary layers. *J. Turbul.* **2017**, *18*, 186–202. [CrossRef]
39. Jiménez, J. Cascades in Wall-Bounded Turbulence. *Annu. Rev. Fluid Mech.* **2012**, *44*, 27–45. [CrossRef]
40. Salesky, S.T.; Katul, G.G.; Chamecki, M. Buoyancy effects on the integral lengthscales and mean velocity profile in atmospheric surface layer flows. *Phys. Fluids* **2013**, *25*, 105101. [CrossRef]
41. Liu, H.; Wang, G.; Zheng, X. Amplitude modulation between multi-scale turbulent motions in high-Reynolds-number atmospheric surface layers. *J. Fluid Mech.* **2019**, *861*, 585–607. [CrossRef]
42. Liu, H.; Zheng, X. Large-scale structures of wall-bounded turbulence in single- and two-phase flows: Advancing understanding of the atmospheric surface layer during sandstorms. *Flow* **2021**, *1*, E5. [CrossRef]
43. Högström, U. Nondimensional wind and temperature profiles in the atmospheric boundary layer: A re-evaluation. *Bound.-Layer Meteorol.* **1988**, *42*, 55–78. [CrossRef]
44. Li, X.; Zimmerman, N.; Princevac, M. Local Imbalance of Turbulent Kinetic Energy in the Surface Layer. *Bound.-Layer Meteorol.* **2008**, *129*, 115–136. [CrossRef]

MDPI AG
Grosspeteranlage 5
4052 Basel
Switzerland
Tel.: +41 61 683 77 34

Symmetry Editorial Office
E-mail: symmetry@mdpi.com
www.mdpi.com/journal/symmetry

www.ingramcontent.com/pod-product-compliance
Lightning Source LLC
Chambersburg PA
CBHW040825250926
51850CB00007BA/1258
9783725863280